WITHDRAWN

PRINCIPLES OF ELECTRONIC INSTRUMENTATION AND MEASUREMENT

Howard M. Berlin / Frank C. Getz, Jr.

MERRILL PUBLISHING COMPANY
A Bell & Howell Information Company
Columbus / Toronto / London / Melbourne

To Our Families

Cover Art: Jolie Muren

Published by Merrill Publishing Company
A Bell & Howell Information Company
Columbus, Ohio 43216

This book was set in Century Schoolbook and Helvetica

Administrative Editor: Steve Helba
Developmental Editor: Don Thompson
Production Coordinator: Rex Davidson
Art Coordinator: Pat Welch

Library of Congress Catalog Card Number: 87–62816
International Standard Book Number: 0–675–20449–6
Printed in the United States of America
1 2 3 4 5 6 7 8 9—92 91 90 89 88

MERRILL'S INTERNATIONAL SERIES IN ELECTRICAL AND ELECTRONICS TECHNOLOGY

ADAMSON	*Structured BASIC Applied to Technology,* 20772-X
BATESTON	*Introduction to Control System Technology, Second Edition,* 8255-2
BEACH/JUSTICE	*DC/AC Circuit Essentials,* 20193-4
BERLIN	*Experiments in Electronic Devices, Second Edition,* 20881-5 *The Illustrated Electronic Dictionary,* 20451-8
BERLIN/GETZ	*Principles of Electronic Instrumentation and Measurement,* 20449-6 *Experiments in Instrumentation and Measurement,* 20450-X
BOGART	*Electronic Devices and Circuits,* 20317-1
BOGART/BROWN	*Experiments in Electronic Devices and Circuits,* 20488-7
BOYLESTAD	*Introductory Circuit Analysis, Fifth Edition,* 20631-6
BOYLESTAD/ KOUSOUROU	*Experiments in Circuit Analysis, Fifth Edition,* 20743-6
BREY	*Microprocessors and Peripherals: Hardware, Software, Interfacing, and Applications, Second Edition,* 20883-X *8086/8088 Microprocessor: Architecture, Programming and Interfacing,* 20443-7
BUCHLA	*Experiments in Electric Circuits Fundamentals,* 20836-X *Experiments in Electronics Fundamentals: Circuits, Devices and Applications,* 20736-3 *Digital Experiments: Emphasizing Systems and Design,* 20562-X
COX	*Digital Experiments: Emphasizing Troubleshooting,* 20518-2

FLOYD	*Electric Circuits Fundamentals*, 20756-8 *Electronics Fundamentals: Circuits, Devices and Applications*, 20714-2 *Digital Fundamentals, Third Edition*, 20517-4 *Electronic Devices, Second Edition*, 20883-1 *Essentials of Electronic Devices*, 20062-8 *Principles of Electric Circuits, Second Edition*, 20402-X *Electric Circuits, Electron Flow Version*, 20037-7
GAONKAR	*Microprocessor Architecture, Programming, and Applications with the 8085/8080A*, 20159-4 *The Z80 Microprocessor: Architecture, Interfacing, Programming, and Design*, 20540-9
GILLIES	*Instrumentation and Measurements for Electronic Technicians*, 20432-1
HUMPHRIES	*Motors and Controls*, 20235-3
KULATHINAL	*Transform Analysis and Electronic Networks with Applications*, 20765-7
LAMIT/LLOYD	*Drafting for Electronics*, 20200-0
LAMIT/WAHLER/HIGGINS	*Workbook in Drafting for Electronics*, 20417-8
MARUGGI	*Technical Graphics: Electronics Worktext*, 20161-6
MILLER	*The 68000 Microprocessor: Architecture, Programming, and Applications*, 20522-0
ROSENBLATT/FRIEDMAN	*Direct and Alternating Current Machinery, Second Edition, 20160-8*
SCHOENBECK	*Electronic Communications: Modulation and Transmission*, 20473-9
SCHWARTZ	*Survey of Electronics, Third Edition*, 20162-4
STANLEY, B. H.	*Experiments in Electric Circuits, Second Edition*, 20403-8
STANLEY, W. D.	*Operational Amplifiers with Linear Integrated Circuits*, 20090-3
TOCCI	*Fundamentals of Electronic Devices, Third Edition*, 9887-4 *Electronic Devices, Third Edition, Conventional Flow Version*, 20063-6 *Fundamentals of Pulse and Digital Circuits, Third Edition*, 20033-4 *Introduction to Electric Circuit Analysis, Second Edition*, 20002-4
WEBB	*Programmable Controllers: Principles and Applications*, 20452-6
YOUNG	*Electronic Communication Techniques*, 20202-7

PREFACE

When you can measure what you are speaking about, and express it in numbers, you know something about it.

Lord Baron William Thompson Kelvin

Lord Kelvin, the British physicist and mathematician, perhaps said it best when remarking that scientific progress is virtually impossible without the ability to measure. This text is intentionally kept at a basic level. The purpose is to teach students how to select and use instruments for various applications based on an understanding of how the instruments interconnect and their capabilities, as well as to learn the fundamentals of how they work.

This text is directed towards many audiences. It is primarily intended for a single-term course in instrumentation and measurement at either a two-year technology school or a standard electrical engineering technology program. For these programs, it is assumed that the student has had previous training in basic AC and DC circuits, as well as fundamentals of solid-state devices such as diodes, transistors, and operational amplifiers. Although not mandatory, some knowledge of digital devices would help in understanding the portions of several chapters, such as chapters 7, 12, 15, and 16. This text can also be used for courses in areas other than electronics that use basic electronic instruments for measurements, such as physics, chemistry, and physiology. Such courses can disregard the advanced details of circuit operation. Finally, this book can be used as a reference work outside and beyond the classroom, providing a refresher for those who have completed their formal training. Since many topics taught in the classroom are often effectively reinforced and further amplified in the laboratory, we have written a companion laboratory workbook for this purpose, covering a wide range of interesting experiments.

Even though this textbook may not match exactly the intended course outline at a particular learning institution, we nevertheless feel that we have covered sufficient material on the desired basic concepts. Many of the chapters are self-sufficient, so that the chapters that are not normally taught can be passed over without fear of missing prerequisite information needed in later chapters.

This book is written in a style that is useful to both the teacher and the student. At the beginning of each chapter is a list of instructional objectives that preview the concepts that the student is expected to master by the end of the chapter. At the end of each chapter, there is both a brief summary of the material presented and a glossary of the important terms presented. When appropriate, numerical examples are worked out as an aid to understanding how many of the equations and concepts are used. Furthermore, numerical problems are placed at the end of most chapters, and the answers to the odd-numbered problems are found at the back of the book.

The 16 chapters of this text cover the gamut from measurement errors, symbols and units, to measurement standards, grounding, shielding, noise, DC/AC analog and digital indicating instruments, potentiometers and recording systems, oscilloscopes and probes, frequency and time measurements, waveform generators and analyzers, RF and fiber optic measurements, logic analyzers, and the interfacing of complex instrumentation systems. While the trend in many books today is to include numerous BASIC computer programs throughout, we have included only two computer programs which appear in the first chapter. In order to focus the students' attention on the primary topic at hand, instrumentation, these two programs serve solely as examples of how computers can be used to perform lengthy calculations. Otherwise, most calculations are reduced to a single formula, requiring only a simple scientific pocket calculator.

ACKNOWLEDGMENTS

There are many individuals and corporations who have helped make this book possible. The following reviewers have provided helpful suggestions and deserve our thanks: William Jackson, Delaware Technical and Community College; Alexander Avtgis, Wentworth Institute of Technology; John P. Brady, Jr., Orange Coast College; Mark Corrao, Vermont Technical College; David Delker, Kansas State University; Donald Ingram, DeVry Institute of Technology—Chicago; and Ulrich Zeisler, Salt Lake Community College. In addition, the following corporate concerns have graciously provided material and advice: National Bureau of Standards, Amprobe Instruments, Biddle Instruments, John Fluke Manufacturing Company, Inc., Gould Instruments, Heath Company, Hewlett-Packard, Interplex Electronics, Inc., Leader Instruments Corporation, Leeds and Northrup, Simpson Electric Company, Tektronix, Inc., Westinghouse Electric Corporation, and Weston/Solarton Electronics, Inc. Finally, we acknowledge the patience and understanding of our families who have put up with the trials and tribulations that writers undergo.

Howard M. Berlin

Frank C. Getz, Jr.

CONTENTS

1

MEASUREMENT FUNDAMENTALS

1–1 INSTRUCTIONAL OBJECTIVES

After completing this chapter, you will be able to:

- Define the following terms: *fundamental unit, derived unit, accuracy, precision, sensitivity, resolution, error, range,* and *span.*
- Round off a number to the proper number of significant digits.
- Differentiate among systematic, random, and gross errors, giving examples of each.
- Determine the accuracy and tolerance of commonly encountered components and measurement systems.
- Determine the average value, deviation from the mean, average deviation, and standard deviation of a series of measured values.
- Determine the least squares regression line, correlation coefficient, and standard error of estimate between two variables that are linearly related.

1–2 INTRODUCTION

Modern engineering depends on the application of mathematics to physical principles. The quantification of physical values depends on a uniform and repeatable system of measurements, which involves determining the size of a physical parameter in terms of a standard unit of magnitude. Specifically, electrical measurements involve the use of some type of instrument to convert the measured value to a form suitable for easy and accurate interpretation by one of the human senses—usually vision.

Instruments used to measure electrical parameters can range from relatively simple meter movements that use technology that has been known since the earliest days of the electrical industry, to highly sophisticated computer-controlled systems that were not possible only a few years ago. Although the spectrum of electrical and electronic measuring instruments is vast and ever-growing, the knowledge of certain common physical principles and techniques will enable the technician or engineer to select and use the proper instrument for a given application.

1–3 THE LANGUAGE OF MEASUREMENT

The science of instrumentation and its associated measurements is referred to as **metrology.** Although measured quantities exist for a wide range of the physical as well as the social sciences, the terms used have been standardized, and it is important that anyone who is seriously involved in instrumentation be familiar with these terms. The most important of these terms are *fundamental unit, derived unit, accuracy, precision, sensitivity, resolution, error, range,* and *span.*

Fundamental Unit– A relatively small number of basic quantities may be used to describe all other units of measure. These quantities are called **fundamental units.** The six fundamental physical quantities are length, mass, time, electric current, temperature, and luminous (light) intensity.

Derived Unit– All units that are not fundamental units are called **derived units.** For example, although electric charge has the units of coulombs, it is actually defined by the amount of current flowing and the time involved. Although the exact relationship is of minor importance with regards to current, the derived unit of coulomb is made up of two fundamental units, i.e., electric current and time.

Accuracy – **Accuracy** is a measure of the closeness with which an instrument measures the true value of a quantity.

Precision – **Precision** is a measure of the consistency or repeatability of a series of measurements. Although accuracy implies precision, precision does not necessarily imply accuracy. A precise instrument can be very inaccurate. As a quantitative measure, the precision of a given measurement can be expressed mathematically as:

$$\text{precision} = 1 - \left| \frac{x_i - \overline{X}}{x_i} \right| \tag{1–1}$$

where:

x_i = the value of the ith measurement

$\overline{X}$ = the average value of n measurements

In addition to *consistency,* precision also takes into account the number of **significant figures** to which a particular measurement is made. As an example, suppose an oscillator has an actual output frequency of 999,764 Hz. Suppose you use a frequency counter and repeatedly measure the frequency to be 1.0 MHz. Although you consistently read 1.0 MHz, you must now realize that consistency alone is not enough. Due to the small number of significant figures displayed by the frequency counter, the result may not be accurate. The lesson that should be learned from this example is that you should not regard instrument readings without question.

Sensitivity – **Sensitivity** is a measure of the change in reading of an instrument for a given change in the measured quantity.

Resolution – **Resolution** is the smallest change in the measured quantity that will produce a detectable change in the instrument reading. For example, if a digital voltmeter can measure to three decimal places, the voltmeter is then said to have a resolution of 0.001 V, or 1 mV.

Error – **Error** is the deviation from the true value of the measured quantity. In experimental work, error can be expressed as absolute quantity or as a percentage. *Absolute error* is most often defined as the difference between the expected value of a quantity and the measured value, or

$$\boxed{\text{absolute error} = X_e - X_m} \qquad (1\text{–}2)$$

where

X_e = expected value

X_m = measured value

On the other hand, *percent error* is defined as

$$\boxed{\text{percent error} = \left|\frac{X_e - X_m}{X_e}\right|} \qquad (1\text{–}3)$$

Range – The **range** of an instrument describes the limits of magnitude over which a quantity may be measured. It is normally specified by stating its lower and upper limits. For example, an ammeter whose scale reads from 0 to 1 mA is said to have a range from 0 to 1 mA.

Span – The **span** of an instrument is the algebraic difference between the upper and lower limits of the instrument's range. Using a −10-mA to +10-mA galvanometer, the span is then 10 mA − (−10 mA), or 20 mA.

1–4 SIGNIFICANT FIGURES

In stating the numerical results of a measurement, only those figures that are meaningful should be recorded. Since the number of significant figures indicates the *precision* of a measurement, it is then important to include only those figures that are justified by the precision of the equipment being used. For example, if a voltmeter reading is 10.7 V, this implies that the actual value must be closer to 10.7 V than to either 10.6 or 10.8 V.

Retaining Digits

When recording values of known precision, the convention is to retain the last digit that is not uncertain by more than ten units. For example, a reading of 10.4 A has an uncertainty in the digit that is to the right of the decimal point. Consequently, the true value could then be between 10.3 and 10.5 A. At worst, it could be between 9.4 and 11.4 A.

Rounding Off

Significant Figures–After the number of figures to be retained is determined, the last retained figure is increased by one unit if the first dropped figure (i.e., the next number to the right) is greater than 5, or is a 5 followed by any other number other than zero. The last retained digit is not changed when the first dropped digit is less than 5. As an example, the number 28.651 rounded to three significant figures is 28.7. Since the first *dropped* figure is a 5 and is followed by a non-zero number (i.e., a 1), the last retained figure is then increased by one from 6 to 7 so that the decimal number 28.651 is written as 28.7 when rounded to three significant figures.

On the other hand, the number 87.349 rounded to three significant digits is 87.3. The last retained figure is the number 3. Since the first dropped figure is a 4 and is less than 5, the last retained figure is then unchanged so that the decimal number 87.349 is written as 87.3 when rounded to three significant figures.

In rounding the number 14.650 to three significant figures (i.e., one decimal place), we have a problem since 14.650 is just as close to 14.6 as it is to 14.7. When the first dropped digit is a 5 and no figures other than zeros follow it, it is the convention to round off to the *even integer* preceding the 5. If the integer preceding the 5 is even, there is no change; if it is odd, the digit is increased by 1. This procedure works well when numbers are averaged, since the rounding will cause an increase to the higher digit just about as often as it causes a decrease to the lower digit, as well as minimizing *cumulative rounding errors*.

EXAMPLE 1–1

Round off the following numbers to the nearest tenth of a decimal place.

(a) 4.4499 (d) 4.55
(b) 4.5499 (e) 4.4501
(c) 4.45 (f) 4.5501

Solution:

Using the rules discussed, the numbers rounded off to the nearest tenth are

(a) 4.4499 becomes 4.4 (d) 4.55 becomes 4.6
(b) 4.5499 becomes 4.5 (e) 4.4501 becomes 4.5
(c) 4.45 becomes 4.4 (f) 4.5501 becomes 4.6

Significant Zeros – To avoid giving an indication of a higher degree of precision than is warranted, zeros that are not significant, but which serve only to indicate the position of the decimal point, should not be used in writing a value. It is then the preferred practice to write the number in *scientific notation:* that is, as the number of significant digits multiplied by the appropriate power of ten. For example, the number 33,000 is somewhat ambiguous, having two significant figures. The same number, written as either 33×10^3 or 3.3×10^4, would then indicate two significant figures without all the trailing zeros. If it were written as 33.000×10^3 or 3.3000×10^4, it would indicate five significant figures.

Another acceptable method is to give the value with its *range of error*. For example, $29{,}300 \pm 50$ implies three significant figures and an error range of 50 units. Consequently, the actual value lies somewhere between 29,250 and 29,350.

Addition and Subtraction – In adding or subtracting numbers, the answer should not contain more decimal places than are present in the number with the fewest number of decimal places. For example, the sum of 12.2, 5.34, and 2.101 would be written as 19.6 when rounded to one decimal place, since one of the three numbers is measured to only one decimal place. Additional digits to the right of the decimal point would be meaningless. The result then is only as accurate as the least accurate number entering into the computation.

EXAMPLE 1–2

Determine the result of the following operations, rounding the answer off to the proper number of decimal places.

Solution:

			19.32
	52.143	37.6	7.251
	+ 5.6	− 5.41	+ 3.21
actual answer	57.743	32.19	29.781
properly rounded answer	57.7	32.2	29.78

When figures involving a range of possible errors are involved in a calculation, the worst-case calculation is performed, as shown by Example 1–3.

EXAMPLE 1–3

Add 15.04 ± 0.03 and 7.32 ± 0.01. Subtract 7.32 ± 0.01 from 15.04 ± 0.03.

Solution:

$$\begin{array}{r} 15.04 \pm 0.03 \\ +\ 7.32 \pm 0.01 \\ \hline 22.36 \pm 0.04 \end{array} \qquad \begin{array}{r} 15.04 \pm 0.03 \\ -\ 7.32 \pm 0.01 \\ \hline 7.72 \pm 0.04 \end{array}$$

When figures involving a range of possible errors expressed as a percentage are to be added or subtracted, convert the percentage error to an actual value and proceed as before.

EXAMPLE 1–4

Add 40 ± 5% and 10 ± 10%.

Solution:

$$\begin{array}{r} 40 \pm 5\% \\ +\ 10 \pm 10\% \\ \hline \end{array} \qquad \begin{array}{r} 40 \pm 2 \\ 10 \pm 1 \\ \hline 50 \pm 3 \end{array} \quad \text{or} \quad 50 \pm 6\%$$

Multiplication and Division–In multiplication and division, each factor should retain only those digits that will give it the *same percentage uncertainty* as the number having the fewest significant figures used to compute the answer. For example, the product of 98.4 and 9.9 would be written as 98 × 9.9.

Since 9.9 is known only to about 1%, the other number, 98.4, is considered only to 1%.

When multiplying two numbers that involve a range of possible errors, perform the following steps:

1. Multiply the worst-case values using the *positive* limits.
2. Multiply the original numbers.
3. The final result is the product of the original numbers, whereas the possible range of error is the difference between the product of steps 1 and 2.

EXAMPLE 1–5

Multiply 10.1 ± 0.1 by 5.3 ± 0.2.

Solution:

(a) Multiplication of worst-case values using positive limits:

$10.2 \times 5.5 = 56.1$

(b) Multiplication of the original two numbers:

$10.1 \times 5.3 = 53.5$

(c) The range of error is the difference of steps (a) and (b):

$56.1 - 53.5 = 2.6$

(d) The final result is then

53.5 ± 2.6

When dividing two numbers (A by B), the following steps are performed:

1. Divide number A plus its *positive* tolerance by number B less its *negative tolerance* to get a worst-case answer.
2. Perform the division using the original numbers.
3. The final answer is the quotient obtained using the original numbers, while the range of error is the difference between the two quotients obtained in steps 1 and 2.

EXAMPLE 1–6

Divide 10.1 ± 0.1 by 5.3 ± 0.2.

Solution:

(a) Dividend plus its *positive tolerance* divided by the divisor less its *negative tolerance:*

$$10.2/5.1 = 2.0$$

(b) Division using the original numbers:

$$10.1/5.3 = 1.9$$

(c) The possible range of error is the difference between the answers obtained in steps (a) and (b):

$$2.0 - 1.9 = 0.1$$

(d) The final result is then

$$1.9 \pm 0.1$$

When multiplying or dividing numbers that involve *percentage errors,* it is generally sufficient to multiply or divide the numbers and then *add* the percentage errors.

EXAMPLE 1–7

Multiply 100 ± 10% by 10 ± 5%. Divide 100 ± 10% by 10 ± 5%.

Solution:

(a)

$$\begin{array}{r} 100 \pm 10\% \\ \times\ 10 \pm \ \ 5\% \\ \hline 1000 \pm 15\% \end{array}$$

(b)

$$(100 \pm 10\%)/(10 \pm 5\%) = 10 \pm 15\%$$

1–5 ERRORS

All physical measurements involve some degree of uncertainty. Whenever measurements are taken, a degree of error must always be assumed, for, in reality, no measurement can ever be made with pinpoint accuracy. A major skill in taking measurements is the ability to interpret results in terms of possible errors. No matter how carefully the measurements are taken, and no matter how accurate the instruments that are used, *some error will always be present.* A measurement is not truly useful unless the size and type of the expected error is known.

One of the first steps in analyzing any measured value is to consider both the magnitude and type of error that is possible. The three major types of error to be discussed here are systematic, random, and gross.

Systematic Errors

Systematic errors are those that remain constant with repeated measurements. Since they arise from inaccuracies in the manufacture of an instrument or from improper adjustment or application of an instrument, they are observed as those errors that do not change (or change very slowly) with time. Consequently, systematic errors can be measured and compensated for. Some of the more common systematic errors are

- **Zero error:** All readings are in error by the same amount. An example would be a voltmeter that reads 1 V low on all readings, including zero.
- Scale error: This depends on the magnitude of the reading. An example would be a voltmeter that reads 1 V high at 10 V, 2 V high at 20 V, and so on.
- **Response time error:** This is due to the instrument's inability to follow dynamic changes in the measured quantity. An example would be the error resulting from the application of a meter having a high inertia movement (or damping) to measure a signal whose value changes rapidly.
- Loading error: The instrument extracts sufficient energy from the system under measurement so that the value of the measured parameter is changed. An example of the loading error would be the incorrect reading caused by using a low-impedance voltmeter to measure the voltage of a high-impedance source.

In addition to these four types, systematic errors due solely to instrument calibration are normally removed by adjusting the instrument if the design permits. Otherwise, a correction factor may be applied to each reading, or a correction table or graph may be constructed. Other systematic errors can be minimized by selecting the proper instrument for the particular application and frequently verifying its calibration against a suitable standard.

Random Errors

Random errors are those errors due to unknown causes and are observed when the magnitude and polarity of a measurement fluctuate in an unpredictable manner. Some of the more common random errors are

- **Rounding error:** This occurs when readings are between scale graduations, and the reading is rounded up or down to the nearest graduation.
- **Periodic error:** This occurs when an analog meter reading swings or fluctuates about the correct reading. In addition, the meter reading quickly changes in the immediate vicinity of the correct value, but changes slowly at the extremes of the swing. Since it would be easier to read the meter when it is slowly changing, the correct value would be less likely read than an incorrect value.
- Noise: The sensitivity of the instrument is changed, or the reading is altered by outside interference.
- Backlash: The reading either lags or leads the correct value because of mechanical play, friction, or damping.
- Ambient influences: Due to conditions external to the measuring system, such as variation in temperature, humidity, or atmospheric pressure.

Random errors cannot normally be predicted or corrected, but they can be minimized by a skilled observer using a well-maintained quality instrument.

Gross Errors

Gross errors, as the name implies are usually quite large and can be divided into two major categories:

- Human error: This occurs when the operator makes a mistake, such as reading the wrong scale or value.
- Equipment faults: This error source can be large and sometimes erratic.

Gross errors cannot be eliminated, but they can be significantly minimized by (1) careful operator attention and cross-checking of results, and (2) frequent equipment calibration.

1–6 ACCURACIES AND TOLERANCES

Instrument Errors

Instruments having analog meters are usually guaranteed to be accurate within certain percentage limits, called **limiting errors,** or *guarantee errors.* For example, an analog voltmeter may have a tolerance of 3% of full-scale reading. This means that the *full-scale* reading is within 3% of an accurate reading.

The magnitude of the limiting error is determined using the accuracy (expressed as a decimal) and the full-scale reading so that

$$\text{limiting error} = \text{accuracy} \times \text{full-scale value} \tag{1–4}$$

On the other hand, the *percentage error* of the actual meter reading is computed from

$$\text{percentage error} = \frac{\text{maximum error}}{\text{scale reading}} \times 100\% \tag{1–5}$$

EXAMPLE 1–8

An analog ammeter with a 0–100-A range and a stated accuracy of 3% of full scale is presently reading 30 A. Determine the magnitude of the limiting error and the percentage error of the reading.

Solution:

(a) From Equation 1–4, the magnitude of the limiting error is

$$\begin{aligned}\text{limiting error} &= 0.03 \times 100 \text{ A} \\ &= 3 \text{ A}\end{aligned}$$

(b) From Equation 1–5, the percentage error is

$$\begin{aligned}\text{percentage error} &= \frac{3 \text{ A}}{30 \text{ A}} \times 100\% \\ &= 10\%\end{aligned}$$

From Example 1–8, you can see that a reading taken on the lower portion of an analog meter scale will always have a larger possible percentage error (e.g., 10%) than a reading taken near the upper scale limit (e.g., 3% at the full-scale reading). Since the limiting error is a fixed quantity that depends on the full-scale meter reading, meter measurements should be taken so that the reading is as close as possible to the full-scale reading.

On the other hand, the accuracy of digital meters is usually specified as a percentage of the reading plus or minus a given number of counts, as illustrated by the following example.

EXAMPLE 1–9

A 4-digit (digital) voltmeter is specified as having an accuracy of ±0.02% of the reading, plus one count. Determine the maximum error and the percentage error when the meter reads 100.0 V.

Solution:

(a) The worst-case limiting error is

$$\text{limiting error} = (0.02 \times 100.0\ \text{V}) + 0.1\ \text{V}$$
$$= 2.1\ \text{V}$$

(b) The maximum percentage error is then

$$\text{percentage error} = \frac{2.1\ \text{V}}{100\ \text{V}} \times 100\%$$
$$= 2.1\%$$

Note that since the voltmeter can be read to within 0.1 V, this then equals the *value of one count*. On the other hand, if the 4-digit voltmeter reads 5.00 V, then the value of one count equals 0.01 V.

Besides a percentage error, the error from the expected value can be expressed in other forms. For example, the frequency drift in oscillators is often specified on the basis of *parts per million* (ppm) at a specified frequency. This is equivalent to 1 Hz per 1 MHz, or as a percentage, 0.0001%. The following example illustrates this type of specification.

EXAMPLE 1–10

An oscillator is guaranteed to have a maximum frequency drift of ±100 ppm at a frequency of 5 MHz. Determine the maximum frequency drift and the maximum percentage drift.

Solution:

(a) The maximum frequency drift is

$$\text{drift} = \frac{100\ \text{Hz}}{1\ \text{MHz}} \times 5\ \text{MHz}$$
$$= 500\ \text{Hz}$$

Therefore, the output frequency of the oscillator can be anywhere between 4.9995 and 5.0005 MHz.

(b) The percentage drift is

$$\text{drift} = \frac{500\ \text{Hz}}{5\ \text{MHz}} \times 100\%$$
$$= 0.01\%$$

Component Tolerances

Electronic components such as resistors and capacitors normally have a tolerance specified as a percentage of the marked, or nominal, value. For example, a 100-Ω resistor with a tolerance of ±10% could have an actual resistance anywhere from 90 to 110 Ω. Unlike most capacitors, some have a different value for the positive tolerance than for the negative tolerance. As an example, a typical tolerance for an electrolytic capacitor having a nominal value of 4.7 μF might be −10% and +20%. When the value of a component having different positive and negative tolerances is used in a calculation, it is best to use the tolerance that will give the *worst-case* answer.

EXAMPLE 1–11

The DC voltage across a resistor with a tolerance of ±10% is measured to an accuracy of ±3%. What would be the maximum percentage error in determining the power dissipated by the resistor?

Solution:

The power dissipated by a resistor is

$$P = \frac{V^2}{R}$$

Using the rules discussed in Section 1–4, we can add percentages when multiplying or dividing numbers. Since the voltage term in the numerator of the equation is squared; that is, the number times itself, the percent error is *doubled* (i.e., 3% + 3%). According to the rules for multiplication and division in Section 1–4, the maximum percentage error is

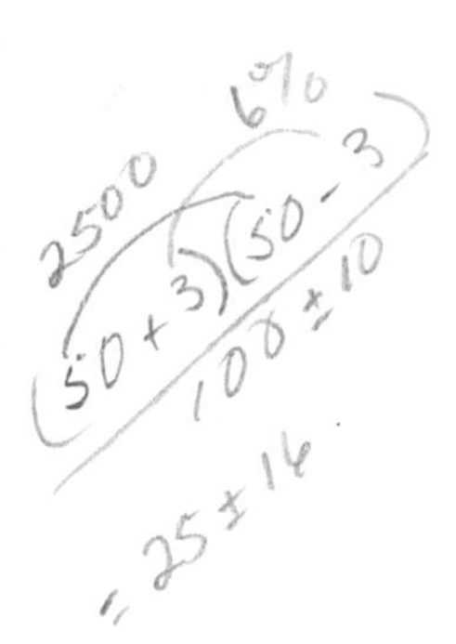

$$\text{maximum percentage error} = (2 \times 3\%) + 10\%$$
$$= 16\%$$

If the measured DC voltage across a 100-Ω resistor is 50 V, the nominal power dissipated by the resistor is 25 W. However, since the maximum percentage error based on the resistor tolerance and meter accuracy is 16%, the actual power may lie between 21 and 29 W.

1–7 PROBABILITY OF ERROR

Statistical methods are frequently used to find the most probable value from a group of readings taken from a given experiment. It is also possible to determine the probable error in one reading and the degree of uncertainty in the most probable value. The laws of probability operate only on *random* errors, not on systematic errors. For this reason, you should be cautioned that statistical analysis does not improve the accuracy of a measurement. For the results of statistical evaluations to be meaningful, however, systematic errors must be small compared to random errors.

Although computational formulas for the various statistical measures are given in this section, many of the low-cost scientific calculators now have provisions for performing these routine calculations easily and accurately. Furthermore, personal computers using BASIC are easily programmed to perform these computations, as illustrated in Section 1–9.

Average or Arithmetic Mean Value

The **average value,** or *arithmetic mean value,* is the most likely value obtained from a series of readings of a given quantity. As a general rule, the more readings, the more closely the computed average represents the most likely value. The average value, $\overline{X}$, is calculated by taking the sum of all of the readings and dividing by the number of readings, so that

$$\overline{X} = \frac{\Sigma x_i}{n} \tag{1–6a}$$

$$= \frac{x_1 + x_2 + \cdots + x_n}{n} \tag{1–6b}$$

where

$\overline{X}$ = average value, or arithmetic mean

x_i = value of the ith reading

n = number of readings

EXAMPLE 1–12

The measurement of five resistors all marked 100 Ω results in the following values: 99.2 Ω, 98.1 Ω, 100.3 Ω, 100.4 Ω, and 100.1 Ω. Determine the average value of these five readings.

Solution:

From Equation 1–6, the average value is

$$\overline{X} = \frac{99.2 + 98.1 + 100.3 + 100.4 + 100.1}{5}$$

$$= 99.6\ \Omega$$

Deviation from the Average Value

The **deviation** from the average value is a measure of how far each measured value departs from the average value. It may be either positive or negative. For a value, x_i, from a group of values having an average value, $\overline{X}$, the deviation, d, of x_i from $\overline{X}$ is expressed as

$$d_i = x_i - \overline{X} \tag{1–7}$$

Note that for n observations, or readings, the algebraic sum of all n deviations is zero.

EXAMPLE 1–13

For the data of Example 1–12, determine the deviation for each reading.

Solution:

Since the average value is 99.6 Ω, the five deviations are then

$$d_1 = 99.2 - 99.2 = -0.4\ \Omega$$
$$d_2 = 98.1 - 99.6 = -1.5\ \Omega$$
$$d_3 = 100.3 - 99.6 = 0.7\ \Omega$$
$$d_4 = 100.4 - 99.6 = 0.8\ \Omega$$
$$d_5 = 100.1 - 99.6 = 0.5\ \Omega$$

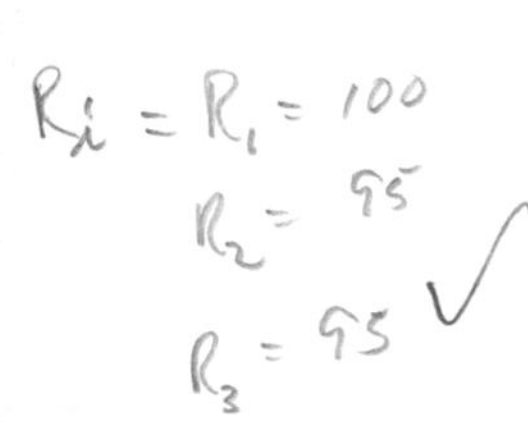

Average Deviation

The *mean*, or *average deviation*, is a measure of how much the data is *dispersed*, or varies from the average value. The mean, $\overline{D}$, is calculated by adding all the absolute values of the deviations of a set of measured values and dividing this sum by the number of observations n, so that

$$\overline{D} = \frac{|d_1| + |d_2| + \cdots + |d_n|}{n} \tag{1–8a}$$

$$= \frac{\Sigma |d_i|}{n} \tag{1–8b}$$

The result, always a positive number, is an indication of the precision of the measurement. A large average deviation compared to the average value of the sample is an indication that the readings varied widely and the measurements as a group were not very precise. In most experimental work, however, the average deviation is not as useful as the *standard deviation,* which is explained in the following section.

EXAMPLE 1–14

Determine the average deviation for the data of Example 1–12.

Solution:

From Example 1–13, the average deviation is then

$$\overline{D} = \frac{0.4 + 1.5 + 0.7 + 0.8 + 0.5}{5}$$

$$= 0.8\ \Omega$$

Standard Deviation

The **standard deviation,** or *root mean square deviation,* of a sample is both mathematically more convenient and statistically more meaningful for analyzing grouped data than is the average deviation. By definition, the standard deviation, s, of a sample is given by

$$s = \sqrt{\frac{\Sigma(\overline{X} - x_i)^2}{n}} \tag{1–9a}$$

$$= \sqrt{\frac{\Sigma d_i^2}{n}} \tag{1–9b}$$

However, as a general practice, the denominator, n, in Equation 1–9 is often replaced by $n - 1$, when the number of observations is less than 20 so that a more accurate value for the standard deviation can be determined. In this case

$$s = \sqrt{\frac{\Sigma(\overline{X} - x_i)^2}{n - 1}} \tag{1–10}$$

As a practical rule, the standard deviation of a set of readings should be less than 10% of the average value of these readings to ensure that the average value is a good indication of the true average value, as shown by the following example.

EXAMPLE 1–15

Using the data of Example 1–12, determine the standard deviation of the five measured resistance values.

Solution:

In terms of the deviations found in Example 1–13 and using Equation 1–10, the standard deviation is

$$s = \sqrt{\frac{(0.4)^2 + (1.5)^2 + (0.7)^2 + (0.8)^2 + (0.5)^2}{4}}$$

$$= 0.97\ \Omega$$

Note that since the number of observations was less than 20, the standard deviation was computed using Equation 1–10. Since the standard deviation of 0.97/99.6 is less than 10% of the average value of 99.6 Ω, this average value can then be considered a good indication of the true average value of this sample.

Gaussian (Normal) Distribution of Error

It is often useful to graph a large number of readings versus the number of times each reading appears in the form of a *histogram,* or bar graph. If all readings are taken with equal care, and if all errors are random, the graph will assume what is known as a *Gaussian distribution,* or *normal distribution,* as shown in Figure 1–1. If enough readings are taken, a smooth curve drawn through the tops of the bars will be *bell-shaped* and will peak at or near the true

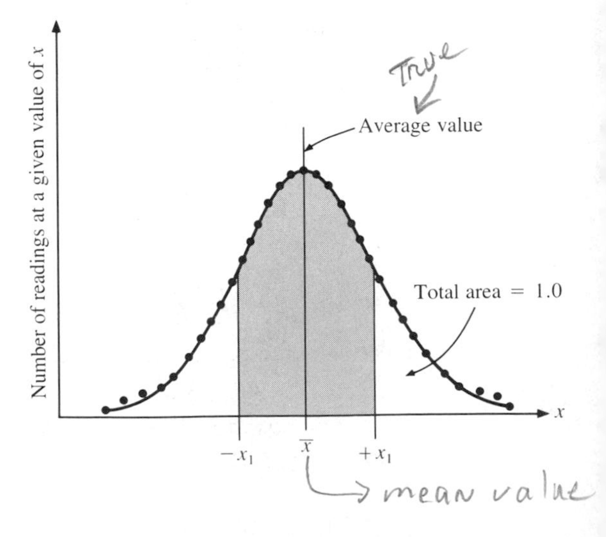

Figure 1–1 Normal curve of error probability. The shaded area represents the probability that a given value will lie within $\pm X_1$ of the mean value.

value. The narrower the bell, the more probable that the center of the bell represents the true value.

Considering a normal distribution of *random* errors, mathematical analysis of the Gaussian curve reveals that 68.3% of all readings will lie within one standard deviation, $\pm 1s$, of the mean value, i.e., within the range $\overline{X} \pm s$. If the curve is normalized in a manner so that the area underneath the curve equals 1.0 (i.e., a probability of 100%), the probability that a value will lie between $+x_1$ and $-x_1$ is given by the area under the curve between $+x_1$ and $-x_1$. In this manner, the area between $-x_1$ and $+x_1$ will translate directly to the probability that a measured value is between two given limits. Table 1–1 gives the probabilities for a reading falling between multiples and submultiples of s. Example 1–16 illustrates how the Gaussian curve is used.

Table 1–1 Probability that a Reading Will Fall between Multiples of the Standard Deviation, s

Deviation s	% Probability (% area under normal curve)
0.5	38.3
0.8	57.6
1.0	68.3
1.5	86.6
2.0	95.5
2.5	98.8
3.0	99.7

EXAMPLE 1–16

The following lists the measured values of 20 resistors having the same marked value. Draw the histogram and determine the average value and the standard deviation for this data.

Reading	*Number of times reading occurred*
92 Ω	1
93 Ω	1
94 Ω	2
95 Ω	3
96 Ω	4
97 Ω	3
98 Ω	3
99 Ω	2
100 Ω	1

Solution:

(a) The histogram is shown in Figure 1–2.

(b) The average value is

$$\overline{X} = \frac{\Sigma x_i}{n} = \frac{1925}{20} = 96.3\ \Omega$$

(c) The standard deviation, after computing the 20 deviations is

$$s = \sqrt{\frac{\Sigma d_i^2}{n-1}} = \sqrt{\frac{83.9}{19}} = 2.1\ \Omega$$

Figure 1–2 Histogram for Example 1–12.

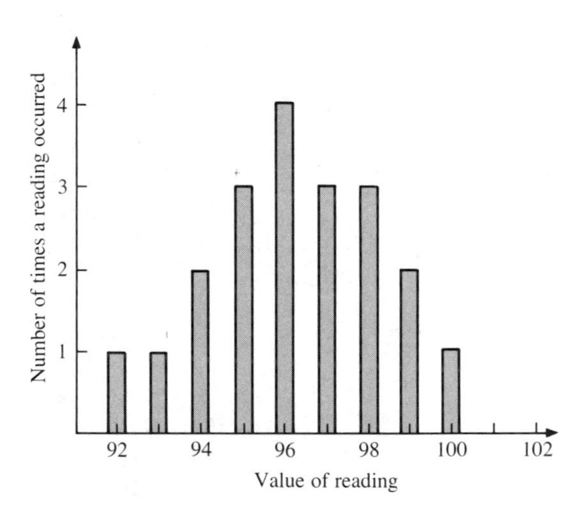

Table 1–1 shows that there is the probability that if the errors of measurement are considered random, then 68.3% (approximately 14 out of 20) of the resistors measured will have values that lie within the range of ±2.1 Ω, or one standard deviation about the mean value of 96.3 Ω. Therefore, if one resistor is selected at random from this original group of 20, there will be a 68.3% chance that its value is between 94.2 and 98.4 Ω. On the other hand, if three standard deviations are considered, there will be a 99.7% chance that a single resistor selected at random from this group will lie within the range of 96.3 ± 0.3 Ω, or between 90.0 and 102.6 Ω.

1–8 CORRELATION OF DATA

In the process of making measurements, it is often evident that there is a relationship between two or more parameters. For example, from Ohm's law we know that there is a relationship between the voltage across a resistor and the current through it. In some cases, we may know the relationship beforehand; in others, we are not sure if one exists. Suppose we measure the voltage developed across a thermocouple as a function of the temperature surrounding it. Such a method may be a good way to measure temperature indirectly by measuring voltage. If we assume that there is a relationship between thermocouple voltage and temperature, we must determine the equation describing the relationship so that we may then be able to calculate any temperature simply by knowing the voltage across the thermocouple. Determining the relationship between two or more variables is accomplished by the method of *least squares regression.* How well these variables are related is quantified by the *correlation coefficient* and the *standard error of estimate.*

Least Squares Regression Line

As an aid in determining a relationship between two (or more) measured parameters, you should graph the data on a rectangular coordinate system. The resulting set of points is sometimes referred to as a *scatter diagram.* From such a graph it is often possible to visualize a smooth line or *approximating curve.* As shown in the scatter diagram of Figure 1–3, a straight line, or *linear,* relationship appears to exist between the dependent variable, Y, and the independent variable, X.

Figure 1–3 Scatter diagram.

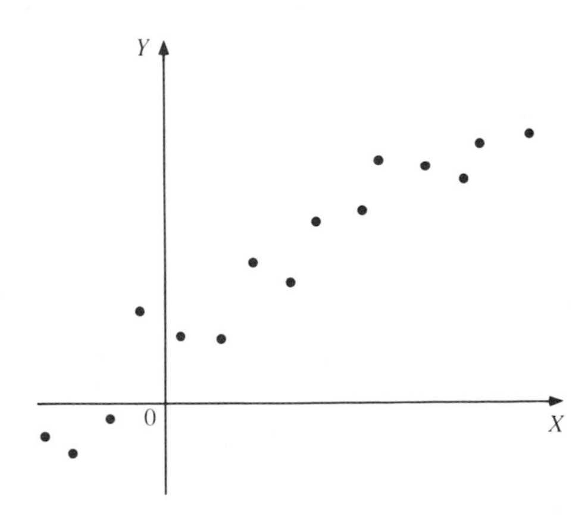

Individual judgment can often be used to draw such an approximating line to fit a set of measured data. This method, called *curve fitting,* can be highly subjective, as two or more people may not draw exactly the same best-fit approximating line. To avoid these individual judgments, an analytical method known as the method of **least squares regression** is often used to describe the best-fit relation between measured variables.

For the majority of measured data, the data constitutes a straight line in the form of

$$Y = mX + b \tag{1–11}$$

where

Y = dependent variable

m = slope of the line

X = independent variable

b = Y-axis intercept, when $X = 0$

For a series of data points (X_1,Y_1), (X_2,Y_2), . . . , (X_n,Y_n), the constants, m and b, are determined from

$$m = \frac{n\Sigma(XY) - \Sigma X\Sigma Y}{n\Sigma(X^2) - (\Sigma X)^2} \tag{1–12}$$

$$b = \frac{\Sigma Y(\Sigma X^2) - \Sigma X\Sigma(XY)}{n\Sigma(X^2) - (\Sigma X)^2} \tag{1–13}$$

In many books on statistics, the equation for the Y-axis intercept is frequently written as

$$b = \frac{\Sigma Y - m\Sigma X}{n} \tag{1–14}$$

which requires that the slope, m, be calculated first. The use of these equations is illustrated by the following example.

EXAMPLE 1–17

An experiment measured the following 10 voltages across a given thermocouple at various temperatures.

Voltage	*Temperature (°F)*
−2.58	−100
−1.11	−20
−0.64	2
−0.44	13
−0.16	25
−0.04	30
0.00	32
1.09	80
1.54	100
3.71	190

Assuming that there is a direct linear relationship, determine the relationship between thermocouple voltage and temperature in the form of an equation of a straight line.

Solution:

Since the measured voltage is dependent on the applied temperature, voltage is then the *dependent variable, Y,* whereas temperature is the *independent variable, X.*

(a) The slope of the line, m, is computed by constructing a table of the various summations as shown in Table 1–2. Using Equation 1–12, we then obtain

Table 1–2

	Y	X	X^2	Y^2	XY
	−2.58	−100	10,000	6.6564	258.00
	−1.11	−20	400	1.2321	22.20
	−0.64	2	4	0.4096	−1.28
	−0.44	13	169	0.1936	−5.72
	−0.16	25	625	0.0256	−4.00
	−0.04	30	900	0.0016	−1.20
	0	32	1024	0	0
	1.09	80	6400	1.1881	87.20
	1.54	100	10,000	2.3716	154.00
	3.71	190	36,100	13.7641	704.90
Total (Σ)	1.37	352	65,622	25.8427	1214.10

$$m = \frac{(10)(1214.1) - (352)(1.37)}{(10)(65{,}622) - 123{,}904}$$

$$= 0.0219 \text{ V/°F}$$

(b) The Y-axis intercept may be determined from either Equation 1–13 or 1–14. Using Equation 1–14 as an example,

$$b = \frac{1.37 - (0.0219)(352)}{10}$$

$$= -0.6339 \text{ V}$$

The least squares equation of the line describing the measured data can then be written as

$$Y = 0.0219X - 0.6339$$

or, in terms of the actual measured parameters:

$$V = (0.0219)°\text{F} - 0.6339$$

There are other computational methods for determining the least squares equation of other relationships in the form of hyperbolas and parabolas, as well as exponential and geometric curves. These, however, will not be discussed as they are beyond the intent and scope of this book.

Once the relationship between X and Y is determined, we are able to *estimate* the value of Y that corresponds to a given value of X without actually having to measure it. This is called a *regression curve of Y on X,* since Y is estimated by knowing the value of X, and is frequently referred to as *interpolation.* If the curve is found to correspond to a straight line between some minimum and maximum values of X for which a straight-line equation was found, then we may be tempted to extend or *extrapolate* the line in either direction and *assume* that points outside the measured data range also behave in a similar manner. While this may be true most of the time, we cannot be *absolutely* sure of this unless it is confirmed by measured data. Since the regression (or *best-fit*) curve was computed over a given range of X-values, the value of X that is used to estimate Y should not be outside this range, as illustrated by the following example.

EXAMPLE 1–18

From the results of Example 1–17, estimate the thermocouple voltage if the measured temperatures are 40 and 200 °F.

Solution:

(a) Using the regression equation determined in Example 1–17, the thermocouple voltage is estimated at 40 °F to be

$$V = (0.0219\ \text{V/°F})\ (40\ \text{°F}) - 0.6339$$
$$= 0.2421\ \text{V}$$

(b) Since a thermocouple temperature of 200 °F is outside the range of the temperatures that were used to determine the equation of the regression line in Example 1–17, we are not able to estimate the value of thermocouple voltage at 200 °F, but only for temperatures between −100 and 190 °F.

Coefficient of Linear Correlation

Although it is possible to determine a least squares regression line for given measured data, it is also important to know how well the data fits the regression line. If we were to estimate the thermocouple voltage at a temperature of 80 °F using the equation determined in Example 1–17, we obtain 1.12 V. However, we know that the thermocouple voltage is 1.09 V at 80 °F, because it was a measured data point. Consequently, even though the measured data is used to determine the best-fit line, not all of the actual measured data points, if any, will actually lie on the line.

Of course, a scatter diagram can be used to visualize whether most of the data points are in the vicinity of the resultant least squares regression line. However, this again is only a subjective conclusion. If Y tends to increase as X increases (as shown in Figure 1–4a), the correlation is considered *positive* or *direct correlation.* If Y tends to decrease as X increases (Figure 1–4b), the correlation is called *negative* or *inverse correlation.* On the other hand, (as shown in Figure 1–4c), there appears to be no correlation between X and Y.

How well a given least squares regression line describes the relationship between measured variables can be determined by the *linear correlation coefficient,* or **coefficient of linear correlation,*** r, as

$$r = \frac{n\Sigma(XY) - \Sigma X\ \Sigma Y}{\{[n\Sigma X^2 - (\Sigma x)^2]\ [n\Sigma(Y^2) - (\Sigma Y)^2]\}^{1/2}} \qquad (1\text{–}15)$$

*Also known as the *Pearson product-moment coefficient of correlation,* after the English statistician Karl Pearson (1857–1936).

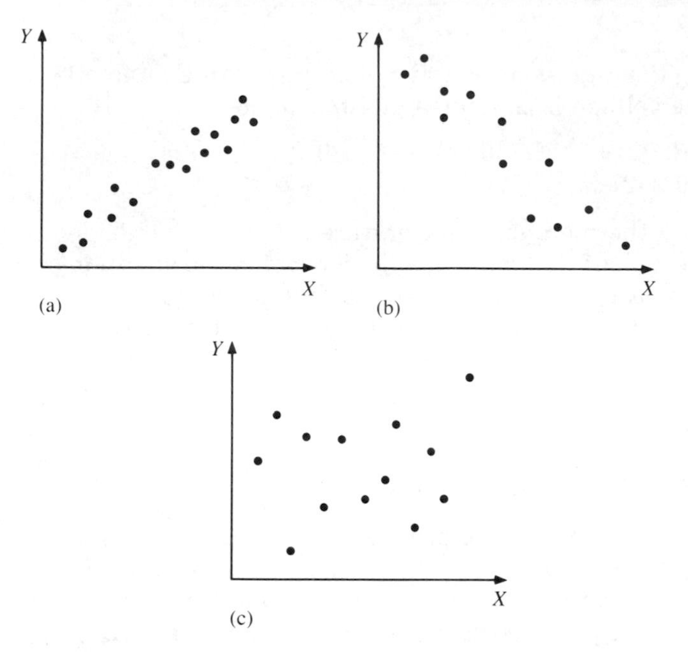

Figure 1–4 Correlation from scatter diagrams. (a) Direct correlation. (b) Negative correlation. (c) No correlation.

which is a dimensionless quantity that varies between -1 and $+1$. If r is positive (Figure 1–4a), then the correlation is said to be positive, as the slope of the line is also positive. If r is negative (Figure 1–4b), both the slope and correlation are negative. The closer the value is to either -1 or $+1$, the more highly the data is correlated. For example, data having a correlation coefficient of 0.956 has a higher correlation between the X and Y data values than data having a correlation coefficient of 0.852. If r is close to zero, then there is virtually no correlation between X and Y, such as the scatter diagram of Figure 1–4c.

Another measure of correlation is the square of the correlation coefficient, r^2, called the **coefficient of determination.** If the correlation coefficient is 0.90, then the coefficient of determination is $(0.90)^2$, or 0.81. Statistically, this tells us that 0.81, or 81% of the variation of the Y-values can be attributed to its straight-line, or linear, relationship with the independent variable, X. Thus, 19% is still unaccounted for. For this reason, a correlation coefficient of 0.90 is not twice as good as 0.45. In fact, a correlation coefficient of 0.90 ($r^2 = 0.81$) is *four times* better than 0.45 ($r^2 = 0.2025$).

EXAMPLE 1–19

For the data of Example 1–17, determine both the correlation coefficient and coefficient of determination.

Solution:

(a) Using Equation 1–15 and the results of Table 1–2, the correlation coefficient is

$$r = \frac{(10)\,(1241.1) - (352)\,(1.37)}{\{[(10)\,(65{,}622) - (123{,}904)]\,[(10)\,(25.8427) - (667.85)]\}^{1/2}}$$

$$= 0.998$$

Since r is positive, the slope of the line is also positive. Furthermore, since r is very close to 1, the ten X-Y pairs of values show a very high degree of correlation. The conclusion is that there is a *definite linear relationship* between the thermocouple temperature and voltage.

(b) The coefficient of determination is then

$$r^2 = (0.998)^2$$
$$= 0.996$$

which means that 99.6% of the variation in the Y-values (i.e., thermocouple voltage) of the data can be attributed to its linear relation to X (i.e., temperature) in Example 1–18.

1–9 USING A COMPUTER

Although computational formulas for the various statistical measures have been given, many of the low-cost scientific calculators now can perform these routine calculations easily and accurately. Note, however, that the built-in routine to compute the standard deviation in many of these calculators uses n instead of $n-1$. On the other hand, personal computers using BASIC are easily programmed to perform these computations. Figure 1–5 illustrates a simple BASIC program for determining the arithmetic mean, average deviation, and standard deviation for n values. Figure 1–6 illustrates a BASIC program that determines the least squares regression for a straight line, correlation coefficient, and coefficient of determination.

```
100 REM ----- PROGRAM TO DETERMINE THE MEAN, AVERAGE DEVIATION
110 REM ----- & STANDARD DEVIATION OF A SERIES OF OBSERVATIONS
120 REM
130 DIM N(100)
140 INPUT "ENTER NUMBER OF OBSERVATIONS"; N
150 AD = 0
160 AV = 0
170 SX = 0
180 V = 0
190 XQ = 0
200 FOR I = 1 TO N
210 PRINT "ENTER VALUE #"; I;
220 INPUT X
230 N(I) = X
240 SX = SX + X
250 XQ = XQ + X*X
260 NEXT I
270 AV = SX/N
280 V = (XQ - N*AV*AV)/(N - 1)
290 SD = SQR(V)
300 FOR I = 1 TO N
310 D = D + ABS(N(I) - AV)
320 NEXT I
330 AD = D/N
340 PRINT
350 PRINT "MEAN ="; AV
360 PRINT "AVERAGE DEVIATION ="; AD
370 PRINT "STANDARD DEVIATION ="; SD
380 END
```

```
ENTER NUMBER OF OBSERVATIONS? 5
ENTER VALUE # 1 ? 99.2
ENTER VALUE # 2 ? 98.1
ENTER VALUE # 3 ? 100.3
ENTER VALUE # 4 ? 100.4
ENTER VALUE # 5 ? 100.1

MEAN = 99.62
AVERAGE DEVIATION = .776001
STANDARD DEVIATION = .973777
```

Figure 1–5 BASIC program for computing the arithmetic mean, average deviation, and standard deviation for n values of a measured sample.

```
100 REM ----- PROGRAM FOR THE LEAST SQUARES REGRESSION AND
110 REM ----- INTERPOLATION OF A STRAIGHT LINE
120 REM
130 DIM X(50), Y(50)
140 SX = 0
150 SY = 0
160 XQ = 0
170 YQ = 0
180 XY = 0
190 INPUT "ENTER NUMBER OF X,Y DATA POINTS"; N
200 FOR I = 1 TO N
210 PRINT "ENTER POINT #"; I; "(X,Y)";
220 INPUT X(I), Y(I)
230 NEXT I
240 FOR J = 1 TO N
250 SX = SX + X(J)
260 SY = SY + Y(J)
270 XQ = XQ + X(J)*X(J)
280 YQ = YQ + Y(J)*Y(J)
290 XY = XY + X(J)*Y(J)
300 NEXT J
310 M = (N*XY - SX*SY)/(N*XQ - SX*SX) : 'SLOPE OF LINE
320 B = (SY - M*SX)/N : 'Y-AXIS INTERCEPT
330 A1 = N*XY - SX*SY
340 A2 = N*XQ - SX*SX
350 A3 = N*YQ - SY*SY
360 R = A1/SQR(A2*A3) : 'CORRELATION COEFFICIENT
370 PRINT
380 PRINT"FOR THE LINE, Y = MX + B:"
390 PRINT "Y-INTERCEPT (B) =";B
400 PRINT "SLOPE (M) =";M
410 PRINT "CORRELATION COEFFICIENT (R) ="; R
420 PRINT "COEFFICIENT OF DETERMINATION ="; R*R
430 PRINT
440 INPUT "DO YOU WANT TO INTERPOLATE (Y/N)"; A$
450 IF A$ = "Y" THEN 460  ELSE 530
460 GOSUB 540
470 INPUT "ENTER VALUE FOR X"; X
480 IF X<X1 OR X>X2 THEN 470  ELSE 490
490 Y = B + M*X
500 PRINT "Y-ESTIMATE ="; Y
510 INPUT"ANOTHER INTERPOLATION (Y/N)"; B$
520 IF B$ = "Y" THEN 470  ELSE 530
530 END
540 REM ----- SUBROUTINE TO DETERMINE MIN/MAX VALUES OF X(I)
550 REM ----- RETURNS X1 = MINIMUM VALUE; X2 = MAXIMUM VALUE
560 X1 = X(1)
570 X2 = X(1)
580 FOR I = 2 TO N
590 IF (X1 - X(I)) <= 0 THEN 620   ELSE 600
600 X1 = X(I)
610 GOTO 640
620 IF (X2 - X(I)) < 0 THEN 630   ELSE 640
630 X2 = X(I)
640 NEXT I
650 RETURN
```

Figure 1–6

```
ENTER NUMBER OF X,Y DATA POINTS? 10
ENTER POINT # 1 (X,Y)? -100, -2.58
ENTER POINT # 2 (X,Y)? -20, -1.11
ENTER POINT # 3 (X,Y)? 2, -0.64
ENTER POINT # 4 (X,Y)? 13, -0.44
ENTER POINT # 5 (X,Y)? 25, -0.16
ENTER POINT # 6 (X,Y)? 30, -0.04
ENTER POINT # 7 (X,Y)? 32, 0
ENTER POINT # 8 (X,Y)? 80, 1.09
ENTER POINT # 9 (X,Y)? 100, 1.54
ENTER POINT # 10 (X,Y)? 190, 3.71

FOR THE LINE, Y = MX + B:
Y-INTERCEPT (B) =-.633949
SLOPE (M) = .021902
CORRELATION COEFFICIENT (R) = .997657
COEFFICIENT OF DETERMINATION = .99532

DO YOU WANT TO INTERPOLATE (Y/N)? Y
ENTER VALUE FOR X? 40
Y-ESTIMATE = .242129
ANOTHER INTERPOLATION (Y/N)? Y
ENTER VALUE FOR X? 200
ENTER VALUE FOR X?
```

Figure 1–6 BASIC program for computing least squares regression of a straight line.

1–10 SUMMARY

After reading Chapter 1, you should understand the common terms used in the language of measurement. These include *accuracy, precision, sensitivity, resolution, error, range,* and *span.* Some of these terms have been mathematically defined. In addition, you have seen how to round off numbers properly and been shown that errors can be either systematic, random, or gross. You have seen how to compute the accuracy and tolerances of electronic components and measurement systems. Finally, when making a series of similar measurements, you can now express grouped data in terms of its average value and standard deviation and determine if two variables are related to each other.

1–11 GLOSSARY

accuracy The measure of the closeness with which an instrument measures the true value of a quantity.

average value The most likely value obtained from a series of independent measurements of a given quantity. Average value is calculated by summing all the readings and dividing by the number of readings.

coefficient of determination A measure of what percentage of the variation of the dependent variable can be attributed to the linear relation with the corresponding independent variable. Numerically, it is equal to the square of the coefficient of linear correlation.

coefficient of linear correlation A measure of how well two or more variables are related. A value of 1.0 indicates a perfect degree of correlation, while zero indicates no correlation.

derived unit All units that are not fundamental units.

deviation A measure of how much each measured value differs from the average value.

error The deviation from the true value of the measured quantity.

fundamental unit A relatively small number of basic quantities may be used to describe all other units of measure. The six fundamental quantities are length, mass, time, electric current, temperature, and luminous (light) intensity.

least squares regression An analytical method of determining the relationship between two or more variables, if a relationship exists.

limiting error The percent accuracy of an analog meter. Also called guarantee error.

metrology The science of instrumentation and its associated measurements.

periodic error The error that occurs when an analog meter reading swings or fluctuates about the correct reading. In addition, the meter reading quickly changes in the immediate vicinity of the correct value, but changes slowly at the extremes of the swing.

precision A measure of the consistency of repeatability of a series of independent measurements.

random errors Errors due to unknown causes and observed when the magnitude and polarity of a measurement fluctuate in an unpredictable manner.

range The limits of magnitude over which a quantity may be measured. Range is normally specified by stating its lower and upper limits.

resolution The smallest change in the measured quantity that will produce a detectable change in the instrument reading.

response time error An error due to the instrument's inability to follow dynamic changes in the measured quantity.

rounding error The error that occurs when readings are between scale graduations, and the reading is rounded up or down to the nearest graduation.

sensitivity A measure of the change in reading of an instrument for a given change in the measured quantity.

significant figures A series of digits from consecutive columns beginning with the leftmost (most important) digit different from zero and ending with the least significant digit whose value is either known or assumed to be relevant.

span Equal to the algebraic difference between the highest and lowest scale values of an instrument. For instruments whose scale starts at zero, the span equals the range.

standard (root mean square) deviation A statistical measure for analyzing grouped data by quantifying the dispersion of individual data points about their mean value.

systematic errors Errors that remain essentially constant with repeated measurements, such as zero, scale, response time, and loading errors.

zero error An error in which all readings are in error by the same amount.

1–12 PROBLEMS

1. Analysis of the current through a resistor yields an ideal value of 14.7 mA. However, when measured, the current is 14.5 mA. Determine the absolute error and the percentage error of the measurement.
2. Determine the number of significant figures in each of the following numbers:
   (a) 14.25
   (b) 52×10^3
   (c) 0.707
   (d) 10,000
   (e) 0.0022
   (f) 348

3. Round off the following numbers to the nearest tenth of a decimal place:
   (a) 28.8251 (d) 28.55
   (b) 28.8591 (e) 28.549
   (c) 28.85 (f) 28.556
4. The following three resistors are placed in series: 10.5 Ω, 2.71 Ω, and 25.008 Ω. Determine the total resistance with the proper number of significant figures.
5. Two resistors are selected: 1000 Ω ± 10% and 2200 Ω ± 10%.
   (a) Determine the value of the series combination with its range of error.
   (b) Determine the value of the parallel combination with its range of error.

   In both cases, express the range of error in terms of an absolute value as well as a percentage.
6. Three resistors have the following marked values: 2.2 kΩ ± 10%, 47 Ω ± 5%, and 560 kΩ ± 20%. Determine the magnitude of error for each resistor.
7. A voltmeter and ammeter, both having full-scale accuracies of ±2%, are used together to measure resistance by Ohm's law. The voltmeter reads 3.8 V on the 5-V scale, while the ammeter reads 62.8 mA on the 100-mA scale. Determine the resistance and the limits within which the result can be guaranteed.
8. A voltmeter has a full-scale accuracy of ±3%. If it reads 152 V on the 250-V scale, determine the absolute error and the percent error of the reading.
9. The maximum drift of a crystal-controlled oscillator is found to be 250 Hz at a frequency of 7.032 MHz.
   (a) Determine the drift in parts per million.
   (b) Determine the maximum percentage drift.
10. The output frequency of an oscillator is controlled by an 8.2-kΩ ± 5% resistor and a 560-pF ± 10% capacitor. Determine the expected output frequency, f, with its absolute range of error, if $f = 1/2\pi RC$.
11. Five students measured the center frequency of a bandpass filter using the same method with the following results:

Student	Measured Frequency
1	652 Hz
2	650 Hz
3	648 Hz
4	654 Hz
5	649 Hz

    Which of the five measurements is the most precise?
12. Each of six resistors selected at random from a supply bin has a nominal value of 10 Ω. These were measured and found to have the following values: 9.8 Ω, 10.1 Ω, 10.2 Ω, 9.8 Ω, 9.9 Ω, and 9.7 Ω.
    (a) Determine the average value, average deviation, and standard deviation of these six resistors.
    (b) If the resistance measurement errors are considered random, over what resistance range will there be a 68.3% chance that a single resistor selected from the group of resistors will be within one standard deviation?
13. An experiment measured a transistor's collector current, I_C, as a function of its collector-emitter voltage, V_{CE}, as follows

V_{CE} (V)	I_C (mA)
0	10.5
2	7.6
4	6.8
6	4.5
8	2.4
10	0.4

    (a) Assuming there is to be a linear relationship between I_C and V_{CE}, determine the slope and Y-axis intercept of the regression line.

(b) From the regression line, determine the collector current in milliamperes for a collector-emitter voltage of 5 V.

(c) Determine the correlation coefficient. Is there a direct, inverse, or no correlation between I_C and V_{CE}?

(d) Determine the percentage of the variation of the values of I_C that can be attributed to the straight-line relationship with V_{CE}.

2

MEASUREMENT UNITS

2–1 INSTRUCTIONAL OBJECTIVES

After completing this chapter, you will be able to

- Distinguish between fundamental and derived units.
- Be familiar with many of the quantity and unit symbols used in electricity and electronics.

- Define fundamental electrical units, such as volt, ampere, ohm, farad, henry, and hertz.
- Determine the logarithmic response of a system in terms of decibels, nepers, and decibel references such as volume unit (VU) and decibel-milliwatt (dBm).
- Convert quantities from one unit system to another.

2–2 UNITS AND DIMENSIONS

Fundamental and Derived Units

All quantities used in the formulation of physical laws and principles must be defined in terms of a *magnitude* as well as a descriptive *unit of measure.* A small number of physical quantities were selected arbitrarily and termed **fundamental quantities.** The intent was to measure these fundamental quantities in terms of arbitrary but internationally accepted base units, but not to equate them to other units. All other physical quantities were then to be defined in terms of fundamental quantities and are known as **derived quantities.** It should be noted that a **unit** is a defined sample of a quantity.

In the physical world there are only six fundamental quantities. The branch of physics known as mechanics gives us the fundamental quantities of length, mass, and time. Thermodynamics adds the fundamental quantity of temperature difference. From electricity and magnetism we get the fundamental quantity of electric current. Finally, from the study of optics, we have the quantity of light intensity. From these six fundamental quantities, all other physical (i.e., derived) quantities are defined.

SI Units

In 1898, the definitions of the basic units were established by an international organization known as the General Conference of Weights and Measures, to which all major countries of the world sent representatives. The definitions, unit symbols, and dimensions of the basic system of units have been refined and updated, and since 1960 the system is now known officially as the **International System** but often referred to as the **SI** units, from the French equivalent, *Système International d'Unités.* Table 2–1 summarizes the six fundamental SI physical quantities. Although SI units are now the official international standard, it should be noted that older unit systems are still currently being used throughout the world. This is particularly true of the **English System,** used predominantly in the United States and Great Britain. Consequently, it is often necessary to convert from one system to the other, as discussed later in Section 2–5. As a general rule, the symbols and units used throughout this book conform to the SI convention.

Table 2–1 Fundamental Physical Quantities

Fundamental Quantity	SI Unit	SI Unit Symbol	Dimension
length	meter	m	L
mass	kilogram	kg	M
time	second	s	T
electric current	ampere	A	I
temperature	kelvin	K	Θ
light intensity	candela	cd	

Letter Symbols for Quantities and Units

Letter symbols fall into two categories: symbols for quantities (quantity symbols) and symbols for the units, or dimensions in which these quantities are measured (unit symbols). Table 2–2 lists the quantities and their unit symbols frequently encountered in the study of electricity and electronics.

A **quantity symbol** is generally a single letter (Roman or Greek) with optional subscripts or superscripts used to identify a measured variable. For example, the letters L and f are the quantity symbols used in equations to indicate inductance and frequency, respectively. A *dimensional symbol,* or **unit symbol,** is generally one or two letters used as an abbreviation for the name of the unit. For example, the unit of current, the *ampere,* or simply *amp,* is abbreviated by the unit symbol A.

Dimensions of derived units are the algebraic combination of fundamental units that enter the measurement of the derived unit. Some derived units, however, are described only in terms of fundamental units, such as A/m (ampere/meter), which are the units associated with magnetic field strength. On the other hand, most SI units have specific names assigned to them, predominantly the last names of those scientists and pioneers closely associated with the measured phenomenon. Although the normal convention in writing is to capitalize the first letter of a person's last name, the last name, when used as a unit of measure, is *not capitalized.* For example, the unit of magnetic flux is named in honor of William Weber. As part of a sentence, for example, it is correctly written as "magnetic flux of 2.49 webers." On the other hand, it is frequently convenient to use the unit symbol, or abbreviation for a given quantity. For the case of those symbols named after people, *the first letter of the abbreviation is capitalized.* As examples of those units named after individuals, the weber is abbreviated Wb, and the siemen is S.

In the SI or any other system, multiples or submultiples of a unit are easily introduced once the fundamental unit is defined. These additional units are related to the fundamental units by multiples or submultiples of 10. The names of additional units are always derived by adding a **multiplier prefix** to

Table 2–2 Standard Unit Symbols Frequently Used in the Study of Electricity and Electronics

Unit	Symbol	Dimension	Unit	Symbol	Dimension
ampere	A	I	kelvin	K	θ
ampere-hour	Ah	IT	lambert	L	
ampere-turn	At	I	lumen	lm	
baud	Bd	$1/T$	lux	lx	
bel	B		maxwell	Mx	I/M^2
coulomb	C	IT	meter	m	M
decibel	dB		mho	mho	T^3I^2/L^2M
degree (angle)	...°		neper	Np	
degree (temperature)		θ	newton	N	LM/T
degree Celsius	°C	θ	nit	nt	
degree Fahrenheit	°F	θ	oersted	Oe	I/M
kelvin	K	θ	ohm	Ω	L^2M/T^3I^2
dyne	dyn	LM/T	radian	rad	
electronvolt	eV	L^2M/T^3I	second	s	T
farad	F	T^4I^2/L^2M	siemen	S	T^3I^2/L^2M
foot candle	fc		tesla	T	M/T^2I
gauss	G	L^2/T^2IM	var	var	L^2M/T^3
gilbert	Gb	I	volt	V	L^2M/T^3I
henry	H	L^2I^2M/T	voltampere	VA	L^2MI/T^3
hertz	Hz	$1/T$	watt	W	L^2M/T^3
horsepower	hp	L^2M/T^3	watt-hour	Wh	L^2M/T^2
hour	h	T	weber	Wb	L^2M/T^2I
joule	J	L^2M/T^2			

the name of the base unit, either spelled out, or in conjunction with a unit symbol. Table 2–3 summarizes the prefixes generally used by scientists and engineers. With the exception of the prefix symbol for kilo, all prefix symbols for multiplier values greater than one are *capitalized.* For multiplier prefix symbol values less than one, the symbol is written in *lowercase* letters. As examples:

1,800,000 ohms = 1.8 megohms = 1.8 MΩ

5426 volts = 5.426 kilovolts = 5.426 kV

0.073 amperes = 73 milliamperes = 73 mA

0.0000000047 farads = 4.7 nanofarads = 4.7 nF

Note that the combination of the prefix *mega-* and the unit *ohm* is spelled *megohm* (without the *a*). Such is similar with the combination *kilo-* and *ohm,* or

Table 2–3 Multiplier Prefixes

Prefix	Symbol	Multiplier	Decimal Value
atto-	a	10^{-18}	0.000 000 000 000 000 001
femto-	f	10^{-15}	0.000 000 000 000 001
pico-	p	10^{-12}	0.000 000 000 001
nano-	n	10^{-9}	0.000 000 001
micro-	μ	10^{-6}	0.000 001
milli-	m	10^{-3}	0.001
centi-	c	10^{-2}	0.01
deci-	d	10^{-1}	0.1
deca-	da	10^{1}	10
hecto-	h	10^{2}	100
kilo-	k	10^{3}	1000
mega-	M	10^{6}	1,000,000
giga-	G	10^{9}	1,000,000,000
tera-	T	10^{12}	1,000,000,000,000
peta-	P	10^{15}	1,000,000,000,000,000
exa-	E	10^{18}	1,000,000,000,000,000,000

kilohm. The proper spellings of the common combined forms are given in Appendix A.

2–3 COMMONLY ENCOUNTERED ELECTRICAL UNITS

In the study of electrical measurements, a number of those quantities listed in Table 2–2 are used more frequently than others. Although these are generally treated in Chapter 3, they deserve a brief introduction here. These are the units of voltage, current, resistance, capacitance, inductance, and frequency.

Volt

The *volt*, named for Alessandro Conte Volta, is the SI unit of *potential difference* and is analogous to the pressure difference in a hydraulic system. Just as a pressure difference forces fluid to flow in a hydraulic system, potential difference is the force that causes charges to move through an electric circuit. On the basis of moving charges, a difference of one volt exists between two points in a circuit if one joule of energy is required to move a 1-coulomb charge from the point of lower potential to the point to higher potential. From this definition, voltage than has the dimensions of joules per coulomb, so that

$$1 \text{ volt} = \frac{1 \text{ joule}}{1 \text{ coulomb}} \tag{2–1}$$

In addition to this definition, the volt has been defined as the potential difference across a resistance dissipating one watt when one ampere is flowing through it, or

$$1 \text{ volt} = \frac{1 \text{ watt}}{1 \text{ ampere}} \tag{2–2}$$

From the fundamental SI units given in Table 2–1, voltage is then dimensionally expressed in units of L^2M/T^3I.

The **quantity symbol** for the potential difference between two points in a circuit is commonly specified by writing the letter V* followed by two subscripts, which correspond to these two points. For example, the potential of point a with respect to point b of a given circuit would be written as V_{ab}. If point a is at a higher potential (i.e., more positive) than b, V_{ab} would then have a *positive* value. If point a is at a lower potential (less positive) than b, V_{ab} would represent a negative value. In this case, point b is the *reference,* which itself in turn may be at either a positive or negative potential with respect to some other point.

In most systems, it is convenient to have a common reference for all voltage measurements, which is assigned the value of zero volts. Potentials above this reference are said to be positive, while potentials below this level are said to be negative. Since this reference point is often connected to the earth for reasons of safety and noise reduction, it is commonly called *ground,*** whether there is an earth connection or not.

Many instruments used to measure potential difference have one lead at ground potential; thus, it is necessary to make all voltage measurements with respect to ground. As a result, most circuit diagrams and equipment manuals show voltage values with circuit ground being the *implied reference.*

Ampere

The *ampere,* named for André Marie Ampere, is the SI unit of *electric current* and is a measure of the rate at which electrical charge passes a given point in a circuit. It is analogous to fluid flow in a hydraulic system. A current of one

*The quantity symbol convention used throughout this text is as follows: Uppercase variables are used for amplitudes of DC and sinusoidal quantities. Lowercase variables are used for generalized time-varying quantities.

**The British term for *ground* is *earth.*

ampere will cause a charge of one coulomb to pass a given point in a circuit in one second. Thus, current flow has the dimensions of coulombs per second.

$$1 \text{ ampere} = \frac{1 \text{ coulomb}}{1 \text{ second}} \tag{2–3}$$

The quantity symbol for current is represented by the letter I, after the French word for current, *intensité*. The direction of current flow is often indicated on a schematic diagram by an arrow pointing in the direction of flow. Two conventions are used to describe the direction of current flow: *electron flow* and *conventional flow*. Electron flow shows the direction that the actual electrons move through the circuit: from negative to positive. On the other hand, conventional current flow shows an arrow in the opposite direction, so that it is assumed the current leaves the positive terminal of an electromotive source and returns to the negative terminal. Of these two conventions, the conventional flow is the more commonly used. There is no inconsistency between the two conventions, however. One assumes a flow of negative charge in one direction; the other assumes a flow of positive charge in the opposite direction, so that the net result is equivalent. In this text, we adhere strictly to *conventional* current flow.

Ohm

The *ohm,* named for Georg Simon Ohm, is the SI unit of *electrical resistance.* Resistance is analogous to friction in a hydraulic system. It is a measure of the opposition to a steady current flow due to the molecular properties of the conductor. The resistance of a circuit element is, by definition, the voltage across the element divided by the *in-phase* current flowing through the element:

$$1 \text{ ohm} = \frac{1 \text{ volt}}{1 \text{ ampere}} \tag{2–4}$$

Resistance may be thought of as a proportionality constant that linearly relates the potential difference between two points in a circuit and the current that results from the potential difference between them (Ohm's law). From the fundamental SI units given in Table 2–1, resistance is dimensionally expressed in units of L^2M/T^3I^2. Although the quantity symbol for resistance is indicated by the letter R, the unit symbol for ohm is marked by the uppercase Greek letter *omega* (Ω).

Farad

The *farad,* named for Michael Faraday, is the SI unit of *capacitance.* Capacitance is a measure of the charge that is stored on a capacitor as a function of the applied voltage, and is equal to the stored charge divided by the voltage across the capacitor so that

$$1 \text{ farad} = \frac{1 \text{ coulomb}}{1 \text{ volt}} \tag{2–5}$$

From the fundamental SI units given in Table 2–1, capacitance is dimensionally expressed in units of T^4I^2/L^2M. The quantity symbol for capacitance is indicated by the letter C and should not be confused with the unit symbol for *coulombs,* represented by the letter C. The unit symbol for farads is represented by the letter F.

Dimensionally, the farad is an extremely large unit that is seldom used in practical work. In its place, capacitance units are most commonly expressed in terms of either *microfarads* (μF) or *picofarads* (pF).

Henry

The *henry,* named for Joseph Henry, is the SI unit of *inductance.* It is a proportionality factor that relates induced voltage in a coil of wire to the rate of change of current with time, so that

$$1 \text{ henry} = \frac{1 \text{ volt}}{1 \text{ ampere/second}} \tag{2–6}$$

From the fundamental SI units given in Table 2–1, inductance is dimensionally expressed in units of L^2M/T^2I^2. The quantity symbol for inductance is indicated by the letter L. It should not be confused with the quantity symbol for *luminance* (in *lamberts*), represented by the letter L. The unit symbol for henry is represented by the letter H.

Hertz

The *hertz,* named for Heinrich Rudolph Hertz, is the SI unit of *frequency.* A frequency of one hertz is defined as one cycle of the waveform per one second, or

$$1 \text{ hertz} = \frac{1 \text{ cycle}}{1 \text{ second}} \tag{2–7}$$

From the fundamental SI units given in Table 2–1, frequency is dimensionally expressed in units of $1/T$. The quantity symbol for frequency is generally indicated by the letter f, whereas the unit symbol for hertz is Hz.

2–4 LOGARITHMIC RESPONSE UNITS

When measurements are made on individual electronic components, the measured values are usually the absolute values of the parameters associated with the components. However, when such components are connected to create *linear* systems, such as amplifiers, filters, or attenuators, the measure of the system's performance is usually expressed as a ratio of the output signal to the input signal rather than the absolute magnitude of the signal level. When the output level is greater than the corresponding input level, this ratio is referred to as the *gain* of the system. Conversely, if the output level is less than the input, the ratio is termed *loss,* or *attenuation.* Ratios such as the ratio of output to input voltage can be quite large. To simplify calculations, it is frequently more convenient to express this ratio in terms of a *logarithmic response unit.* Several logarithmic units are used extensively in science and engineering, primarily to designate power ratios. In electronic instrumentation, the most common are the decibel and neper.

Decibel

The decibel, abbreviated dB, is a unit that simplifies calculations involving power or voltage gain calculations with cascaded stages or systems in both audio and radio-frequency work. Although the *bel,** abbreviated B, is the base response unit, the smaller and more convenient unit for engineering purposes is the *decibel,* or 1/10 bel.

As was originally defined in terms of the input and output power levels of a system, the dB gain or loss is based on taking the *common logarithm* of the power ratio, so that

$$A_{dB} = 10 \log_{10}\left(\frac{P_{out}}{P_{in}}\right) \qquad (2\text{–}8)$$

The factor of 10 in the right-hand side of Equation 2–8 scales the dimension from bel to decibel. If the output power is greater than the input power, the dB gain is positive. If the output power is less than the input level, the dB gain is negative, which implies a power loss. Thus, gain is associated with positive decibel values, and loss is measured by negative decibel values.

The use of the *common* (Briggsian) logarithm, or *base-10* logarithm, in Equation 2–8 gives us two advantages. First, when several systems are connected in cascade, as shown in Figure 2–1, the gain of the overall system is the product of the power gains of the individual systems, or

*Named for Alexander Graham Bell, the inventor of the telephone.

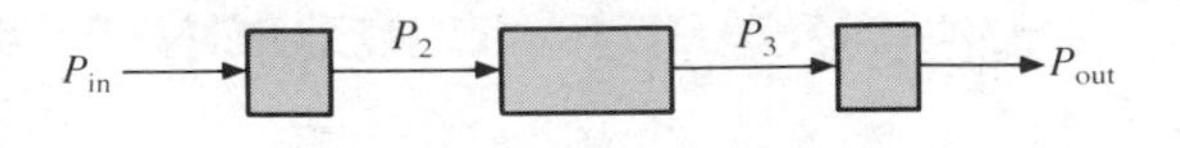

Figure 2–1 Three systems connected in cascade. The power gain of the overall system is the product of the power gains of the three individual systems.

$$A_P = \frac{P_2}{P_{in}} \times \frac{P_3}{P_2} \times \frac{P_{out}}{P_3} \tag{2–9}$$

However, if the individual system power gains are expressed in decibels, the overall gain or loss is obtained by simple addition of the decibel gain of each system rather than by multiplication:

$$\begin{aligned} A_{dB} &= 10 \log_{10}\left(\frac{P_{out}}{P_{in}}\right) \\ &= 10 \log_{10}\left(\frac{P_2}{P_{in}}\right) + 10 \log_{10}\left(\frac{P_3}{P_2}\right) + 10 \log_{10}\left(\frac{P_{out}}{P_3}\right) \end{aligned}$$

The second advantage is that system gains or losses may range from very small to very large numbers. By using a logarithmic scale, this wide range is then compressed to a more convenient range, which generally is less than 100 dB for an entire system.

It should be recalled that Equation 2–8 is the *basic* definition of the decibel. However, electrical signals are more frequently measured in terms of voltage or current than power, especially in audio-frequency applications. Thus, it is more appropriate to express the decibel gain in terms of voltage or current ratios. Using Figure 2–2 as a guide, Equation 2–8 can be rewritten in terms of the input and output voltages as

$$A_{dB} = 20 \log_{10}\left(\frac{V_{out}}{V_{in}}\right) + 10 \log_{10}\left(\frac{R_{in}}{R_L}\right) \tag{2–10}$$

by letting $P = V^2/R$. Resistances R_{in} and R_L are the resistive input and load impedances, respectively, of the system.

For the special case when the load and input impedances are equal, then $\log_{10}(R_{in}/R_L) = 0$ and Equation 2–10 reduces to

$$A_{dB} = 20 \log_{10}\left(\frac{V_{out}}{V_{in}}\right) \tag{2–11}$$

Figure 2–2 Generalized two-port network with input and output parameters.

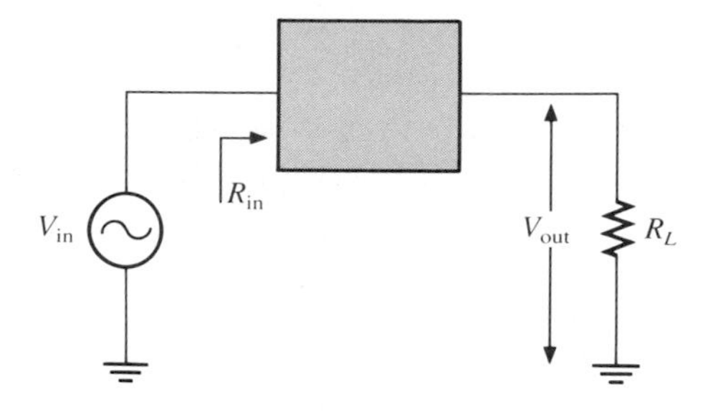

which is frequently used as the expression for voltage gain expressed in decibels. However, it should be noted that Equation 2–11 only applies when R_{in} and R_L are equal. Because of the convenience in using the decibel, it has been a common practice, although incorrect, to express a voltage ratio in terms of Equation 2–11 without regard for the impedance across which the input and output voltages are developed. Consequently, any information about the true power gain is lost. As a general rule, when voltage gain is discussed in terms of decibels, it is generally assumed that Equation 2–11 is being used rather than Equation 2–10.

EXAMPLE 2–1

An amplifier has a voltage gain of 10,000. Find the gain in decibels assuming $R_{in} = R_L$.

Solution:

$$\begin{aligned} A_{dB} &= 20 \log_{10} (10{,}000) \\ &= 20 \times 4 \\ &= +80 \text{ dB} \end{aligned}$$

EXAMPLE 2–2

An amplifier has a voltage gain of 100. The input impedance is 1600 Ω, and the load impedance is 400 Ω. Find the voltage gain in dB.

Solution:

$$\begin{aligned} A_{dB} &= 20 \log_{10}(100) + 10 \log_{10}(4) \\ &= +46 \text{ dB} \end{aligned}$$

EXAMPLE 2–3

A signal path has the following elements in series: a coupler with a 2-dB loss, a transmission line with a 5-dB loss, and an amplifier with a 10-dB gain. Find the total path gain or loss in dB.

Solution:

$$\begin{aligned} A_{dB}(\text{total}) &= -2 \text{ dB} + (-5 \text{ dB}) + (+10 \text{ dB}) \\ &= +3 \text{ dB (gain)} \end{aligned}$$

Since the decibel is not an absolute level, it may be used only to state the ratio of a signal to a reference level. For this reason, decibel units with specific reference levels have been established. Commonly used references are listed in Table 2–4.

Of these decibel references, two are worthy of special attention: the dBm and VU. The VU, or volume unit, reference equals 1 mW across a 600-Ω load, so that

$$\text{VU} = 10 \log_{10}\left(\frac{P_{out}}{1 \text{ mW}}\right) \tag{2–12}$$

which can be simplified to

$$\text{VU} = 30 + 10 \log_{10}(P_{out}) \tag{2–13}$$

From Table 2–4, both the dBm (m = *milliwatt*) and the VU have the same definition. The dBm scale is generally used when the signal is a sinewave,

Table 2–4 Decibel References

Unit	Reference
dBk	1 killowatt
dBm	1 milliwatt, 600 Ω (sine wave, typically 1 kHz)
dBv	1 volt
dBw	1 watt
dBvg	voltage gain
dBrap	10^{-16} watts acoustical power
VU	1 milliwatt, 600 Ω (complex waveforms)

normally 1 kHz, whereas the VU is used in conjunction with "complex" audio signals, such as speech or music. In either case, a 0-dBm or 0-VU level is equal to a voltage of 0.775 V across a 600-Ω load.

Neper

Like the decibel the neper,* abbreviated Np, is used to measure differences in power levels. Unlike the decibel, the neper is based on Napierian (natural), or *base-e* ($e = 2.718281$) logarithms. By definition, the gain expressed in nepers is

$$A_{\mathrm{Np}} = 0.5 \log_e\left(\frac{P_{\mathrm{out}}}{P_{\mathrm{in}}}\right) \tag{2–14}$$

In addition, the gain or loss of a system based on voltage ratios in terms of nepers is

$$A_{\mathrm{Np}} = \log_e\left(\frac{V_{\mathrm{out}}}{V_{\mathrm{in}}}\right) \tag{2–15}$$

In either case, 1 Np = 8.686 dB, or conversely, 1 dB = 0.1151 Np. It should be noted that most calculators generally use Ln to mean $\log_e$.

2–5 CONVERSION OF UNITS

Very often, quantities expressed in one type of unit system must be converted to another type. This is particularly true in the United States and other countries that use the English (foot-pound-second) system. As a general rule, however, SI units in all measurements are preferred so that all scientists and engineers are consistent.

Appendix C lists the conversion factors for the most commonly encountered units. Although a *multiplier factor* is given in each case, it is perhaps more appropriate to perform the conversion in terms of an equation. The desired units are written on the left-hand side of the equation, while the given units and the conversion factors are written on the right-hand side. In this manner, units are treated like algebraic symbols. To perform any conversion successfully, you should routinely include the units of *all* physical quantities in any calculation. Doing this makes it easy to determine if the result is *dimensionally* correct, as illustrated by the following example.

---

*Named for the Scottish mathematician John Napier, who discovered the logarithm.

EXAMPLE 2–4

Convert 3.62 ergs to joules.

Solution:

Since 1 erg = 10^{-7} joule, the conversion from ergs to joules then becomes

$$3.62 \text{ ergs} = (3.62 \cancel{\text{ergs}})\left(\frac{10^{-7}\text{ joule}}{\cancel{\text{erg}}}\right)$$
$$= 3.62 \times 10^{-7} \text{ joules}$$

In Example 2–4, note that an algebraic equation was written such that the right-hand side was then multiplied by the appropriate conversion factor relating ergs to joules. Since the units are treated just like algebraic symbols, the unit, erg, cancels, leaving only the joule unit. Many times, however, it is necessary to convert more than one unit in the same equation, as shown in Example 2–5.

EXAMPLE 2–5

The speed of an automobile is 48 miles per hour. Convert this to an equivalent number of meters per second.

Solution:

$$48 \text{ mi/hr} = \frac{48 \cancel{\text{mi}}}{\cancel{\text{hr}}} \times \frac{1 \cancel{\text{hr}}}{3600 \text{ s}} \times \frac{5280 \cancel{\text{ft}}}{1 \cancel{\text{mi}}} \times \frac{1 \text{ m}}{3.048 \cancel{\text{ft}}}$$
$$= 21.46 \text{ m/s}$$

2–6 SUMMARY

In this chapter, a distinction was made between fundamental and derived quantities, as well as quantity and unit symbols. The six fundamental quantities of length, mass, time, electric current, temperature, and light intensity are internationally defined in what has now become the International System of units. From these, commonly encountered units of volt, ohm, farad, henry, and hertz were defined. Logarithmic response was defined in terms of the decibel and neper.

2–7 GLOSSARY

derived quantity A physical quantity defined in terms of fundamental quantities.

English system A system of measure based on the foot, pound, and second, which is predominantly used in the United States and Great Britain.

fundamental quantity Arbitrarily selected quantities by which all other quantities are derived. The six fundamental quantities are length, mass, time, electric current, temperature, and light intensity.

International System (SI) The official international system of units since 1960. The units of the six fundamental quantities are meter (length), kilogram (mass), second (time), kelvin (temperature), ampere (electric current), and candela (light intensity).

metric system A measurement system based on the meter, gram, and second. The metric system is used by the majority of the world's scientists and engineers.

multiplier prefix A prefix that designates either a greater or smaller unit than the original. The multipliers are multiples or submultiples of 10.

quantity symbol A single Roman or Greek letter with optional subscripts used to identify a measured variable.

unit A defined sample of a quantity.

unit symbol One or two letters used as an abbreviation for the name of a unit of measure. Also called a dimensional symbol.

2–8 PROBLEMS

1. If 100 mV is applied to an amplifier having an input impedance of 75 Ω and whose output is connected to a 300-Ω load, determine the output voltage if the overall gain is to be 12 dB.
2. A system with an input impedance of 500 Ω and an input voltage level of 10 mV produces an output voltage of 200 mV across a 100-Ω load. Determine the gain or loss in terms of
   (a) decibels
   (b) nepers
3. An amplifier has an output voltage of 1.5 V across a 600-Ω load. Determine the output level in VU.
4. Convert
   (a) 560 pF to μF
   (b) 22 MΩ to kΩ
   (c) 4300 μV to mV
   (d) 18,125 kHz to MHz
   (e) 39 mH to μH
   (f) 0.028 mA to nA
   (g) 330 μS to S
   (h) 4.7 Np to dB
   (i) 18 °C to °F
5. The speed of light in a vacuum is 3×10^8 m/s. Convert this to ft/hr.
6. The average current in a conductor flowing for 2 minutes and 15 seconds is 25 mA. Calculate the electric charge required in coulombs.
7. A magnetic flux of 52 lines was measured over a 100-ft^2 area. Determine the flux density in maxwells/meter^2 (Mx/m^2).
8. If 1000 joules of energy is measured over exactly a 2-day period, determine the power in milliwatts.
9. The capacitance in farads of two identical parallel plates is given by

$$C = \varepsilon_o \frac{A}{d}$$

where the permittivity (ε_o) of the air between the plates is 8.85×10^{-12} F/m, A is the area of one plate, and d is the distance between them. For the dimensions shown in Figure 2–3, determine the capacitance in picofarads.

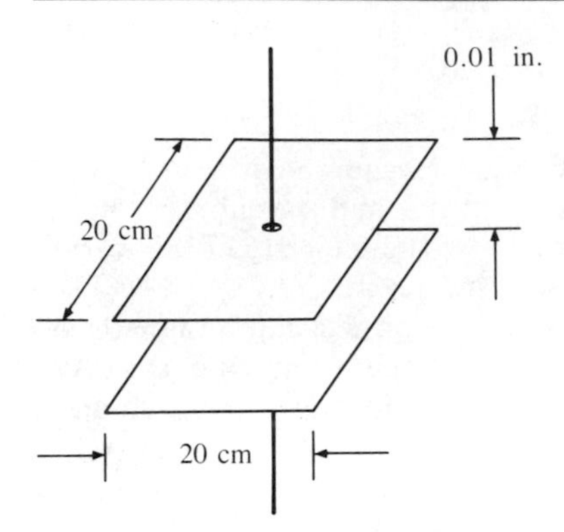

Figure 2–3 Diagram for Problem 9.

10. The resistance of ohms of a conductor with uniform cross-sectional area, A, and length, L, is given by

$$R = \rho \frac{L}{A}$$

where ρ is the resistivity of the conductor material in CM-Ω/ft (CM = circular mil, 0.001 in. = 1 CM, $\pi/4$ mil = 1 CM). Determine the resistance of a 50-m length of copper wire (ρ = 10.37 CM-Ω/ft) having a diameter of 0.02 in.

11. Show that the units of the R-C time constant (ohm-farads) can be expressed in units of time.

12. Show that the units of inductance divided by resistance (henry/ohm) are equivalent to units of time.

13. Show that $1/2\pi\sqrt{LC}$ dimensionally equals units of frequency.

14. Express the value of the Bohr magnetron constant, $\mu_o = 9.2731 \times 10^{-20}$ erg/gauss (erg/G), in terms of joule/tesla (J/T).

3

STANDARDS

3–1 INSTRUCTIONAL OBJECTIVES

After completing this chapter you will be able to

- Describe the construction of and the differences between the saturated and unsaturated Weston standard cell.
- Describe how voltage standards are based upon the Josephson effect.
- Describe how a temperature-controlled zener diode is used as a voltage standard.
- Appreciate the accuracy of and the methods used by the National Bureau of Standards to maintain the following standards:

  time and frequency
  resistance

current
capacitance
inductance

3–2 INTRODUCTION

A **standard** is a physical device having stable, precisely defined characteristics, that is used as a reference for a unit of measurement. For the most part, internationally accepted standards have been established for all electrical and magnetic units. International agreement on electrical units and standards begins with the International Advisory Committee on Electricity, which, in turn, passes its recommendations on to the International Committee on Weights and Measures. The General Conference is then convened every few years.

The International Bureau of Weights and Measures, located in Sevres, France, serves as the laboratory facility for this international system that has evolved over nearly a 100-year period.

A number of national standards laboratories such as the National Bureau of Standards (NBS, United States); the National Physics Laboratory (England); and the Physikalisch-Technische Bundesanstalt (West Germany) cooperate in this international program.

The mission of the NBS in the United States is to develop and maintain the national standards of measure and furnish the essential services that lead to accurate and uniform physical measurements throughout the nation. It has evolved a hierarchy of standards that will show *traceability* to a common reference—the NBS. The general hierarchy of references in use in the United States is as follows

Echelon I

1. International Standards
2. Primary Standards (National Standards)
3. Secondary Standards (NBS reference standards)
4. Working Standards (NBS calibration facilities)

Echelon II

1. Reference Standards—secondary standards maintained by private laboratories and industries.
2. Working Standards—standards of lower order used to calibrate and check general laboratory instruments.

Echelon III

General instrumentation for production, maintenance, and field test.

3–3 TIME AND FREQUENCY

The unit of time now has two definitions depending upon the user's needs. In what is known as *ephemeris time,* the basic unit is a second that is 1/31,556,925.9747 of the tropical year 1900 January 0 at 12 hrs ET (December 31, 1899 at 12 Noon). This particular and unusual time scale is known as *UT1* and is inferred from astronomical observations. Despite its apparent awkwardness, it nevertheless has attained a universal acceptance and is used in applications such as precise navigation, celestial mechanics, and satellite tracking, which must be referred to the slightly varying speed of the Earth's rotation. The frequently used, internationally accepted time scale known as *UTC,* or *Coordinated Universal Time,* has a second that is based on the radiation of the cesium-133 atom. The *UTC second* was defined by the 13th General Conference of Weights and Measures in October 1967, and is used extensively in radio, instrumentation, and electronics work.

Primary Frequency Standards

The current state-of-the-art in UTC frequency standards is the atomic or cesium beam frequency standard that provides an accuracy of 2 parts in 10^{13} in the laboratory, and 7 parts in 10^{12} in commercial equipment. Such a commercial frequency standard used as the frequency source for a time clock would give an accuracy equivalent to the gain or loss of one second in 4530 years.

Standard Time and Frequency Broadcasts

Both the primary time and frequency standards for the United States are maintained by the NBS. In addition, the NBS maintains two radio transmitting stations, WWV at Fort Collins, Colorado, and WWVH near Kekaha, Kauai (Hawaii), for broadcasting standard time and radio frequency information. These stations provide

Time announcements
Standard time intervals
Standard frequencies
Geophysical alerts
Marine storm warnings
Omega Navigation System status reports
UT1 time corrections
Binary Coded Decimal (BCD) time code

As summarized in Table 3–1, WWV broadcasts on the frequencies of 2.5, 5, 10, 15, and 20 MHz, while WWVH broadcasts on 2.5, 5, 10, and 15 MHz. Atomic frequency standards are used to control the transmitter frequencies to

Table 3–1 High-Frequency NBS Time and Frequency Services

Service	WWV	WWVH
Location	Ft. Collins, Colorado	Kehaha, Kauai (Hawaii)
Transmitting Frequencies		
2.5 (MHz)	X	X
5	X	X
10	X	X
15	X	X
20	X	
Standard Audio Frequencies		
440 (Hz)	X	X
500	X	X
600	X	X
1000	X	
1200		X
1500	X	X
Time Intervals	X	X
Time Signals*	man's voice	woman's voice
Time Code	X	X
UT1 Corrections	X	X
Official Announcements	X	X
Geophysical Alerts	X	
Marine Storm Warnings	X	X

*WWV transmits its signal 15 seconds before the minute; WWVH transmits 7.5 seconds before the minute.

an accuracy of one part in 100 billion at all times. Fluctuations in the carrier frequency as received by the user, however, may be greater than this as a result of Doppler effect, diurnal shifts, or other results of changes in the propagating path. At regular intervals, as shown in Figure 3–1, the carrier frequency is modulated with the standard audio frequencies of 440, 500, and 600 Hz. Voice announcements of the time, in English, are given every minute. One-second time markers, audible *ticks* are transmitted (except on the the 29th and 59th second of each minute) throughout all programs.

A binary coded decimal (BCD) time code is transmitted continuously by both stations on a 100-Hz subcarrier. This code is transmitted serially, and may be used as a standard time base for scientific observations made simultaneously at different locations. It is also used to keep specially equipped time clocks automatically synchronized with WWV or WWVH. The UTC time is occasionally adjusted by adding what are known as *leap seconds* so that the maximum difference between UTC and UT1 never exceeds 0.9 seconds. The broadcasts of WWV may also be heard by telephone by dialing (303) 499-7111,

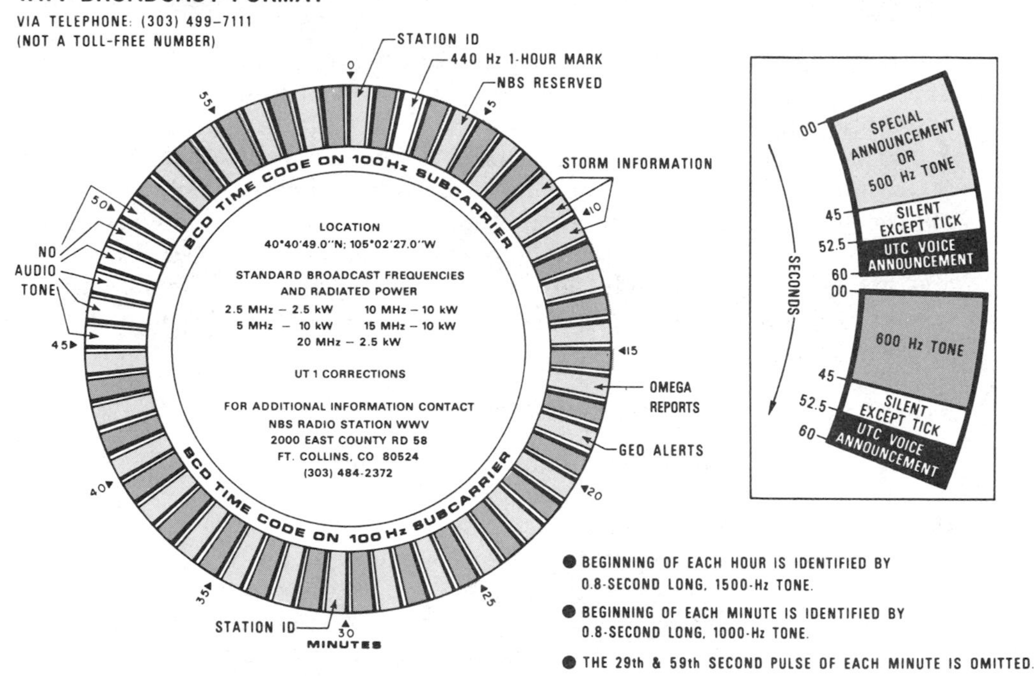

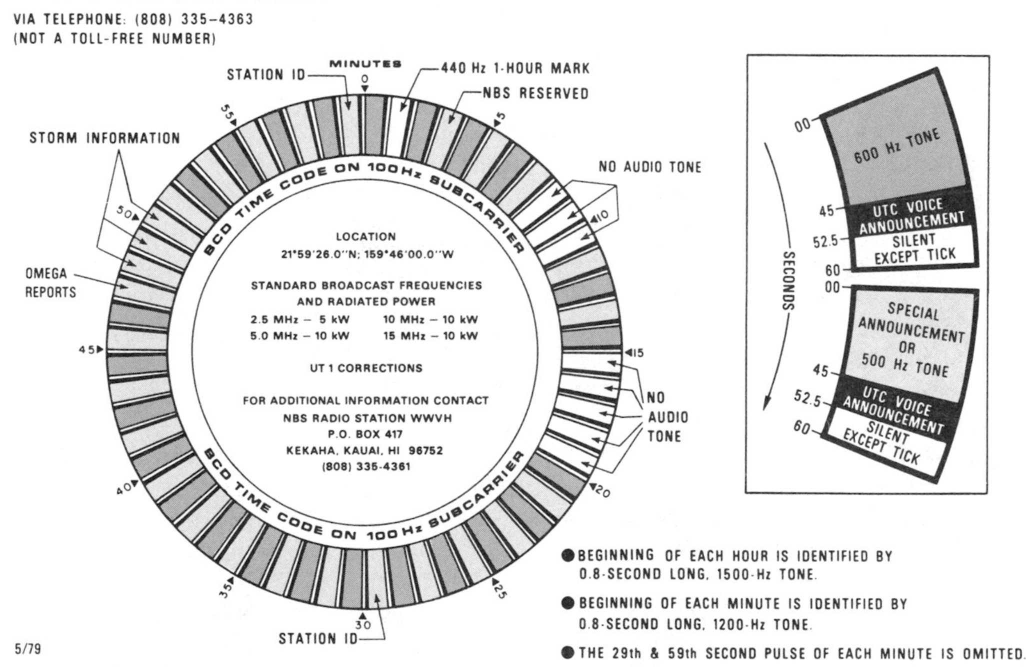

Figure 3–1 WWV broadcast format. Courtesy National Bureau of Standards.

which is located in Boulder, Colorado; or (808) 335-4363 for WWVH in Hawaii. Each call is limited to three minutes. The station identification and time-of-day announcements are prerecorded.*

The Canadian time-signal station, CHU, transmits on 3.330 MHz, 7.335 MHz, and 14.670 MHz. Voice announcements of the time are made each minute in French and English. Each second is marked by a time tick similar to those of WWV. CHU emanates from the Dominion Observatory, Ottawa (Ontario) Canada, and is particularly useful on the east coast of the United States. Throughout the world, there are many other stations that broadcast similar data, as listed in Table 3–2.

Frequency Calibration Using Network Television

Extremely stable rubidium oscillators are used to generate the 3.58-MHz color subcarrier frequency that is transmitted with all color programs originated by major networks in the United States. This color signal may be used as a **transfer standard** for frequency. NBS monitors these network signals and publishes the difference between the network oscillators and the NBS frequency standard in a monthly publication: *NBS Time and Frequency Services Bulletin*. With the proper equipment, it is possible to achieve accuracies of one part in 100 billion.

*Complete information on the services can be found in NBS Special Publication 432, *NBS Frequency and Time Dissemination Services,* available for a small charge from the Superintendent of Documents, U.S. Government Printing Office, Washington, D.C. 20402.

Table 3–2 Worldwide Time and Frequency Broadcasting Stations (High Frequency)

Call Sign	Location Country/City	Carrier Frequency (MHz)	Modulation Frequency (Hz)
ATA	India (New Delhi)	10	1, 1000
CHU	Canada (Ottawa)	3.33, 7.335, 14.671	1
HBN	Switzerland (Neuchatel)	5	1
JJY	Japan (Tokyo)	2.5, 5, 10, 15	1, 440, 1000
LOL	Argentina (Buenos Aires)	5, 10, 15	1, 1000
MSF	England (Rugby)	2.5, 5, 10	1
OMA	Czechoslovakia (Prague)	2.5	1, 1000
RWM/RES	USSR (Moscow)	5, 10, 15	1, 1000
VNG	Australia (Lyndhurst)	5.425, 7.515, 12.005	1, 1000
ZUO	South Africa		
	(Olifantsfontein)	5	1
	(Johannesburg)	10	1

3–4 VOLTAGE STANDARDS

Local primary voltage and resistance standards provide the backbone for the calibration of direct current (DC) and low-frequency alternating current (AC). This calibration makes it possible to achieve accuracy, precision, and interchangeability. The accuracy of local primary standards is ensured by periodically submitting them to the NBS for comparison with their working standards. From the NBS standards, accurate transfer of certified values may be made to calibrators and portable reference standards.

Until recently, the Weston cell was normally considered the standard for voltage. The Weston standard cell, as shown in Figure 3–2, has a positive electrode of mercury, a negative electrode of cadmium-mercury amalgam (10% cadmium), and a cadmium sulfate electrolyte. There are two forms of the cell, saturated and unsaturated. The saturated cell is more stable, but it has a higher temperature coefficient than the unsaturated cell. Temperature must be carefully controlled when using the saturated cell.

In the saturated cell, the electrolyte contains a surplus of cadmium sulfate, and crystals of cadmium sulfate cover the electrodes. A paste of mercurous sulfate and cadmium sulfate cover the mercury electrode as a depolarizer. The unsaturated cell contains cadmium sulfate crystals only at temperatures lower than room temperature. The unsaturated cell is more portable and better suited as a working reference. The saturated Weston cell provides a voltage of approximately 1.01830 volts, and yields a voltage that is within 1 μV of its stated value at 20 °C. The saturated cell has a temperature coefficient of about

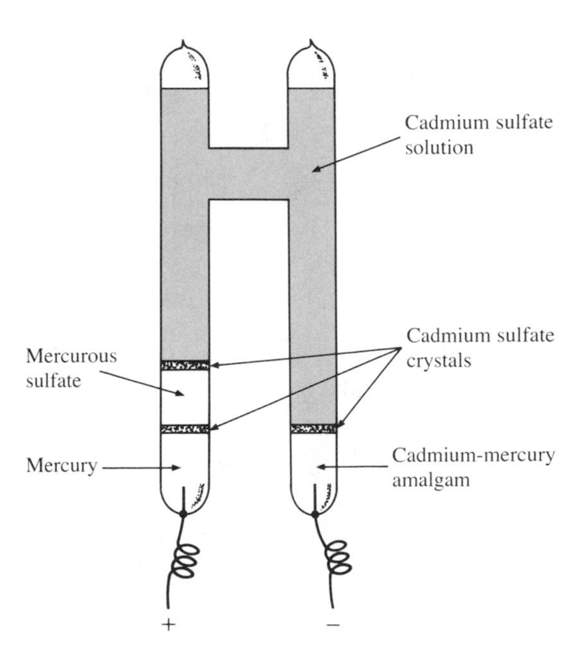

Figure 3–2 Construction of a Weston standard cell.

$-40\ \mu V/°C$; the unsaturated cell exhibits no voltage change with small temperature changes near 20 °C, but its voltage varies with time at about 3 μV per month.

Weston cells must be used only as a reference. They must never be permitted to supply any appreciable current, because current in excess of about 100 μA may permanently change the terminal voltage. Weston cells are also subject to damage by ambient temperatures of less than 4 °C or more than 40 °C. The NBS maintained 40 temperature-controlled, saturated Weston cells as the national primary reference standard and ten saturated cells as the NBS working standards. The primary standards were compared against each other, and the working standards were compared against the primary standards.

Since 1972, the Josephson junction has been used by NBS as the national standard for the United States. It is used to calibrate standard cells to an accuracy of a few parts in 10^8. Resonant, thin-film tunnel junctions of lead with lead oxide insulation barriers on a glass substrate are mounted in a superinsulated helium Dewar system held at a temperature near absolute zero. As illustrated by the block diagram of Figure 3–3, the junction is irradiated at approximately 9 GHz using a microwave source. The microwave source is phase-locked to a high stability, 100-MHz temperature-controlled crystal oscillator. The Josephson junction produces a voltage given by

$$V_J = \frac{nhf}{2e} \times 10^9 \tag{3–1}$$

where

f = microwave frequency, GHz

h = Planck's constant, 6.626×10^{-34} J/Hz

e = charge on an electron, 1.602×10^{-19} C

n = an integer

The advantage here is that voltage becomes solely a function of frequency, and that frequency can be measured to a higher degree of accuracy than any other physical quantity. Such a junction operated on the 250th step (i.e., $n = 250$) will produce a voltage of

$$V_J = \frac{(250)(6.626 \times 10^{-34}\ \text{J/Hz})(9\ \text{GHz})}{(2)(1.602 \times 10^{-19}\ \text{C})}$$
$$= 4.66\ \text{mV}$$

Note that the units of joule/coulomb in Equation 3–1 are dimensionally equal to volts.

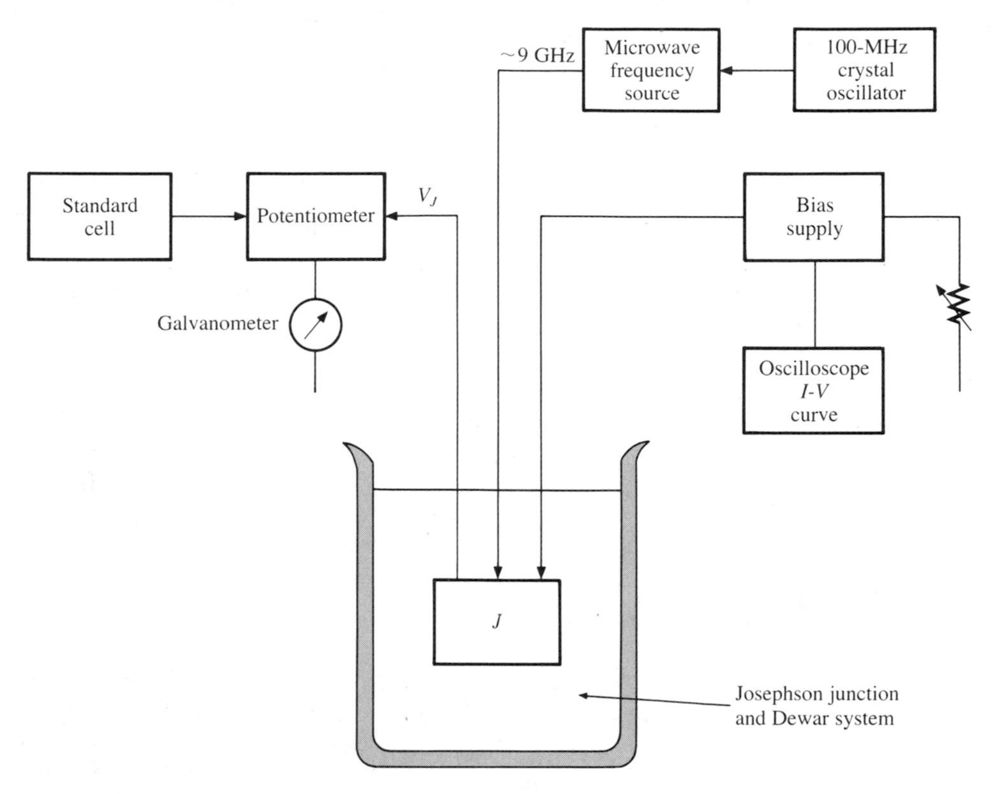

Figure 3–3 Block diagram of Josephson effect-based voltage standard.

The desired value for n is selected by adjusting the bias supply current as the current-voltage characteristics are viewed on an oscilloscope. A potentiometer consisting of a Hamon network (a series-parallel resistor network) and a Kelvin-Varley resistive decade voltage divider is used to compare the junction voltage with the standard cell being calibrated.

In recent years, selected temperature-controlled zener diodes have been used as the reference voltage for local primary reference standards. These higher quality, solid-state standards offer better resolution, lower noise, and simpler operation than standard cells.

The basic voltage-reference circuit of Figure 3–4 is constructed by placing a high-quality zener diode in a temperature-controlled oven. For small changes in source voltage, V

$$\frac{\Delta V_{\mathrm{REF}}}{\Delta V} = \frac{1}{1 + \dfrac{R_1}{R_Z} + \dfrac{R_1}{R_2}} \tag{3–2}$$

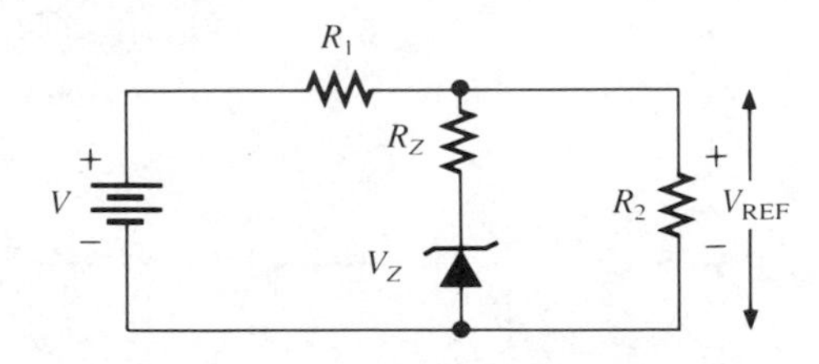

Figure 3–4 Zener diode voltage reference circuit.

If R_1 is made large with respect to the dynamic resistance of the diode (R_Z), then the variation of the voltage reference (ΔV_{REF}) across R_2 will be substantially less than the variation of the source voltage, ΔV.

Zener diodes are as accurate as a secondary standard cell. If properly maintained, the accuracy of the diodes may approach that of a primary cell. The accuracy of these units is directly traceable to the NBS as they come from the manufacturer. Precision voltage dividers are usually included to provide several output voltages.

3–5 RESISTANCE STANDARDS

The NBS maintains ten 1-Ω resistors as the national standard. They are maintained to an accuracy of 0.08 ppm and are compared against each other and against international reference standards to ensure that their values are constant. Additional standard resistors are used as working standards by the NBS.

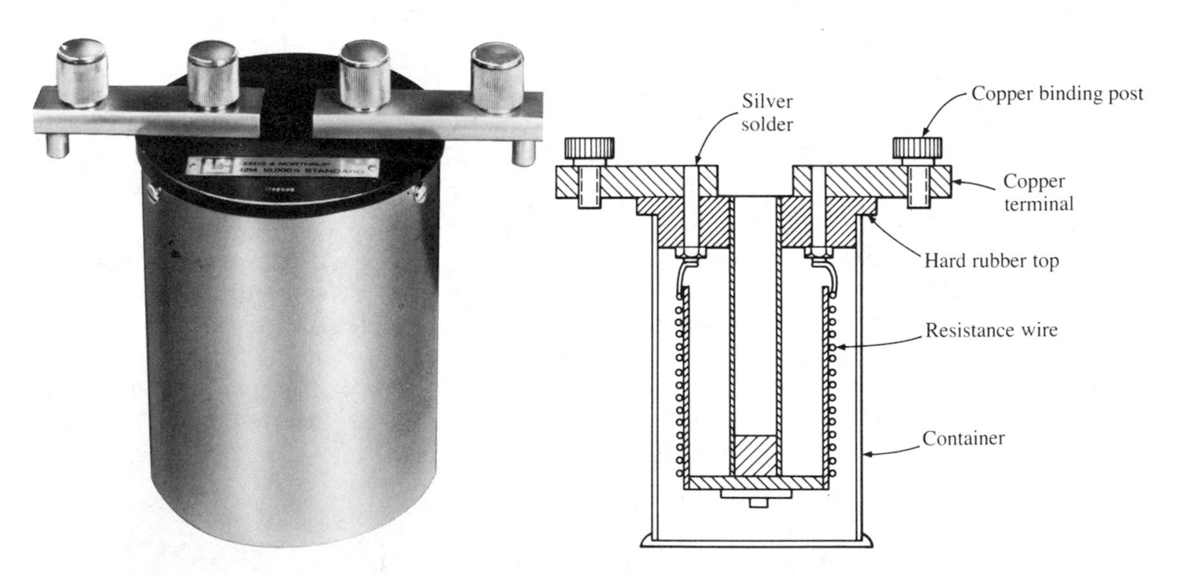

Figure 3–5 Standard resistor. Courtesy Leeds and Northrup.

These are compared to the national standards with a special circuit, known as a *Wenner bridge*.

The standard resistor shown in Figure 3–5 is constructed of a coil of wire that has high resistivity and a very low temperature coefficient of resistance. The wire is fully annealed to remove residual strains and sealed from contact with the air to protect it from humidity and oxidation. *Manganin* is a particularly good material since it has a low temperature coefficient of resistance and a low thermoelectric EMF at junctions with copper. Another material used for high resistance standards is *evanohm*. As summarized in Table 3–3, evanohm has properties similar to manganin but has a higher resistivity.

Table 3–3 Properties of Resistance Wire Standards

Material	Composition	Resistivity Ω-cm	Temperature Coefficient, ppm/°C	Thermal EMF against copper μV/°C
manganin	Ni 4%, Cu 84%, Mn 12%	48	15	3.0
evanohm	Ni 74.5%. Cr 20%, balance Al and Cu	133	20	2.5

3–6 CURRENT STANDARDS

A current standard usually consists of a precision four-terminal resistor (Figure 3–6). One pair of terminals is used to connect a precision potentiometer across the resistor to measure voltage. The other pair of terminals is used to pass the current through the resistor. A standard current is thus derived from a standard voltage and a known resistance using Ohm's law.

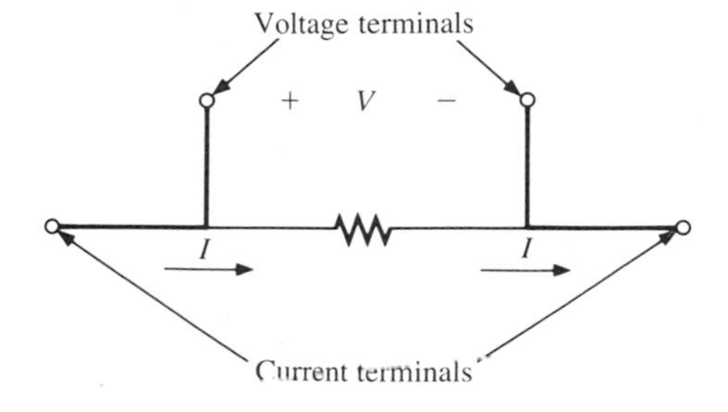

Figure 3–6 Current standard using a 4-terminal resistor.

3–7 CAPACITANCE STANDARDS

The National Bureau of Standards has constructed standard capacitors with values known to 2 parts in 10^8. These follow a design by Thompson and Lampard.* They use four equal, closely spaced cylindrical rods. The NBS primary standard is a 1-pF capacitor that is compared with 10-pF working standards using a special transformer ratio-arm capacitance bridge.

Most commercial standard capacitors with values up to 1000 pF use multiple, parallel-plate construction with dry-air or nitrogen as a dielectric. A low-expansion alloy such as *Invar* is used to minimize the effects of temperature, and mechanical strains are minimized by using fully annealed materials.

With two-terminal construction, capacitors are subjected to the effects of surrounding objects on the value of capacitance. Three-terminal, shielded construction provides a definite value of capacitance that is independent of external objects and fields. A good working standard might be calibrated to an accuracy of ± 5 ppm and have a drift of less than 20 ppm/year.

3–8 INDUCTANCE STANDARDS

The standard unit of inductance is based on the ohm and the farad. The NBS determines inductance by using standard capacitors and standard resistors as part of a Maxwell-Wien bridge.

Commercial standard inductors consist of a single layer solenoid wound on dimensionally stable forms of fused silica (silicon dioxide). Working standards often use multilayer windings on ceramic, marble, or bakelite forms. Toroidal cores are also popular because they are nearly immune to the effects of external magnetic fields. The effective inductance is a function of the applied frequency due to the effects of winding capacitance, and therefore the calibration frequency must be known. Standard inductors are stable to about $\pm 0.01\%$ per year, and are available in a range from 50 μH to 10 H.

3–9 SUMMARY

This chapter introduces the reader to the fundamentals of the who, what, where, when, and why of standards. Without standards, no one could make a single measurement with any degree of accuracy, nor could a measurement be compared with a similar measurement taken in another country. The National

*Lampard, D. G. "A New Theorem in Electrostatics with Application to Calculate Standards of Capacitance." *Proc. Indust. Elect. Engg* (London), 1957, Vol. 104, part C, pp. 271–80. Cutkosky, R. D. "New NBS Measurements of the Absolute Farad and Ohm." *IEEE Trans. on Inst. and Meas.*, December 1974, Vol. IM-23(4), pp. 305–9.

Bureau of Standards maintains the standards for the United States. Those standards that are primarily used for electrical measurements include voltage, resistance, current, capacitance, inductance, time, and frequency.

3–10 GLOSSARY

international standard A standard defined by international agreement and periodically checked by absolute measurements in terms of fundamental units.

primary standard A basic fundamental or derived unit standard maintained by a country's national standards laboratory. These are independently calibrated by absolute measurement by various worldwide national laboratories and compared against each other. In the United States, this is done by the National Bureau of Standards (NBS).

secondary standard A reference standard used in industrial laboratories. It is periodically verified and calibrated against the national laboratory's primary standard.

transfer standard An interlaboratory standard.

UTC Abbreviation for *coordinated universal time,* an internationally accepted time standard based on the cesium-133 atom.

UT1 Mean solar rotation derived from astronomical observations but corrected for period variations.

Weston cell A chemical cell formerly used by the National Bureau of Standards as the primary voltage standard. It has a voltage of 1.01830 V at 20 °C.

working standard A day-to-day standard used to check and calibrate laboratory instruments for accuracy and performance or to perform comparison measurements.

4

GROUNDING, SHIELDING, AND NOISE

4–1 INSTRUCTIONAL OBJECTIVES

At the completion of this chapter, you will be able to

- Give examples of natural and man-made noise sources.
- Discuss the differences between white noise, pink noise, and atmospheric noise and give examples of each.
- Explain thermal noise, shot noise, flicker noise, and burst noise in terms of frequency spectrum and amplitude.
- Calculate S/N ratio, noise figure, and SINAD ratio from noise measurements.
- Describe how interference arises from magnetic (inductive) coupling, electrical (capacitive) coupling, and electromagnetic sources.
- Describe how a 3-wire AC power line is wired and how a ground fault can occur.
- Explain how a GFI works and how 3-wire receptacle analyzers work.
- Describe the physiological effects of contact with an electrical current.
- Explain how shielding and filtering are used to substantially reduce the influence of interference and noise.
- Describe crosstalk and how it occurs.

4–2 INTRODUCTION

Measurements of voltage levels less than 100 mV are frequently required. Such low-level signals rarely escape being affected by sources of externally generated interference or internally generated noise. These unwanted signals are amplified along with the measured signals. If the level of the interference and noise is large with respect to the signal of interest, then the interference tends either to cover up or distort the signal being measured. This chapter discusses the sources of noise and interference and the steps generally used to eliminate or minimize their effects.

4–3 NOISE

Any undesired electrical signal that is present in addition to the voltage or current being measured or processed is termed **noise.** Noise originates either in circuit components or in the atmosphere. Both types can be classified as either *natural* or *man-made.* Natural noise is caused by noise-producing phenomena inherent in the component or the atmosphere. On the other hand, man-made noise has its origins from power supplies, and the electrical discharges from motors, X-ray machines, electrical welders, ignition systems, or fluorescent lamps. No electrical signal, however, can be free of noise because small, random voltages and currents occur naturally in every electrical component whose temperature is above absolute zero.

The types of noise discussed in the following sections can be subdivided into the following categories:

- **White noise:** thermal and shot noise.
- **Pink noise:** flicker and burst noise.
- **Atmospheric noise:** lightning.

Just as white light contains all colors of the visible spectrum, **white noise** has a frequency spectrum composed of all frequencies with amplitudes that are "flat" with frequency (Figure 4–1). The major examples of white noise are *thermal noise* and *shot noise.*

The term *pink noise* is probably derived analogously because pink color contains the lower frequencies of the visible spectrum, and pink noise contains frequency components that have higher amplitudes at lower frequencies. Examples of pink noise are *flicker noise* and *burst noise. Atmospheric noise* is caused mainly by lightning, which causes electromagnetic radiation that travels over great distances. Like pink noise, its amplitude increases as frequency decreases. It is not much of a problem at frequencies greater than 20 MHz.

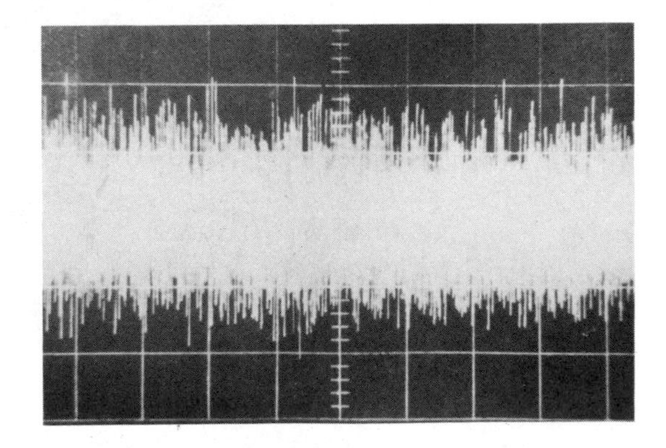

Figure 4–1 Oscilloscope display showing white noise with essentially a constant amplitude.

Thermal Noise

Except at absolute zero (−273 °C), where all molecular motion stops, electrons in any conductor are in random motion. From the basic definitions of electricity, the motion of an electron defines a current traveling in a conductor, which can be thought of as a resistance in series with a noise generator producing **thermal noise** (Figure 4–2). Thermal noise arises from the motion of electrons in a conductor and increases with increasing temperature. Since it is generally associated with carbon and wire-wound resistors, thermal noise is frequently called **resistance noise,** in addition to the terms **Johnson,** or **Nyquist** noise.

In terms of a generated RMS voltage, the level of thermal noise is found from

$$v_{\text{RMS}} = \sqrt{4kTR\,\Delta f} \tag{4–1}$$

where

k = Boltzmann's constant, 1.38×10^{-23} W/K-Hz*
T = absolute temperature (Kelvin, K)
R = resistance (Ω)
Δf = effective noise bandwidth (Hz)

The absolute temperature in Kelvin can be derived from either the Celsius or Fahrenheit temperature by

$$\text{K} = 273 + {}^\circ\text{C} \tag{4–2a}$$

$$= 273 + \frac{5}{9}({}^\circ\text{F} - 32) \tag{4–2b}$$

*Boltzmann's constant is frequently expressed as 1.38×10^{23} J/K, noting that 1 joule equals 1 watt-second, or 1 watt/hertz.

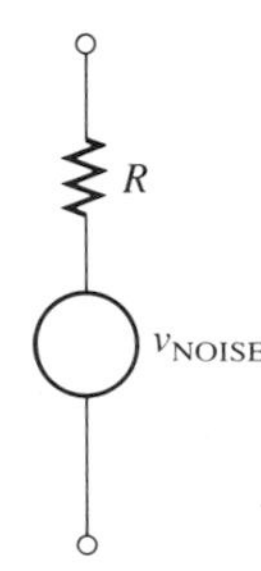

Figure 4–2 Equivalent circuit of a thermal noise source.

From Equation 4–1, the power dissipated in a resistance R is

$$P_N = 4kT\,\Delta f \tag{4–3}$$

The effective noise bandwidth (Δf) in Equations 4–1 and 4–3 should not be confused with the 3-dB bandwidth of a given system. Figure 4–3 illustrates the frequency response of a given circuit, such as a tuned amplifier. The 3-dB bandwidth is the frequency range between the points where the response drops to 3 dB less than its maximum value. In terms of voltage gain, the response decreases to 0.707 of its peak voltage gain; in terms of power, it decreases to 0.5 of the peak power gain:

$$\text{voltage: } -3 \text{ dB} = 20 \log_{10} (0.707)$$

$$\text{power: } -3 \text{ dB} = 10 \log_{10} (0.5)$$

The **effective noise bandwidth** is defined as the width of a rectangle with the same height and area as the frequency response curve. The width of the dashed rectangle in Figure 4–4 shows the effective noise bandwidth, which is greater than the 3-dB bandwidth. Although the computation of the area underneath the frequency response curve is beyond the scope of this text, the effective noise bandwidth for most cases can be approximated by

$$\Delta f \simeq \frac{\pi}{2} \Delta f_{3\text{ dB}} \tag{4–4}$$

The effective noise bandwidth is then approximately 57% greater than the 3-dB bandwidth.

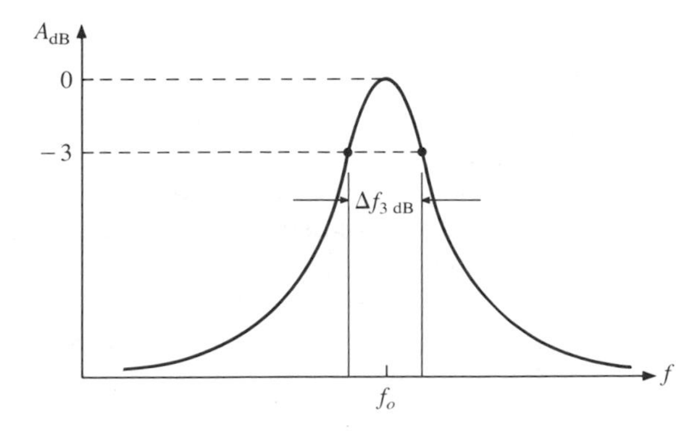

Figure 4–3 Frequency response of a tuned amplifier.

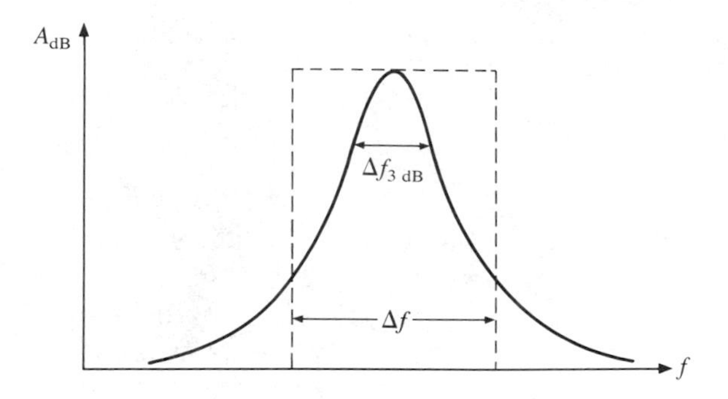

Figure 4–4 Frequency response of a tuned amplifier showing the noise bandwidth.

EXAMPLE 4–1

For a 1-MΩ resistor at room temperature (18 °C), determine the thermal noise voltage and power generated, assuming an effective noise bandwidth of 10 MHz.

Solution:

From Equation 4–1, the thermal noise RMS voltage is

$$v_{RMS} = \sqrt{(4)\,(1.38 \times 10^{-23}\ \text{W/K-Hz})\,(291\ \text{K})\,(1\ \text{M}\Omega)\,(10\ \text{MHz})}$$

$$= 0.4\ \text{mV RMS}$$

The generated noise power is computed from $P = v^2/R$, so that

$$P_N = \frac{(0.4\ \text{mV})^2}{10\ \text{k}\Omega}$$

$$= 16\ \text{pW}$$

From the above example, it should be pointed out that the magnitude of the generated noise, whether in terms of voltage or power, is *independent of the actual current flowing through the resistor,* and regardless of whether the current is DC or AC. From Equation 4–1, thermal noise is dependent only on the resistance, effective noise bandwidth, and absolute temperature. For this reason, you should try to keep impedance levels as low as possible to reduce the magnitude of the noise. For example, Figure 4–5*a* shows an inverting amplifier with a closed-loop gain of 10. The input impedance is approximately 1 MΩ. In Figure 4–5*b,* the gain of the inverting amplifier is also 10, but with an input impedance of 100 kΩ. If both amplifiers are operated under identical conditions, the 1-MΩ feedback resistor of the amplifier of Figure 4–5*b* will contribute less noise than the 10-MΩ feedback resistor of the amplifier of Figure 4–5*a*.

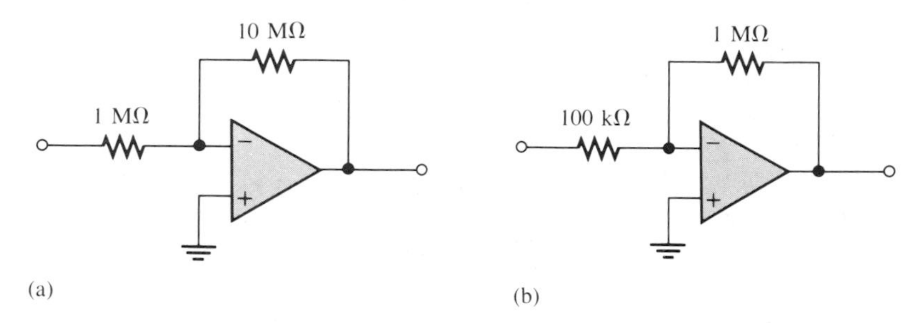

Figure 4–5 Inverting amplifier with a closed loop gain of 10. (a) 1-MΩ input impedance. (b) 100-kΩ input impedance which produces less noise than (a).

If resistance levels cannot be reduced because of design considerations, then special "low-noise" resistors, manufactured from glass or metal-film materials, should be used.

If two or more resistors are connected in parallel or series, then the equivalent resistance of the resistance combination should be used in Equation 4–1. For resistors in series, the *squares* of the individual generated RMS voltages are added

$$(v_{\text{total}})^2 = v_1{}^2 + v_2{}^2 + \ldots + v_n{}^2 \qquad (4\text{–}5)$$

Shot Noise

Shot noise, also called **Schottky, Schrot,** or **partition** noise, is due to the inherent small fluctuations in the instantaneous value of a DC current wherever it flows. Like thermal noise, shot noise is a natural source of white noise. The RMS value of the shot noise current is determined from

$$i_{\text{RMS}} = \sqrt{2q_e I_{\text{DC}}\,\Delta f} \qquad (4\text{–}6)$$

where

q_e = charge of an electron, 1.6×10^{-19} C
I_{DC} = DC current flowing in the device or circuit
Δf = effective noise bandwidth

Flicker and Burst Noise

Two examples of pink noise are **flicker noise** and **burst noise.** *Flicker noise,* frequently called *1/f* noise, is thought to be caused by the variations of the velocity of electrons as a result of defects in semiconductor materials. It decreases as the operating frequency is increased, hence the *1/f* name. Flicker noise is proportional to a DC bias current. Above several kilohertz, however, the noise power is essentially flat.

Burst noise is like *flicker noise* in that it is random and varies inversely with frequency, but in a $1/f^2$ fashion. It therefore is a low-frequency noise source whose sound can be described as similar to that of popcorn popping. For this reason, burst noise is often called **popcorn noise.** The magnitude of both flicker and burst noise depend on the devices in which they occur. This device-dependent variation makes it difficult to predict their magnitude with any degree of certainty, unlike thermal or shot noise sources.

Signal-to-Noise Ratio and Noise Figure

A term quantifying the amount of noise present with a given signal is the **signal-to-noise ratio,** usually abbreviated S/N. Expresed in decibels, it is the ratio of the overall signal power or voltage level, S, to the existing noise power or voltage level, N. In terms of power levels

$$S/N = 10 \log_{10}\left(\frac{S}{N}\right) \tag{4–7}$$

Since power is proportional to V^2, the signal-to-noise ratio in terms of voltage becomes

$$S/N = 20 \log_{10}\left(\frac{S}{N}\right) \tag{4–8}$$

The **noise figure** (NF) is a measure of how much noise a given network or amplifier stage adds to the signal. It is simply the signal-to-noise ratio measured at the circuit's input to the signal-to-noise ratio measured at the output. Expressed in decibels

$$NF = 10 \log_{10}\left[\frac{(S/N)_{\text{input}}}{(S/N)_{\text{output}}}\right] \tag{4–9}$$

Otherwise

$$NF_{\text{dB}} = (S/N \text{ input})_{\text{dB}} - (S/N \text{ output})_{\text{dB}} \tag{4–10}$$

Usually, the lower the noise figure, the better. However, a lower noise figure may not result in the lowest signal-to-noise ratio at the circuit's output.

EXAMPLE 4–2

An op-amp circuit of Figure 4–6 is connected to a signal source with an output impedance of 2 kΩ, while the amplifier is connected to a 500-Ω load. If the input signal level is 0.5 mV RMS while the noise levels contributed by the signal source and the input of the amplifier are 21 μV RMS and 14 μV RMS respectively, determine

(a) the signal-to-noise ratio at the amplifier's input,
(b) the signal-to-noise ratio at the amplifier's output,
(c) the noise figure.

Figure 4–6 Circuit for Example 4–2.

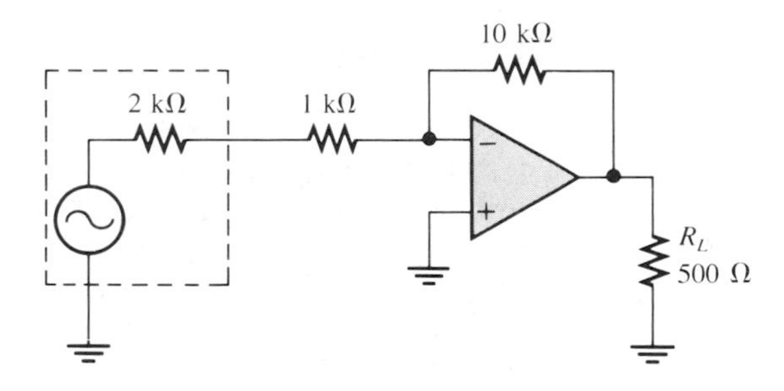

Solution:

(a) The input signal-to-noise ratio in terms of voltage levels is

$$(S/N)_{IN} = \frac{(0.5\ \text{mV})^2}{(21\ \mu\text{V})^2}$$

$$= 566.9\ (27.5\ \text{dB})$$

(b) The total input noise voltage is

$$N_{IN} = \sqrt{(N_{AMP})^2 + (N_{GENERATOR})^2}$$

$$= \sqrt{(21\ \mu\text{V})^2 + (14\ \mu\text{V})^2}$$

$$= 25.2\ \mu\text{V RMS}$$

The closed-loop gain of the inverting amplifier is 10, giving an output voltage without noise of 10 × 0.5 mV, or 5 mV. However, the amplifier looks like a *noninverting amplifier* to noise, since the signal generator is effectively shorted to ground. In this case, the closed-loop gain is 1 + (10 kΩ/1 kΩ) = 11. Therefore, the output RMS noise voltage is 11 × 25.2 μV, or 277.6 μV. The output signal-to-noise ratio is then

$$(S/N)_{\text{OUT}} = \frac{(5 \text{ mV})^2}{(277.6\ \mu\text{V})^2}$$

$$= 324.4 \text{ (25.1 dB)}$$

(c) The noise figure is then

$$NF = 10 \log_{10}\left(\frac{566.9}{324.4}\right)$$

$$= 2.4 \text{ dB}$$

or from Equation 4–10,

$$NF = 27.5 \text{ dB} - 25.1 \text{ dB}$$

$$= 2.4 \text{ dB}$$

SINAD Ratio

The unwanted changes of a waveform's shape as it passes through an amplifier is termed *distortion*. Generally, it is due to the frequency response and nonlinearities of the amplifier. In two-way communications systems, distortion is combined with signal and noise together as a standard performance specification for radio receivers, which is called the **SINAD ratio.** SINAD is an acronym for **signal plus noise plus distortion** and is expressed in decibels so that

$$\text{SINAD} = 10 \log_{10}\left(\frac{\text{signal} + \text{noise} + \text{distortion}}{\text{signal} + \text{noise}}\right) \tag{4–11}$$

SINAD measurements do not attempt to separate either noise or distortion from the signal because it is assumed that a receiver, like any amplifier stage, may introduce some distortion to the signal.

Noise Measurement

In general, noise is not a symmetrical sinusoidal waveform. Thermal noise contains all frequencies and does not have a specific peak value, as was shown in Figure 4–1. As is discussed in Chapter 6, most AC meters that measure RMS voltage are calibrated for *sine waves* only. If an AC meter is used to measure the RMS value of a triangle waveform, an error will result unless a correction factor is supplied.

On the other hand, certain meter types, like the thermocouple meter, are true RMS-indicating instruments and read the correct RMS value regardless of the waveform's shape. This type of meter can be used to measure noise levels.

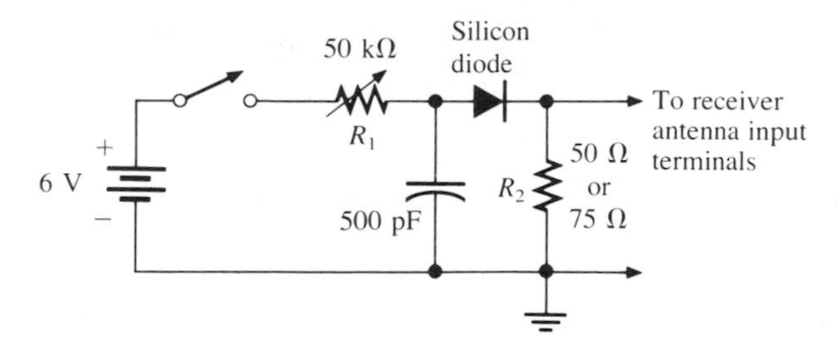

Figure 4–7 Simple diode white noise generator.

However, if a true RMS meter is not available, a d'Arsonval-rectifier meter circuit or most AC meters can be used to reasonably estimate the noise level by multiplying the reading by 1.11 to obtain the approximate "true RMS" noise voltage.

For testing the noise characteristics of amplifiers and radio receivers, a noise generator, like the simple diode circuit of Figure 4–7 is frequently used. The RF, or "hiss"-type, noise is evenly distributed throughout the frequency spectrum. The current in the diode also flows through R_2, which equals the characteristic impedance of the transmission line connected to the receiver. In effect, R_2 then substitutes for the transmission line because the noise generator is connected directly at the receiver's antenna input terminals.

When using the generator for adjusting a receiver's input circuit, or "front end," for an optimum noise figure, set the receiver's RF and audio frequency (AF) gain controls to mid range and turn the automatic gain control (AGC) off. Connect an RMS voltmeter to the receiver's output across either the speaker or headphones. With the generator connected to the receiver's antenna terminals and the switch open, adjust the RF and AF gain controls for an output reading that is far enough below the maximum obtainable to ensure that the receiver is operating linearly. This then is the *noise reference level.* Close the switch, and R_1 is adjusted so that the receiver's output level increases slightly. Note the ratio of the two readings. Make experimental adjustments to the receiver's input stage with the object of obtaining the largest dB or ratio increase in output when the generator is switched on.

This type of generator is especially useful for receiver adjustment when comparing the performance of different receivers checked with the same generator. However, it does not permit the actual measurement of the receiver's noise figure. The commercial noise figure meter is discussed in Section 14–8.

4–4 INTERFERENCE

In many situations, any signal disturbance other than the desired signal is termed **interference.** These extraneous signals, including noise, (1) make the measurement of the desired signals more difficult, (2) assume a variety of forms, and (3) are easily able to enter or exit electronic equipment.

External sources of interference may be divided into the following major types:

1. **Electrically coupled**
2. **Magnetically coupled**
3. **Electromagnetic**
4. **Common-mode**

The basis for these classifications rests solely on the phenomena responsible either for their production or propagation. Of the many types of interference, perhaps the best-known and most hated is 60-Hz power-line interference, commonly called **60-cycle hum.** The first three types are discussed in the following sections, while common-mode interference is discussed in Section 4–6 as part of the ground loop problem.

Electrically Coupled Interference

Electrically coupled, or **capacitive** (high impedance) interference is the situation where the signal of a nearby conductor is capacitively coupled to a nearby wire or cable, as shown in Figure 4–8*a*. An electrostatic field is created between conductors with different potentials. The major sources that contribute to this type of interference are fluorescent light bulbs, and, to some extent, *unconnected* power-line and ceiling light sockets. Since the power line is the primary carrier, the 60-Hz sine-wave voltage (approximately 120 V RMS, or 167 V peak) is present virtually everywhere. The resultant interference is *60-cycle hum* because the 60-Hz signal is an audible low-level buzzing.

From the equivalent circuit of figure 4–8*b,* the voltage that is capacitively coupled from points A to B can be expressed as

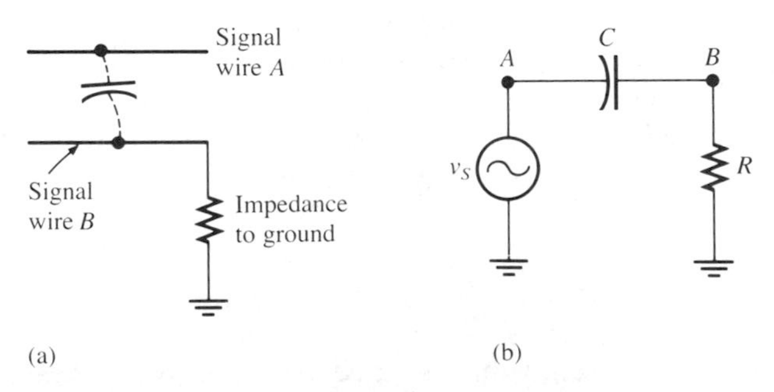

Figure 4–8 Capacitively coupled interference from one wire to another. (a) Physical arrangement. (b) Equivalent circuit.

$$v_B \text{ (RMS)} = \frac{R}{\sqrt{R^2 + X_c^2}} v_s \text{ (RMS)} \tag{4–12}$$

It is difficult to assign exact numbers to the value of the coupling capacitance because the exact configuration must be known. However, for frequencies less than 10 MHz and distances less than two feet, the reactance will be several thousand ohms and decrease with increasing frequency.

From Equation 4–12, the level of the voltage at point B due to capacitively coupled interference increases as

1. The capacitance between the source and the pickup circuit increases.
2. The frequency of the voltage source increases.
3. The level of the voltage source increases.
4. The input impedance of the pickup circuit increases.

With respect to 60-cycle (60-Hz) hum, frequency and level of the voltage source cannot usually be controlled. To prevent or minimize capacitively coupled interference, either reduce the capacitance between the two points or keep the input impedance low as possible. The impedance level comes down to a tradeoff between circuit performance (i.e., instrument loading) or interference reduction. This leaves trying to reduce the amount of capacitance by *shielding* the pickup circuit from the source. Proper shielding methods are discussed in Section 4–8.

Magnetically Coupled Interference

When a current flows in a conductor, a magnetic field surrounds that conductor. If a time-varying current flows in a conductor, then the magnetic field that results will also change. **Magnetically coupled,** or inductive (low-impedance) interference arises from such magnetic fields created from current-carrying conductors as described above. Normal sources are from power transformers, the power cord wires, and large, ground-loop currents. The coupling between the source of interference and any pickup circuit is inversely proportional to both frequency and distance. For frequencies below 1 MHz and distances less than two feet, the coupling impedance can be on the order of a few ohms.

Possible solutions for reducing inductive interference include the use of low-current and high-impedance levels, large spacing, short current-carrying wires, and eliminating low-frequency response. However, many of the possible remedies for inductive interference actually increase the level of capacitive interference. Both types of interference may be reduced by increasing the distance between noise source and pickup point as well as proper shielding. Magnetically coupled interference is more difficult to shield against. For power cord wires, twisting the wire pair minimizes the magnetic field surrounding the wires by reducing the area between the conductors, which reduces the induced current.

Electromagnetic Interference (EMI)

Capacitively coupled interference produces an electric field; magnetically coupled interference creates a magnetic field. In the region near the electric charge or current-carrying conductor, the field variation will be synchronized with the variation in the charge or current. When this is the case, the field is referred to as a *near field,* or *induction field.* At points farther away, the effect of the near field is small, and the electromagnetic field propagation predominates in what is termed *far field,* or *plane-wave propagation.*

Electromagnetic fields are created by the high-frequency energy associated with radio waves, or RF radiation. Besides the obvious situation of a radio station transmitter, RF signals are produced by electric motors and generators, microwave ovens, remote control garage door openers, radar detectors, and oscillator circuits in televisions and radios.

As its name suggests, the electromagnetic field simultaneously contains both an electric field component and a magnetic field component. Since EMI is caused by high-frequency RF sources, many types of circuits are subjected to both capacitive and inductive interference. However, it either the magnetic or electric field component is effectively reduced, then the electromagnetic field is reduced. Generally, metal-wire mesh (Section 4–8) is used as the shield material.

4–5 GROUNDS AND GROUNDING

The electrical ground was originally considered as a low-resistance connection between a given circuit and the earth. A long copper rod, called a *ground rod,* was driven at least four feet into the earth and the ground connection was made to it. Today, however, the term **ground** is loosely used to mean any point having a zero-voltage reference.

The AC Power Line

As shown in Figure 4–9*a,* the 120-V, single-phase power line comprises three lines: *hot, neutral,* and *ground.* Most electricians adopt the following convention for color-coding these wires:

hot line: black
neutral line: white
ground line: green

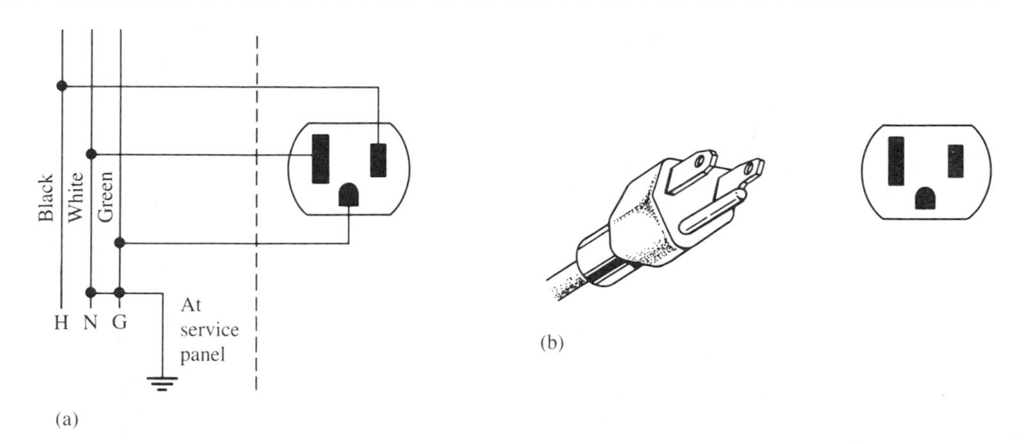

Figure 4–9 The 3-wire power AC power line. (a) Wire connections. (b) Polarized plug and receptacle.

The National Electric Code (NEC) requires that the 120-V, single-phase electrical power-line outlet have a polarized 3-wire socket and plug arrangement as shown in Figure 4–9*b*. The third, or ground, wire is tied to the neutral wire (usually at the distribution or service panel) where the fuses or circuit breakers are located. Ideally, the ground wire is at ground potential, and the potential difference between the ground and neutral wires is zero. However, like any transmission line, there are distributed resistances and capacitances that may create a return path from the hot wire back through the ground lead.

Ground Faults

As used by the NEC, a **ground fault** refers to the return of current to ground by any path other than the neutral wire. There are two possible paths for current flow in the AC power line. As shown in Figure 4–10*a,* the hot line to neutral is the correct and safe path. The supply current enters the hot line,

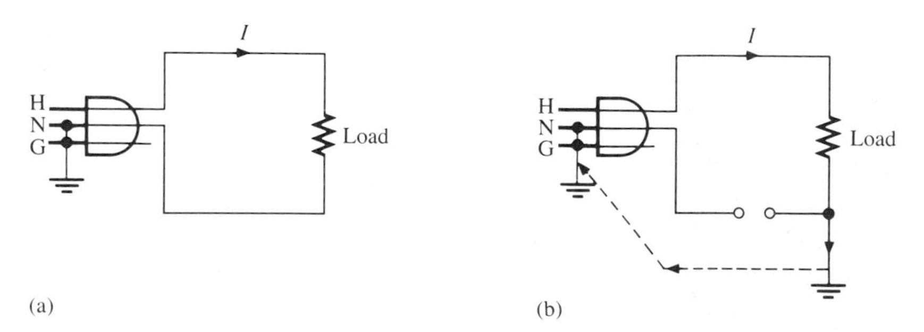

Figure 4–10 Current flowing in the AC power line. (a) Normal current path. (b) Ground fault path.

travels through the load, and returns to the power line via the neutral line. The other path, as shown in Figure 4–10*b*, is from the hot line to ground where the current returns directly to ground via a ground conductor (or any other grounded object) rather than the neutral conductor. This situation, called a ground fault, is potentially dangerous.

The current flowing in the ground wire as a result of a ground fault is termed the **fault current.** Although a given piece of equipment may have a fuse or circuit breaker, the fault current often is not large enough to open the fuse or circuit breaker. The hazardous situation is remedied by the installation of a **ground-fault interrupter** (GFI), which compares currents flowing in the hot and neutral lines supplying a load, and disconnects the supply lines if the currents are not equal. A GFI may be actuated for fault currents as low as 5 mA.

As shown in Figure 4–11, the GFI is nothing more than a transformer connected to a sensitive AC relay. As long as the hot and neutral line currents are equal, the two oppositely connected transformer windings cancel each other's magnetic fields, and no voltage is across the winding connected to the relay coil. A current imbalance then energizes the relay, disconnecting the power.

In Figure 4–12*a*, a **leakage current** flows in the ground wire of a 3-wire power line as a result of the combination of a distributed *R*-*C* network and wire-insulation breakdown near the power transformer inside a piece of equipment. Since the equipment's power cord ground wire is connected internally to the chassis, a potential difference now exists between the chassis and another ground point. If a person were to touch the chassis of an instrument with such a fault condition and another ground point, that person's body then provides a path for current to flow back through the ground and neutral lines of the power line (Figure 4–12*b*). The current may be of sufficient level to seriously injure or even kill a person, as is discussed in Section 4–9.

Besides insulation breakdown, another common ground-fault culprit is the use of 3-wire **"cheater" adapters**, shown in Figure 4–13, to convert 3-wire plugs to the older 2-wire (hot and neutral) outlets. Without any further inspection or testing, it cannot be readily determined which of the two connections is the neutral wire. The ground connection is made at the screw holding the outlet

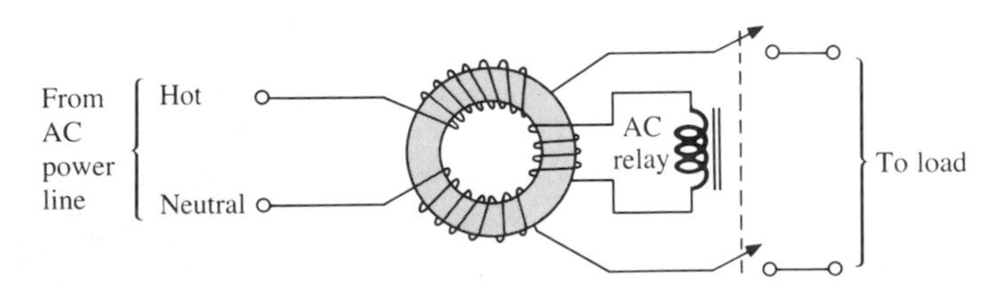

Figure 4–11 Ground-fault interrupter.

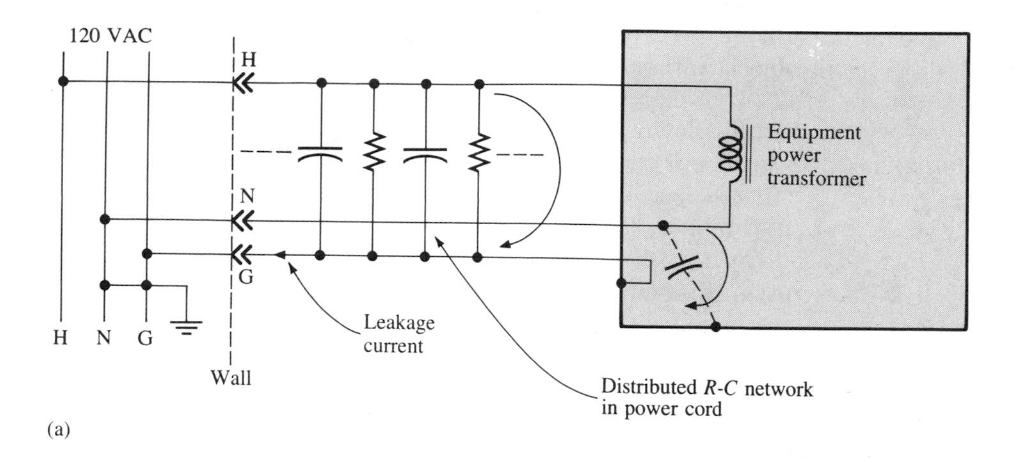

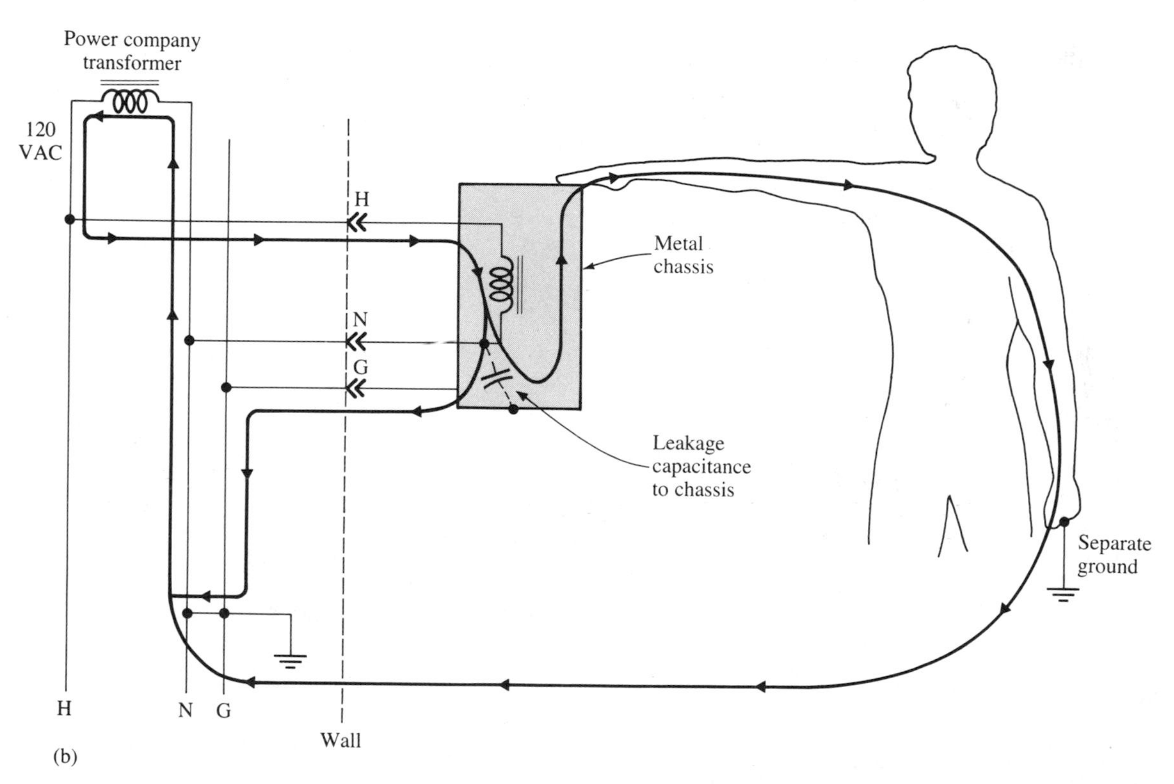

Figure 4–12 Leakage current paths. (a) Path created by distributed R-C network in AC power cord and leakage capacitance to equipment case. (b) Path created by leakage capacitance to equipment case through the human body and to another ground point.

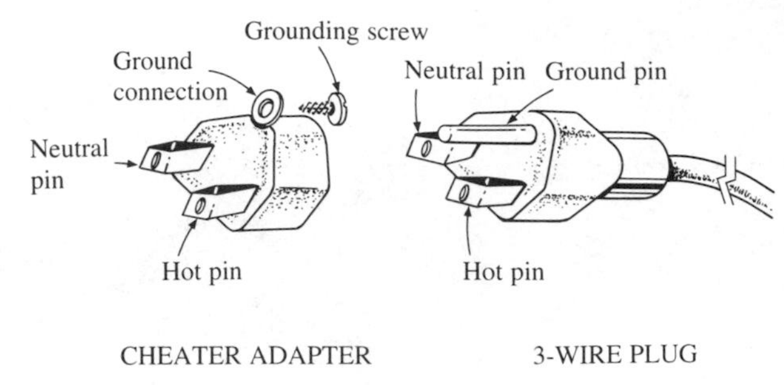

Figure 4–13 Cheater adapter.

cover plate in place. The hazard associated with cheater plugs is perhaps the most serious of all, especially if the ground connection is broken off, or intentionally not made (See Figure 4–14).

Besides correctly converting 3-wire plugs to mate with the older 2-wire sockets, the cheater adapter can be used to convert a single-ended input on an instrument such as an oscilloscope or recorder to a *differential* input. For example, as shown in Figure 4–15*a*, an oscilloscope with a single-ended input cannot correctly measure the voltage across the resistor as the probe measures the voltage at point A with respect to ground. Since the ground clip connected at B

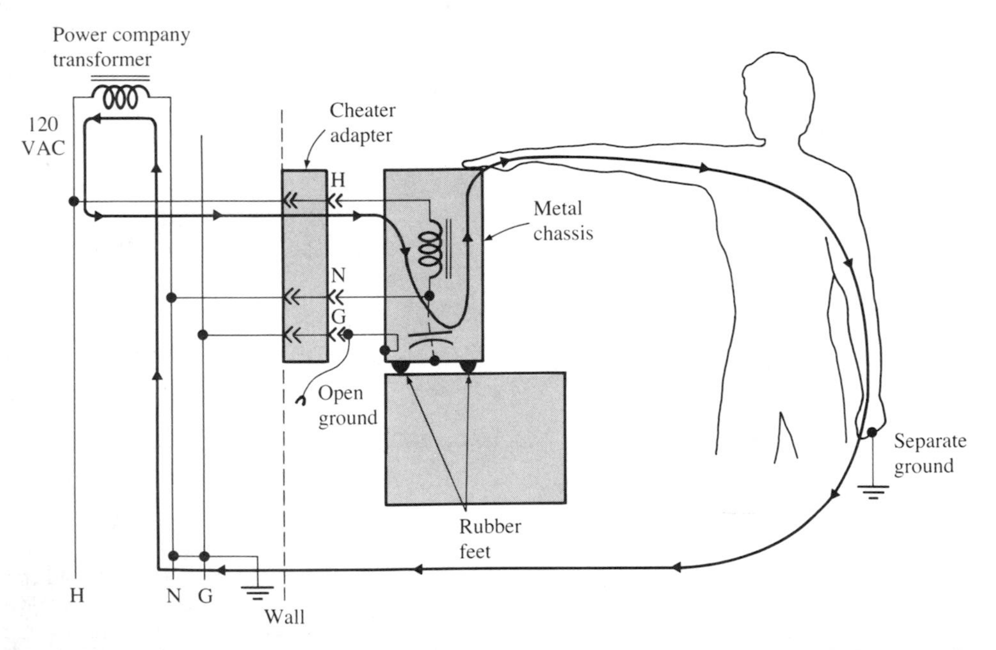

Figure 4–14 Hazardous condition created by leakage capacitance and the use of a cheater adapter without connecting its ground lead.

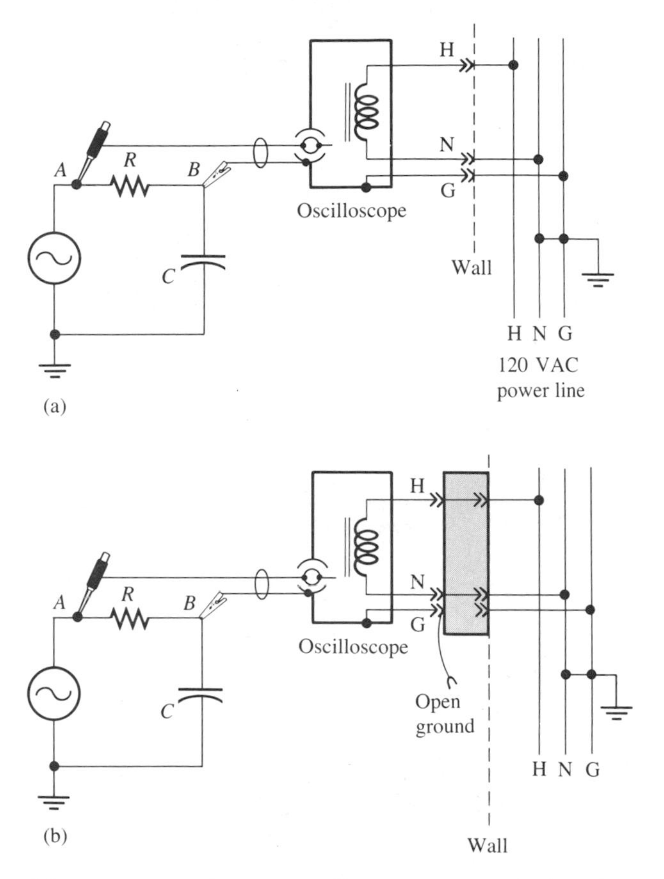

Figure 4–15 Using an ocilloscope to attempt to measure a voltage difference between points A and B. (a) Incorrect method shorts capacitor at B to ground. (b) Correct method by using a cheater adapter to float the oscilloscope's input and case ground.

is already at ground, the capacitor is shorted out. In Figure 4–15*b,* a cheater adapter (because it does not make the ground connection at the 3-wire socket), allows the oscilloscope to "float" and measure the voltage difference between points A and B across the resistor. However, the hazard condition similar to that shown in Figure 4–14 is still possible if a fault occurs within the equipment and the operator touches a ground. If an operator forgets to remove the cheater adapter, he or she leaves an open invitation to danger the next time any one uses the oscilloscope.

Neon and LED 3-Wire Receptacle Analyzers

A simple method of determining if the power-line wiring is faulty is to use an incandescent bulb to test all three lines. If the wiring is correct, as shown in Figure 4–16, the light bulb will light when connected between the hot and

Figure 4–16 Three-step testing of the 3-wire AC power line receptacle with an incandescent bulb tester.

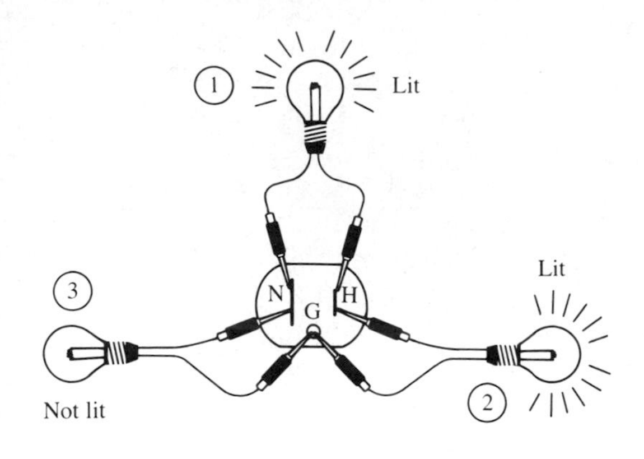

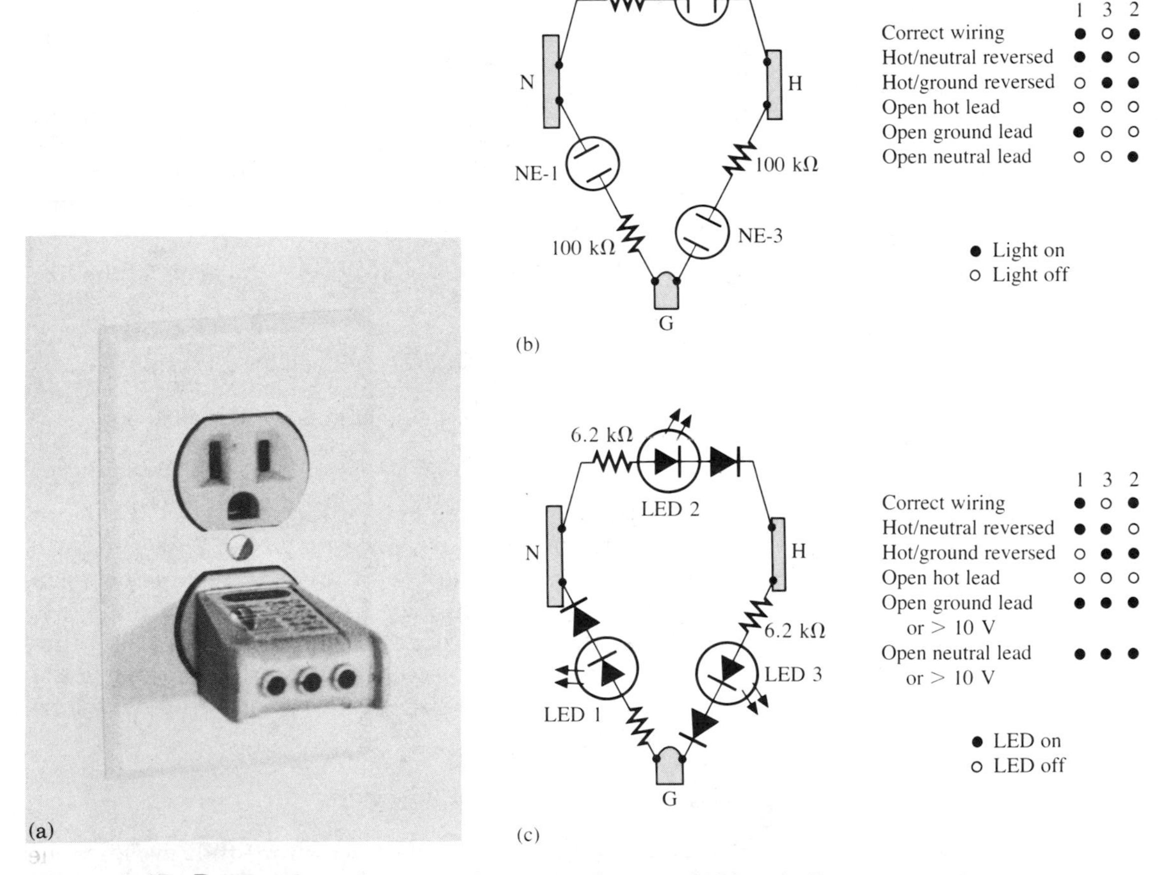

Figure 4–17 Receptacle analyzers. (a) Commercial tester. (b) Neon bulb tester circuit. (c) LED tester circuit.

neutral lines or the hot and ground lines; it should not light when connected between the ground and neutral lines.

The above method requires three separate tests. On the other hand, simple, 3-bulb receptacle analyzers, similar to that shown in Figure 4–17*a* are available. These simple testers, which plug into the three-wire wall socket, come in two basic forms: neon bulb (Figure 4–17*b*) and LED (Figure 4–17*c*). The condition of the wiring of the wall socket is indicated by the status of the neon bulbs or LEDs.

The LED tester is generally superior to the older neon bulb tester.

1. The neon bulb, which is a voltage-controlled device similar to a vacuum tube, requires 70 V to turn on. On the other hand, the LED, which is a current-controlled device similar to transistors, requires only approximately 1.7 V to light.
2. The LED tester has a higher test current capability (20–50 mA) than the neon type, which operates with only a few milliamperes.
3. The neon bulb does not detect polarity, while the LED emits light only when forward biased.
4. The LED tester lights all three LEDs when the ground voltage is approximately 10 V above the neutral voltage, but the neon tester will not give warning until the voltage exceeds 70 to 80 V. This difference can be important when checking for high ground voltage due to a ground fault.
5. The NEC specifies a maximum allowable neutral voltage of 5% of the line voltage to minimize the risk of fire. For a 120-V line, this is 6 V from neutral to ground. The LED tester can detect approximately an 8- to 10-V ground-to-neutral potential, but the neon tester can detect only gross errors of approximately 70 to 80 V.

In general, neither type of tester is to be left plugged in for an extended period of time. Both types inject leakage current into the ground circuit.

Instrument Grounds

There are three basic categories of grounds associated with electronic systems:

1. Power grounds
2. Signal grounds
3. Chassis and shield grounds

The **power ground** is used for a return path for the current that provides the power required to operate the equipment. A **signal ground** is both the reference point and return path for all signal currents that flow. Finally, **chassis**

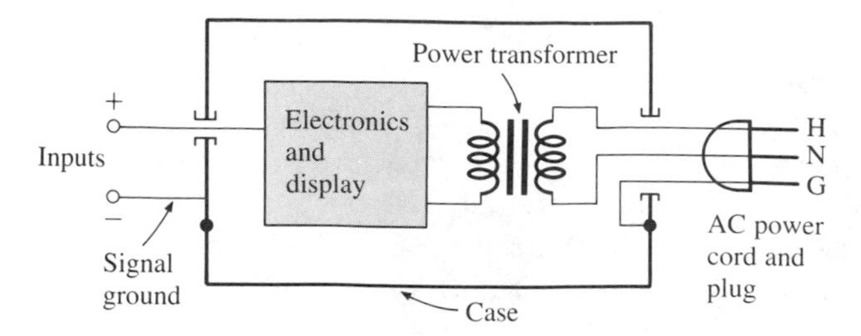

Figure 4–18 Common method of testing for signal input and power grounds.

and **shield grounds** are the connections made at the chassis or outer metal case of an instrument and any shielded cables.

Depending on the type of the instrument, signal ground paths vary depending on the type of instrument involved. As shown in Figure 4–18, a typical situation using an oscilloscope, recorder, or electronic voltmeter takes a given input quantity and displays it in some form to the operator. Some of these instruments have input signals referenced to their chassis ground. The chassis ground is also tied to the earth ground at the building's service panel via the third wire of the AC power line. As long as the ground wire is connected, the input signal ground should not be connected to any point that has a potential difference with respect to (earth) ground.

Some instruments have differential, or *floating* inputs, where the signal ground is isolated from the case or shield ground, as is shown in Figure 4–19. The displayed signal then is the voltage difference between the two inputs. The part of the voltage that is common to both inputs is the *common-mode* voltage and is not displayed. Because both input terminals are isolated, either input can be considered as the signal ground.

4–6 GROUND LOOPS

Of all the causes of noise and interference, perhaps the most common, but most misunderstood, is the **ground loop.** The ground loop is generally considered as a closed electrical path, or loop, made of the ground wires of a system and

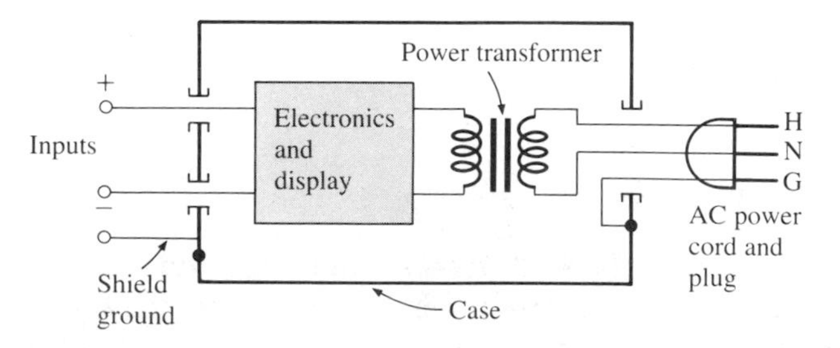

Figure 4–19 Grounding system for devices having differential inputs.

Figure 4–20 The ground loop.

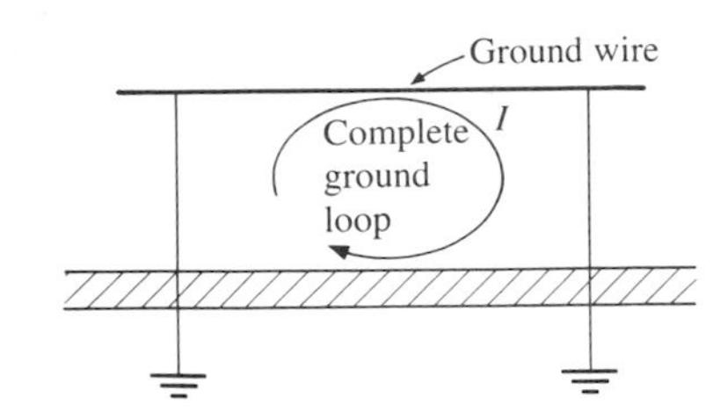

the earth's ground plane, as illustrated in Figure 4–20. Whenever two or more terminals of a ground conductor associated with an electrical system are connected at different points, a ground loop is created. Since ground connections are intended to be low-impedance paths, ground loops are then also low-impedance paths. When a current flows, a voltage, or IR drop, is developed between two different ground connections.

Ground loops can be created by

1. A potential difference between two or more points of a ground plane to which the external grounds are connected.
2. Inductive coupling.
3. Capacitive coupling between a system and ground.
4. Common-mode noise.

The potential difference between two or more points is briefly discussed above. The current that flows is the fault current. It can be produced by many sources.

Inductive Coupling

Inductive coupling, or *inductive pickup,* in a ground loop is similar to the magnetic induction created in a closed loop of wire. As a general rule, because ground loops are low-impedance paths that cover a relatively large area, inductive pickup from surrounding magnetic fields occurs easily.

Capacitive Coupling

Capacitive coupling, as shown in Figure 4–21, occurs when the ungrounded shielded chassis of an amplifier is capacitively coupled to a ground point. The amplifier's circuit ground is capacitively coupled to the chassis, which is coupled, in turn, to ground. Although no resistive path to ground is present, a ground loop now exists as a result of capacitive coupling. The complete loop as shown in Figure 4–21, travels from the input ground through the circuit ground capacitively coupled to the chassis. The chassis is, in turn, coupled to ground. If a potential difference exists between these two external ground points, a current flows in the ground loop.

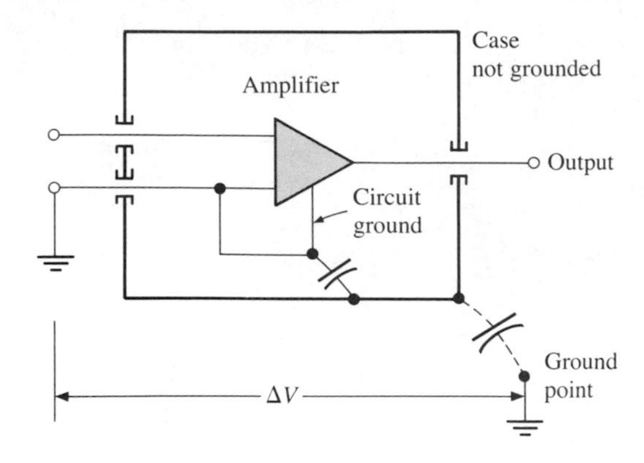

Figure 4–21 Ground loop path created by capacitive coupling to chassis and a ground point.

Common-Mode Noise

As shown in Figure 4–22, a typical instrumentation arrangement is where the signal source is grounded via the AC power line at one point, and the amplifier is grounded at a different point. The two power cords are generally plugged into power outlets at two different locations. The cable shield from the signal source, although grounded to the chassis of the signal source, may not be connected to the chassis ground of the amplifier. If the signal source ground and the amplifier ground are at different potentials, then the input voltage of the amplifier is the sum of the signal source voltage (v_s) plus the potential difference (v_{cm}) that exists between the signal source and amplifier ground points. This voltage is frequently called the **common-mode noise voltage.**

If the cable shield from the signal source is connected to the chassis ground of the amplifier (as in Figure 4–23), a ground loop is formed. Because the common-mode voltage is present in the ground loop system, interference resulting from ground loops is frequently referred to as *common-mode* interference.

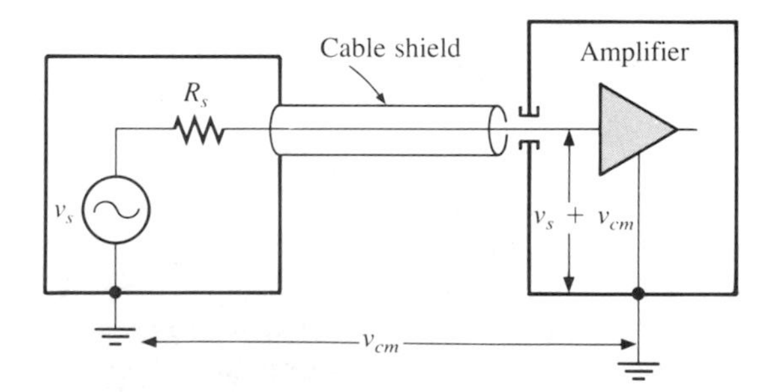

Figure 4–22 Input voltage of an amplifier includes the common-mode voltage when the signal cable shield is not connected to the amplifier case.

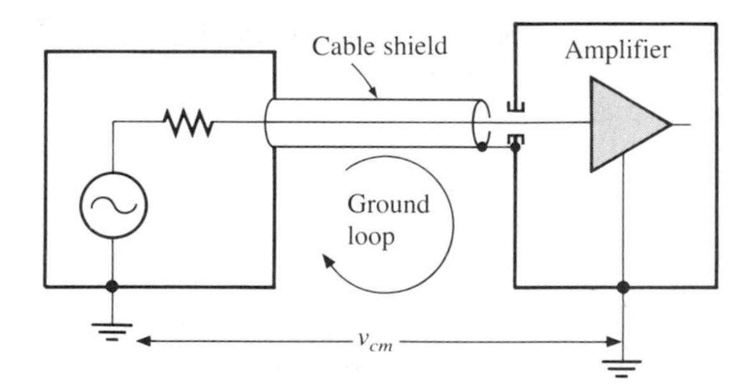

Figure 4–23 Formation of a ground loop by the connection of cable shield between source and amplifier.

The reduction or elimination of ground loops and common-mode interference in equipment powered by power lines is achieved in one of two ways:

1. Single-point grounding
2. Use of differential amplifiers

Single-point grounding is where *all* ground leads are connected to the earth ground at a *single point,* as illustrated in Figure 4–24. One of the easiest ways to achieve a common ground point is to use a multi-outlet power-line outlet. In this way, all the power cords are plugged into what is considered the same point.

Complete single-point grounding is often not possible, especially when the signal source is a significant distance away from the amplifier. When this is the case, a differential amplifier is frequently used as an alternative approach. Figure 4–25 shows a differential amplifier using an operational amplifier.

If the operational amplifier were ideal, the output voltage would be

$$v_o = (v_2 - v_1)A_d \tag{4–13}$$

where

A_d = differential gain, R_2/R_1

The output voltage depends only on the voltages v_1 and v_2. The common-mode voltage, v_{cm}, because it is *common* to both inputs, is not a factor. The actual

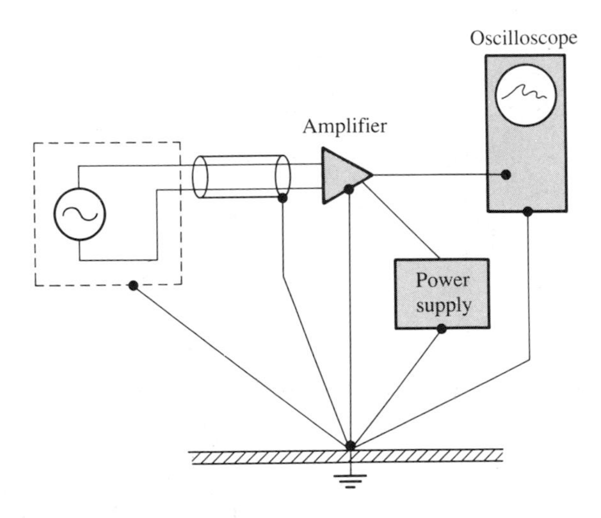

Figure 4–24 Single-point grounding.

input voltages are $v_1 + v_{cm}$ and $v_2 + v_{cm}$. The differential amplifier magnifies the difference of the two input signals $(v_2 + v_{cm}) - (v_1 + v_{cm})$, or $v_2 - v_1$. However, operational amplifiers are not ideal, and some of the common-mode voltage gets amplified in addition to the input difference. The measure of how well an operational amplifier rejects the common-mode interference signal, **common-mode rejection** (CMR), is expressed in decibels. The **common-mode rejection ratio** (CMRR) is the CMR expressed as a simple ratio. The relationship between CMR and CMRR is

$$\boxed{\text{CMR} = 20 \log_{10}(\text{CMRR})} \qquad (4\text{–}14)$$

For example, a CMR of 100 dB means that 1/100,000 of the common-mode input signal is amplified by the differential amplifier. The CMRR is equal to the circuit's differential gain divided by the common-mode gain. The common mode gain is the ratio of the output of the amplifier of the common-mode signal alone to the level of the common-mode input (i.e., $v_2 = v_1$).

EXAMPLE 4–3

For the differential amplifier circuit of Figure 4–25, $R_1 = 1\ \text{k}\Omega$ and $R_2 = 100\ \text{k}\Omega$. If both inputs are tied together and the input common-mode signal of 2 V RMS produced an output voltage of the differential amplifier of 0.08 V RMS, determine the CMR of the operational amplifier.

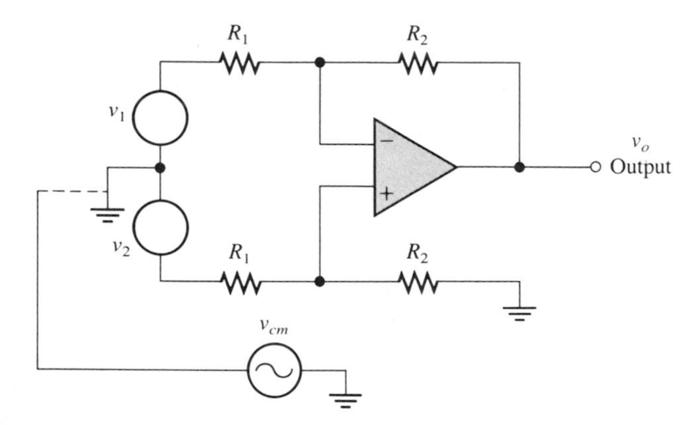

Figure 4–25 The differential amplifier using an operational amplifier.

Solution:

The common-mode gain is

$$A_{\text{cm}} = \frac{0.08\ \text{V}}{2\ \text{V}}$$

$$= 0.04$$

Since the differential gain is 100 kΩ/1 kΩ = 100, the common-mode rejection ratio and common-mode rejection are respectively

$$\text{CMRR} = \frac{2\ \text{V}}{0.08\ \text{V}}(100)$$

$$= 2{,}500$$

$$\text{CMR} = 20\ \log_{10}(2{,}500)$$

$$= 67.9\ \text{dB}$$

When the op-amp's common-mode rejection ratio for the nonideal operational amplifier is taken into consideration, the output voltage for the circuit of Figure 4–25 becomes

$$v_o = (v_2 - v_1)A_d + A_d\left(\frac{v_{\text{cm}}}{\text{CMRR}}\right) \tag{4–15}$$

where

CMRR = op-amp common mode rejection ratio
v_{cm} = common mode voltage

The first term of the right-hand side of Equation 4–15 is the differential output. The second term is the error voltage that results from a nonideal differential amplifier. If the operational amplifier were ideal, the CMRR would be infinite. Equation 4–15 reduces to Equation 4–13.

The CMR of a differential amplifier is seriously affected if the signals to its two inputs do not have the same impedances. For example, Figure 4–26 shows a differential amplifier connected to a signal source with one end grounded. The output impedance of the signal source is R_S. The total resistance connected to the inverting (−) input of the operational amplifier is now $R_1 + R_S$. The resistance of the other input is R_1, so that an imbalance equal to R_S exists. Thus, the CMR will be degraded. If an additional resistance equal to the output impedance of the source is added to the resistance connected to the operational amplifier's noninverting (+) input, then the imbalance will be

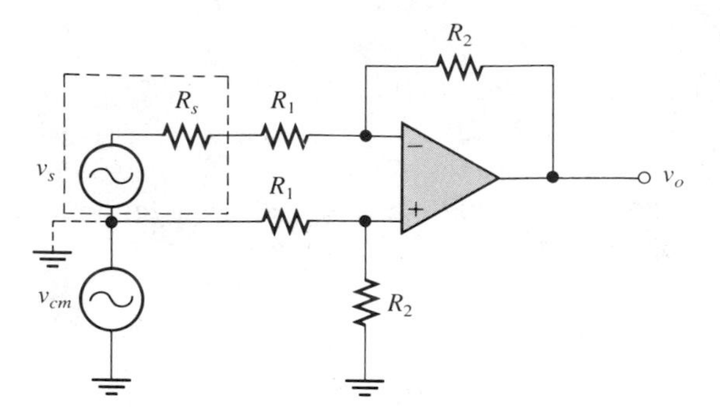

Figure 4–26 Differential amplifier with one input grounded.

eliminated, but the differential gain will be reduced. The voltage at the inverting input is then $v_s + v_{cm}$, while the voltage at the noninverting input is simply v_{cm}. For an ideal amplifier, the output voltage is then

$$\boxed{v_o = v_s A_d} \qquad (4\text{–}16)$$

If the op-amp's CMRR is taken into account, then

$$\boxed{v_o = v_s A_d + A_d \left(\frac{v_{cm}}{\text{CMRR}}\right)} \qquad (4\text{–}17)$$

4–7 CROSSTALK

Multichannel systems, as the name implies, have many signals present simultaneously in different channels, or inputs. The stereo, with its right and left channels is one example; the dual or quad operational amplifier is another. If at any time, the signal from one channel appears in a second adjacent channel, the result is termed **crosstalk.** Often the problem is caused by something in common being shared by the two subsystems, such as a signal ground or power-supply ground. Although the obvious solution is to require complete shielding between the two circuits and the use of separate power supplies, this may not be economically possible. Leads and cables common to both circuits should be avoided whenever possible.

4–8 SHIELDING AND FILTERING

One of the most effective methods of filtering is the use of a shield at the noise source, the pickup point, or both. The basic method is to enclose the device in a shield that reflects and/or absorbs electromagnetic energy. The design of the shield itself generally depends on the frequencies and the type of interference that must be contained or excluded.

Electric-Field and Plane-Wave Shielding

When a plane wave or a high-impedance signal strikes a high-conductivity surface, the signal is completely reflected from the surface. Using this principle, a shield is formed by *completely* surrounding the object in an electrically closed, high-conductive surface. There should be no openings in the enclosed shield. If this is not possible, any opening must be very small in relation to the wavelength of the plane wave or signal being shielded against. Most equipment requires holes throughout the shield for ventilation. As a general rule, all uncovered holes should be less than ⅛ inch in diameter. If larger holes are necessary for the passing of cables or wires through the shield, they should be covered with a fine-wire mesh. To avoid heavy metal enclosures, a metal-wire mesh or screen is often used. The wire mesh itself is often used as a shield, and it is nearly as effective as solid metal.

Where no screen is possible, the opening should be designed with a connecting sleeve to form a waveguide as illustrated in Figure 4–27. The dB attenuation of the waveguide is given by

$$\text{dB attenuation} = 54.5\left(\frac{L}{\lambda_c}\right)\sqrt{1 - \left(\frac{\lambda_c}{\lambda}\right)^2} \quad (4\text{–}18)$$

where

λ_c = waveguide cutoff wavelength, πr
λ = wavelength of the interference signal
L = length of the waveguide sleeve
r = radius of the opening

The waveguide, which acts as a filter for wavelengths greater than its cutoff wavelength, effectively rejects signals below the cutoff frequency of the waveguide. If the sleeve is made three times as long as its diameter, an attenuation of approximately 100 dB is achieved for frequencies less than one-fifth of the cutoff frequency.

EXAMPLE 4–4

Using the waveguide of Figure 4–27, the diameter is 1 cm and the sleeve is 4 cm long. If the interference has a frequency of 1 GHz, determine

(a) the cutoff frequency of the waveguide, and
(b) the dB attenuation.

Figure 4–27 A waveguide to attenuate interference from penetrating a shielded enclosure through an opening.

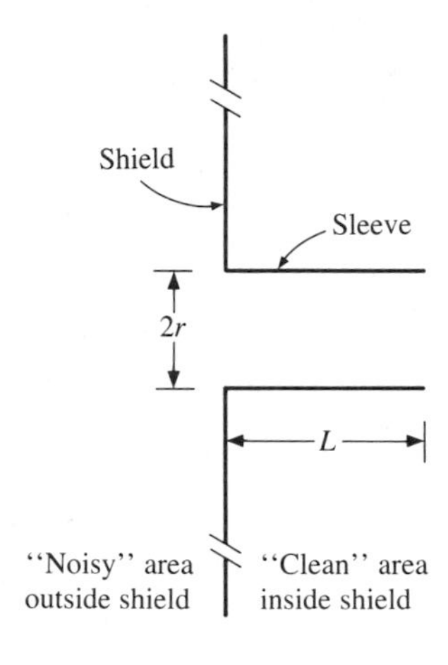

Solution:

(a) The cutoff wavelength of the waveguide is

$$\lambda_c = (\pi)(0.5\text{ cm})$$
$$= 1.57\text{ cm}$$

Since $f = C/\lambda$, where C is the speed of light (3×10^8 m/s), the waveguide cutoff frequency is

$$f_c = \frac{3 \times 10^8\text{ m/s}}{1.57\text{ cm}}$$
$$= 19.1\text{ GHz}$$

(b) In similar fashion, the wavelength of the 1-GHz interference is

$$\lambda = \frac{3 \times 10^8\text{ m/s}}{1\text{ GHz}}$$
$$= 0.3\text{ m, or }30\text{ cm}$$

From Equation 4–18, the dB attenuation is

$$\text{attenuation} = 54.5\left(\frac{4\text{ cm}}{1.57\text{ cm}}\right)\sqrt{1 - \left(\frac{1.57\text{ cm}}{30\text{ cm}}\right)^2}$$
$$= 138.7\text{ dB}$$

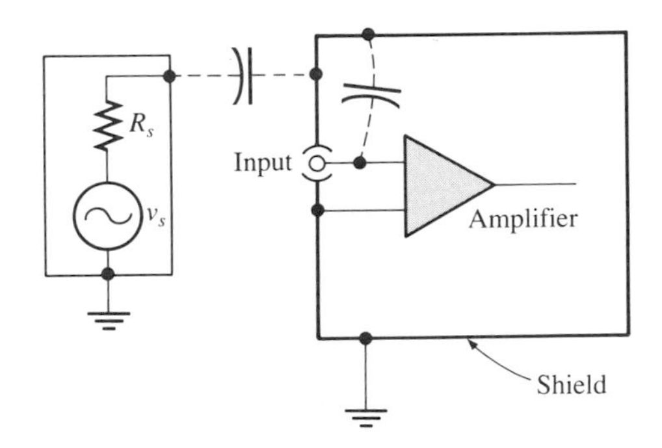

Figure 4–28 Connection of shield enclosure to ground to shunt coupling capacitances.

With the exception of floating inputs, as shown in Figure 4–28, the shield enclosure is connected to the instrument ground, which in turn may be connected to the signal ground. If the shield is at the same potential as the ground point of the input of the amplifier, the interfering signal is shunted out and does not affect the input signal. The input signal is capacitively coupled to both the shield and the amplifier's input. However, as long as the shield is maintained at the same potential as the amplifier's input, no energy flows between the shield and the amplifier input.

Magnetic Shielding

As mentioned in Section 4–4, shielding to eliminate or reduce magnetic fields is more difficult than shielding against other types of interference due to the low-impedance levels that couple the field noise to the pickup point. As a general rule, it is preferable to reduce the interference at its source. Once this is done, it is easier to reduce its effects further by a magnetic shield at the pickup point.

Because magnetic shielding requires a low-reluctance path, a high-permeability metal is required. Transformer iron or mu-metal is often used, as is the case with oscilloscopes (Chapter 11). The common technique is to build up the shield in layers with either an air gap or a non magnetic material between each layer. The lowest-permeability metal is placed toward the source of interference to avoid saturation. Depending on the field strength, many layers may be required.

Filtering

Besides shielding, interference may be reduced by the use of **filters**. The choice of the type of filter is determined primarily by the frequency of the signals to be passed versus the frequency of the interference to be rejected. In all cases, the basic idea is to provide a high-series impedance and a low-impedance shunt for the interference.

Filters fall into four basic types as follows

1. low-pass
2. high-pass
3. bandpass
4. band reject, or notch

For each of these types, filters may be either *passive* or *active.* Passive filters are made solely from passive components: resistor-capacitor or inductor-capacitor combinations. Active filters are generally those that have a resistor-capacitor filter network built around an active device, such as a transistor or operational amplifier. As a general rule, active filters are used primarily to filter signals in the audio range from 20 Hz to 15 kHz. RF signals above the audio range use passive filter networks.

Filters must be installed carefully to ensure their effectiveness in reducing the interference. No part of the wire in a shielded area prior to the filter can be exposed. An exposed wire can seriously reduce the filter's effect by radiating or receiving noise.

The 60-Hz hum, which is probably the most frequently encountered type of interference in audio systems, is effectively eliminated by an active, 60-Hz notch filter, similar to the one shown in Figure 4–29. The filter network itself is a *twin-T R-C* filter combined with an operational amplifier. Since this type of filter has sharp rejection characteristics, it is best suited for use at a single frequency, such as 60 Hz. The frequency at which the rejection is a maximum is called the *notch frequency,* and is given by

$$f_N = \frac{1}{2\pi R_1 C_1} \tag{4–19}$$

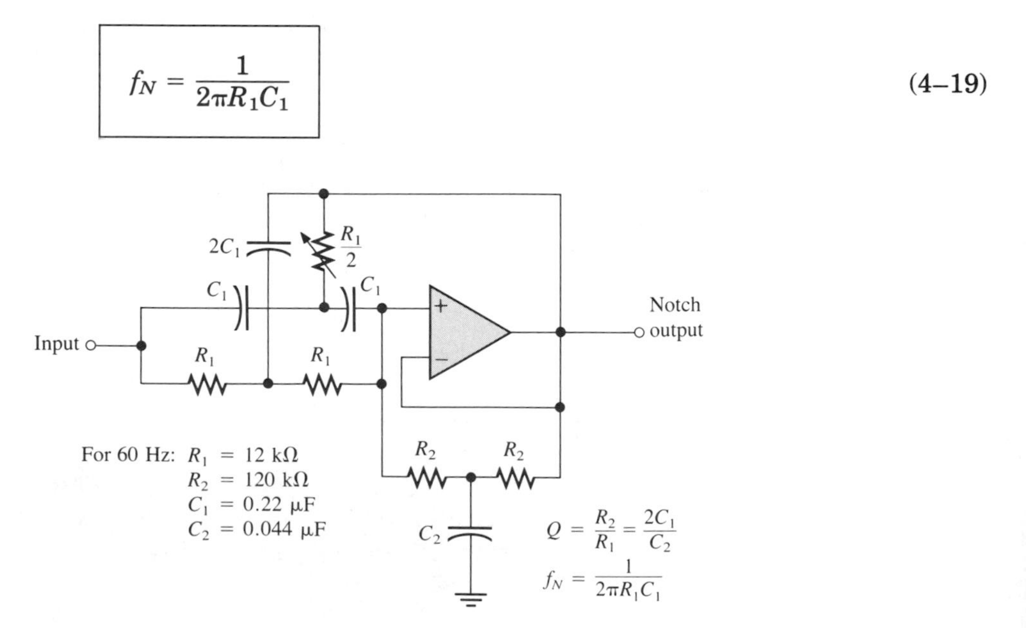

Figure 4–29 60-Hz twin-T active notch filter.

When properly adjusted, the amount of attenuation, called the *null depth,* is generally 60 dB or better.

4–9 PHYSIOLOGICAL EFFECTS

The physiological effects of electricity are primarily a function of the *applied current,* not the voltage. However, voltage cannot be discounted altogether. In most circumstances, the current increases as the voltage level increases. As shown in Table 4–1, a current of approximately 1 mA applied externally to the skin is required before any physical sensation is noted. Pain begins current increases, and there is an *involuntary* contraction of the body's muscles. For the average person, the muscular contractions reach the "can't let go" threshold at 11 to 16 mA. At approximately 100 mA, the heart *fibrillates,* or beats erratically. With currents greater than 100 mA, there is a tendency for muscular contraction to be so rapid and forceful that the affected individual is involuntarily jerked away from contact with the electrical source. As a result of this phenomenon, high current levels are often not as lethal as moderate levels.

Improper stimulation of the heart muscle is caused primarily by the local *current density* (current/area). Since the human body acts as a conductor, current flows from the point of contact with the electrical source to the exit point, usually the soles of the feet. Many studies have been done on the effect of frequency on the ventricular fibrillation threshold in animals, and interestingly enough, the greatest hazard occurs, as shown in Figure 4–30, at the vicinity of 60 Hz, the power-line frequency in the United States.

Table 4–1 Effects of 60-Hz current flowing through the average human body with 1-second contact

Current intensity	Physiological effect
1 mA	Threshold of perception
5 mA	Accepted as maximum harmless current intensity
10–20 mA	"Can't let-go" current before sustained muscular contraction
50 mA	Pain, possible fainting, exhaustion. Heart and respiratory functions continue.
100–300 mA	Ventricular fibrillation will start, but respiratory center remains intact.
6 A	Sustained contraction of heart muscle, followed by normal heart rhythm. Temporary respiratory paralysis. Burns, if current density is high. Death can result depending on voltage levels.

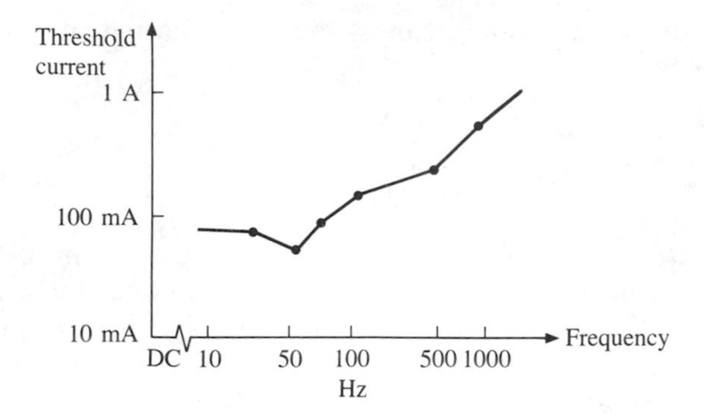

Figure 4–30 Graph of the threshold of sensation of current in the human body as a function of applied frequency.

4–10 SUMMARY

This chapter discussed the causes and source of noise and interference, either internal or external, and natural or man-made. The types of internally generated noise can be classified as thermal, shot, flicker, or burst. In addition, noise measurements in terms of signal-to-noise ratio, noise figure, and SINAD ratio were presented. Interference is noise that is externally generated and is coupled from its source to the pickup point in several ways. Among these are capacitive coupled interference, magnetically coupled interference, as well as electromagnetic and common-mode interference resulting from ground loops. The effects of noise and interference can be minimized either by proper grounding, shielding, filtering, the use of differential amplifiers, or a combination of two or more of these methods.

4–11 GLOSSARY

atmospheric noise Interference, such as that from lightning discharges, that originates in the atmosphere.

burst noise Pink noise, whose frequency spectrum varies inversely proportional to the square of the frequency. It is frequently referred to as "popcorn noise" because of the sound it makes.

capacitively coupled interference Electrically coupled, high-impedance interference whereby the electric field signal of a nearby conductor is capacitively coupled to a wire or cable.

chassis ground A reference point of zero potential which is the metal chassis of an instrument or measurement device.

cheater adapter A 3-wire socket to 2-wire plug adapter for the AC power line.

common-mode interference Interference that results from the presence of a common-mode voltage in a ground loop system.

common-mode noise voltage The input voltage that is applied simultaneously to both inputs of a differential amplifier.

common-mode rejection (CMR) The common-mode rejection ratio of a differential amplifier expressed in decibels.

common-mode rejection ratio (CMRR) The ability of a differential amplifier to cancel a common-mode signal, and is equal to the ratio of the common-mode input voltage to the generated output voltage.

crosstalk Electrical disturbances in one circuit as a result of coupling with other circuits.

effective noise bandwidth The frequency width of a rectangle with the same height and area as a given frequency-response curve, and is approximately 57% greater than the 3-dB bandwidth of a given system or network.

electrically coupled interference High-impedance interference whereby the signal of a conductor is capacitively coupled to a nearby wire or cable.

electromagnetic interference (EMI) A source of interference that contains both a magnetic field and an electric field, and created by the high-frequency energy associated with RF radiation.

fault current The current flowing in the ground wire as a result of a ground fault.

filter A selective electrical network which allows the passage of a band of frequencies while rejecting the passage of all other frequencies.

flicker noise Pink noise caused by the variations of the velocity of electrons from defects in semiconductor materials. Its magnitude is inversely proportional to frequency. It is frequently referred to as *1/f* noise.

ground A reference point considered to be at zero potential and to which all other potentials are referred.

ground fault The condition whereby the current is a 3-wire power-line system returns to ground by any path other than the neutral wire.

ground fault interrupter (GFI) A device that automatically disconnects the hot and neutral power wires from the load when it senses that the current in these two lines is not equal.

ground loop The generation of noise signals within a ground path as a result of a current within this path originating from another voltage source. This frequently results from connecting two separate grounds to a circuit.

interference Extraneous signals, including noise, which tend to make proper measurements difficult.

Johnson noise See **thermal noise.**

leakage current An unwanted (but small) flow of current.

magnetically coupled interference Inductive, low-impedance interference caused by a magnetic field created from a current-carrying conductor.

noise Any undesired electrical signal present in addition to the desired voltage or current being measured or processed.

noise figure (NF) A measure of how much noise a given network or amplifier stage adds to the signal. It is the ratio of the S/N ratio at the input of the network to the S/N ratio at the output.

Nyquist noise See **thermal noise.**

partition noise See **shot noise.**

pink noise Noise whose frequency spectrum decreases with frequency.

popcorn noise See **burst noise.**

power ground A reference point return path for the current that provides the power required to operate a given piece of equipment.

radio frequency interference (RFI) Electromagnetic interference occurring between approximately 10 kHz and 1000 GHz.

random noise An irregular signal whose instantaneous amplitude is distributed randomly with respect to time.

receptacle analyzer A device having three neon bulbs or light-emitting diodes that automatically determine if the 3-wire power-line socket is wired correctly.

resistance noise See **thermal noise.**

60-cycle hum An electrical disturbance at the AC power-line frequency or its harmonics.

Schottky noise See **shot noise.**

Schrot noise See **shot noise.**

shield ground The ground connection made at the braided shield of a shielded cable.

shot noise A natural source of white noise generated as a result of small fluctuations in the instantaneous value of a direct current. Also referred to as Schottky, Schrot, or partition noise.

signal ground The reference point and return path for all signal currents that flow.

signal-to-noise (S/N) ratio The ratio of the overall signal power to the existing noise power level.

SINAD ratio An acronym for **signal plus noise plus distortion.** It is a performance specification for radio receivers, and equals the ratio of the level of the signal plus noise plus distortion to the level of the signal and noise only.

thermal noise Random, radio-frequency white noise generated by thermal motion of electrons in a resistor. Also referred to as either Johnson, Nyquist, or resistance noise.

white noise Random radio-frequency noise, which includes shot and thermal noise, whose amplitude is essentially constant with frequency.

4–12 PROBLEMS

1. Determine the thermal noise voltage across a 1-kΩ resistor at a temperature of 70 °F if the noise bandwidth is 500 kHz.
2. For the resistor of Problem 1, determine the noise power dissipated.
3. If a 470-Ω resistor and an 1.5-kΩ resistor are placed in parallel, determine the total RMS noise voltage for a temperature of 295 K and a noise bandwidth of 1 MHz.
4. Repeat Problem 3 if the two resistors are connected in series.
5. If a 100-Ω resistor has a DC current of 5.7 mA flowing through it, determine the RMS shot noise current for a noise bandwidth of 100 kHz.
6. For the resistance of Problem 5, determine the noise voltage and the power dissipated.
7. An amplifier has an input signal of 30 μV and a noise level of 7.3 μV. Determine the signal-to-noise ratio in decibels.
8. Determine the signal-to-noise ratio in decibels if the signal power level is 16 μW while the noise power is 50 nW.
9. If the signal-to-noise ratio is 15 dB, and the noise power is 10 pW, determine the signal power.
10. Determine the noise power if the signal-to-noise ratio is 12 dB and the signal power is 10 μW.
11. If the input signal-to-noise ratio of an amplifier is 20 dB, and it has an output signal-to-noise ratio of 5 dB, what is the noise factor?
12. An amplifier has an output RMS noise voltage of 20 μV when a noise generator is coupled to the input. Determine the noise figure if, when the noise generator is removed, an RMS noise voltage of 7.2 μV is measured at the amplifier's output.
13. A differential amplifier has a differential gain of 50 and a CMR of 85 dB. What is the output voltage if the input voltages are 60 mV and 32 mV RMS while the 60-Hz common mode voltage is 9.2 V RMS?
14. In Problem 13, what is the minimum common-mode rejection of the differential amplifier if the output voltage is to be no more than 10 mV from its ideal value?
15. For the waveguide shown in Figure 4–27, the radius of the hole is 1.7 cm, while the length of the sleeve is 5.2 cm. Determine the dB attenuation if the interference has a frequency of 842 MHz.
16. If a waveguide like that shown in Figure 4–27 has an opening of 2.4 cm in diameter and a sleeve that is 6.1 cm long, what is the maximum frequency of the interference if the waveguide is to provide 80 dB of attenuation?
17. If the length of the sleeve of the waveguide of Figure 4–27 is made three times larger than its diameter, determine the dB attenuation if the frequency of the interference signal is one-fifth the cutoff frequency of the waveguide.

5

DC INDICATING METERS

5–1 INSTRUCTIONAL OBJECTIVES

At the completion of this chapter you will be able to

- Describe the operation of the d'Arsonval movement as well as the construction and characteristics of a permanent magnet moving coil meter.
- Describe the operation of the hot-wire movement and galvanometer.
- Describe the operation of the DC ammeter to include current shunts, multiple range ammeters, errors due to loading effects, and precautions in using DC ammeters.
- Describe how a DC ammeter is used to form a DC voltmeter and explain voltmeter ohms-per-volt specifications, errors due to loading effects, and precautions in using DC voltmeters.
- Determine the internal meter resistance using both the half-scale and the voltage-current methods.
- Describe how to calibrate DC ammeters and voltmeters using comparison methods.

5–2 INTRODUCTION

Until about 1960, nearly all DC measurements depended upon some type of an electromechanical meter movement. Although digital instruments have made considerable inroads, the analog meter is still very popular. An analog meter in its simplest form uses an electromechanical movement and pointer to display the quantity being measured along a continuous (i.e., analog) scale. This chapter describes the characteristics and construction of DC ammeters and voltmeters, as well as the error each meter introduces to the measurement process.

5–3 THE D'ARSONVAL MOVEMENT

The most common electromechanical meter movement is the **d'Arsonval** or **permanent magnet moving coil** (PMMC) type. It was developed by Jacques d'Arsonval in 1881, and has been in continuous use ever since in all type of electrical measuring instruments.

The meter movement is driven by current and uses the force arising from the interaction of a magnetic field and a current-carrying conductor to rotate a moving coil against the restraining force of a spiral spring. Figure 5–1 shows the direction of the force that is exerted on a single current-carrying conductor that is oriented perpendicular to a magnetic field.

The resulting force is perpendicular to both the magnetic field and to the direction of current flow. The force is proportional to the magnitude of the field strength, the magnitude of the current, and the length of the conductor immersed in the magnetic field. Mathematically this is represented by

$$F = BLI \tag{5–1}$$

where

F = force in newtons
B = magnetic flux density in teslas
I = current in amperes
L = length of conductor in the magnetic field measured in meters.

Figure 5–2 shows how a torque is produced on a single-turn, current-carrying coil when it is immersed in a magnetic field between the two poles of a permanent magnet.

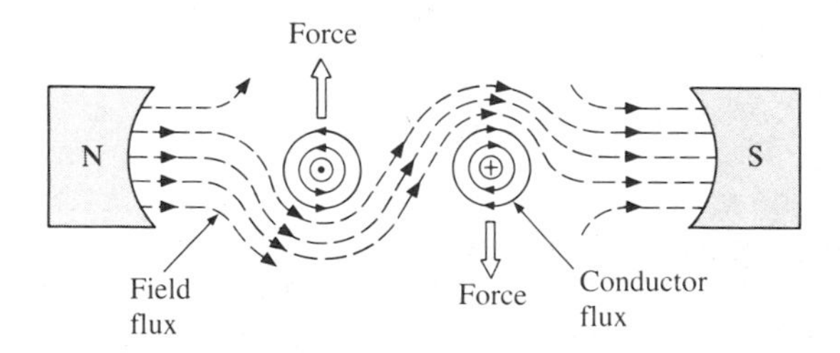

Figure 5–1 The direction of the force exerted on a single current-carrying conductor perpendicular to a magnetic field.

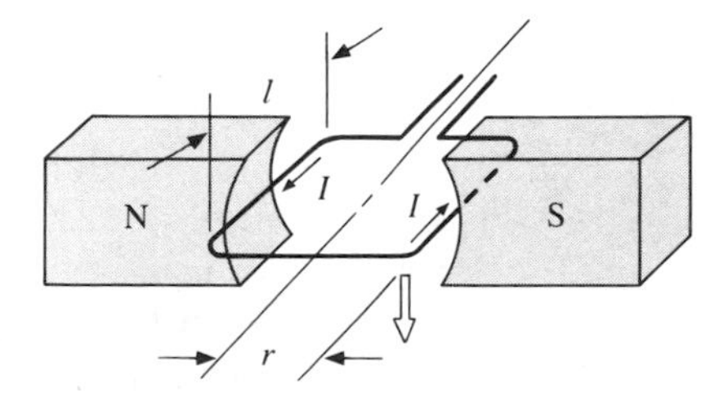

Figure 5–2 Producing a torque on a single-turn loop in a magnetic field.

The force on each conductor acting at a radius, r, from the axis of rotation to produce a torque (τ) is equal to

$$\begin{aligned}\tau &= 2Fr \\ &= 2BLIr\end{aligned} \tag{5–2}$$

Since $2r$ is the width of the coil of length L, the resulting torque may also be expressed as the area (A) of the coil (in the magnetic field) multiplied by the current and the flux density, so that

$$\tau = BIA \tag{5–3}$$

Adding turns to the coil increases the torque proportionally. Thus, the torque of a coil having n turns is given by

$$\tau = nBIA \tag{5–4}$$

Reversing the direction of current flow reverses the direction of the resulting torque.

PMMC Meter Construction

The movement that d'Arsonval patented uses a permanent magnet to supply the stationary magnetic field as well as a moving coil to drive the pointer. Figure 5–3 shows the essential elements of the d'Arsonval movement.

The coil is wound on a rectangular, nonmagnetic, light metal frame, such as aluminum. The frame is free to rotate in a narrow air gap between soft iron pole pieces and a soft iron concentrically mounted cylinder. The geometry of the magnetic path is designed to provide low magnetic reluctance and keep the flux at right angles to the sides of the air gap. This configuration maintains the maximum possible flux density and provides the highest possible torque for a given current, number of turns, and coil area.

From Equation 5–4, the torque may also be increased by adding more turns or by increasing the area of the coil. However, both of these options soon reach their limitations. Increasing the dimensions of the coil means greater inertia, bearing loading, and physical size. Additional turns increase the electrical resistance of the meter.

A modern improvement of the d'Arsonval movement is one that substitutes a magnet for the concentrically mounted cylinder, and completes the magnetic path outside the coil with a low-reluctance yoke that completely sur-

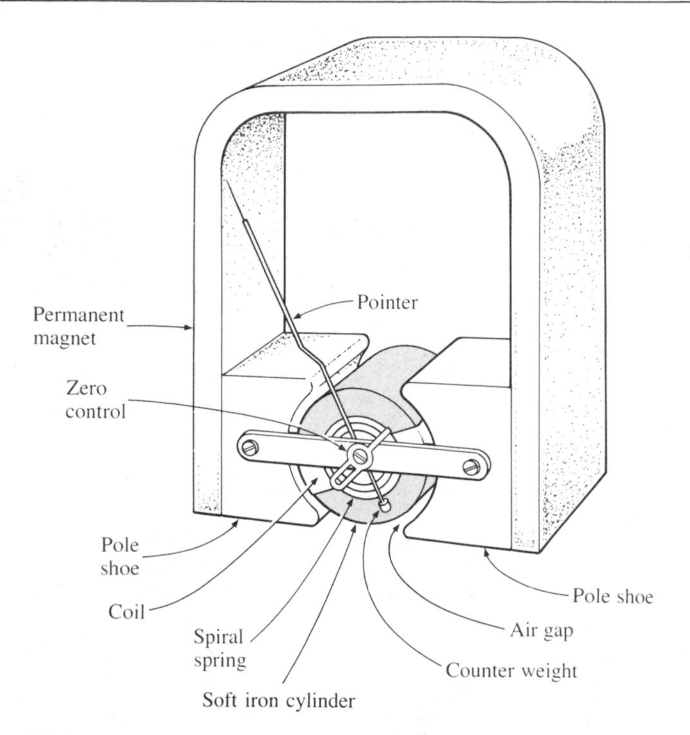

Figure 5–3 Construction details of a d'Arsonval meter movement.

rounds the coil. This type of construction provides enhanced magnetic shielding, which makes the meter less sensitive to stray magnetic fields and the shunting effect of steel panels. This improved arrangement is said to be *self-shielded.* It is illustrated in Figure 5–4.

The torque caused by a current flowing through the coil is opposed by a pair of phosphor bronze spiral springs, which also serve as conducting paths between the ends of the coil and the stationary wires to the meter terminals. The counter-torque due to these springs is a linear function of their angular displacement. When there is no current flowing in the coil, the coil and attached pointer rest at the position of zero spring torque. However, when a current is present, the coil rotates to a position where the magnetic torque is opposed by an equal spring torque. This then gives an angular displacement that is a *linear* function of coil current.

Maximum limits on angular rotation are provided by a pair of mechanical stops, one at each end of travel. The zero position is set by an eccentric pin that is accessible through the front cover. This eccentric pin engages a Y-shaped member that controls the angular position of the front spring anchor point. By rotating the eccentric pin, the rest position (the position with zero coil current) of the pointer may be adjusted over a narrow range about the zero scale mark.

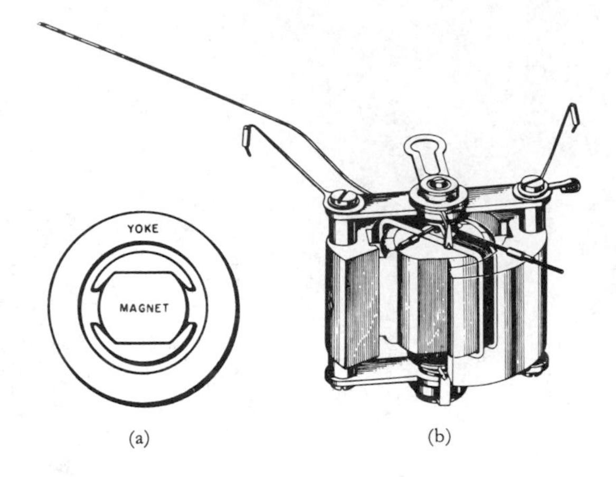

Figure 5–4 Construction of a core-magnet moving coil mechanism. (a) Top view of a permanent magnet surrounded by a yoke acting as an electromagnetic shield. (b) Complete assembly. Courtesy Solartron Electronics Inc., Weston USA Division.

Motion Damping

The aluminum coil frame serves to dampen mechanical oscillations due to sudden current changes by acting as a dynamic brake. The aluminum frame acts as a shorted turn, and rotation of the frame through the magnetic field induces a circulating current in the frame. The magnitude of this circulating current is proportional to the speed of rotation. The resulting magnetic forces are directed in such a manner as to oppose the motion that creates them. Thus, the presence of the coil frame has no influence on the position of the coil when it is at rest, but it opposes any sudden rotational motion and quickly dampens any oscillation tendencies when the movement comes to a new position.

Connecting a resistor across the meter terminals or short circuiting them also effects a dynamic braking. Here, as the meter winding itself cuts the magnetic field, a voltage is induced in the winding, and a circulating current passes through the external path. This current produces a torque that tends to retard motion. A **shorting wire** is often placed between meter terminals during shipment to dampen meter motion due to rough handling.

Balance

Several small, adjustable counter weights are normally provided to statically balance the rotational parts of the movement for all positions of rotation.

Moving-Coil Support System

There are two commonly used methods of supporting the coil shaft. As shown in Figure 5–5a, the older method uses a pair of jeweled bearings with cone-shaped recesses, one at each end of a pointed shaft. The bearings themselves are usually made of sapphire or glass, and the shaft is pointed at the ends to provide the least possible friction. Since the actual contact area is quite small, the bearing loading can be large. It may exceed 10 tons per square inch, despite the fact that the meter coil assembly itself is relatively light. The high bearing load

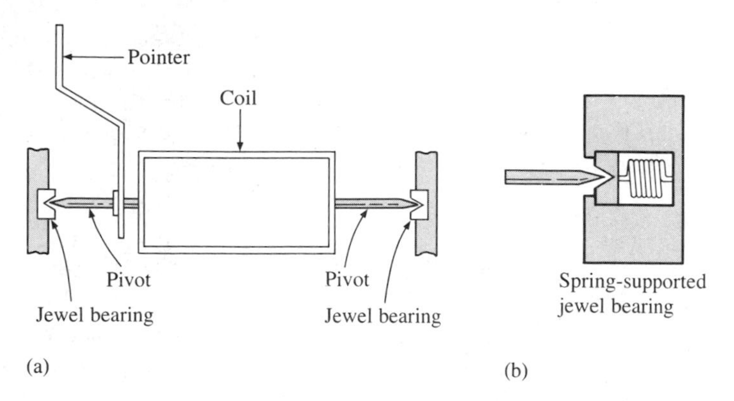

Figure 5–5 Jeweled bearing moving coil support mechanism.

and brittle bearing material combine to make this arrangement susceptible to damage due to mechanical shocks. A slightly better mechanical arrangement uses spring-supported jeweled bearings to better absorb the effects of shock (Figure 5–5b).

A more rugged mounting arrangement, shown in Figure 5–6, consists of two torsion members called **taut bands.** These replace the usual shaft and spiral spring arrangement. Taut bands are flat metal bands, one at each end of the coil frame, held under tension by springs anchored to the meter frame. Any twisting of these bands produces a counter torque that is a linear function of the rotational displacement. The taut band system eliminates the delicate bearings and springs, and greatly reduces friction.

Suspension bands are made of materials such as platinum-nickel that exhibit great tensile strength as well as a small degree of fatigue in elasticity. Since friction is reduced, taut-band meters can be made more sensitive than those types mounted on bearings. Since the suspension tends to cushion the moving parts, this type of meter movement is better able to absorb mechanical shocks.

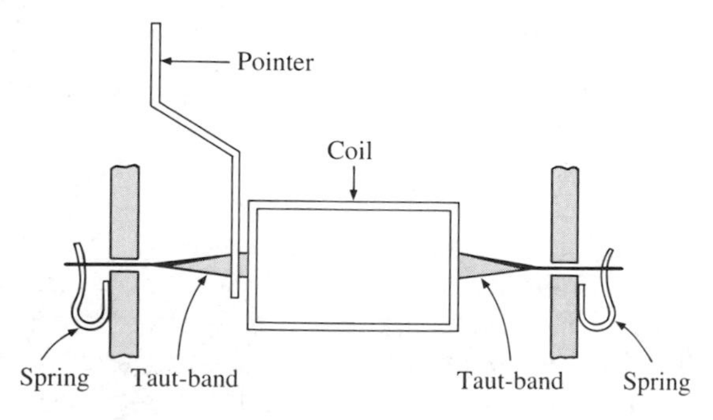

Figure 5–6 Taut-band coil support mechanism.

The taut-band meter construction is suited for most applications, but the use of a meter having a pivot and jewel suspension is preferred in those environments where vibrations cause the pointer to oscillate so much that it is difficult to read the meter. This may occur in automobiles or on control panels connected to machinery such as motors or generators.

d'Arsonval Response

The d'Arsonval meter movement has a certain amount of mechanical inertia as well as a controlled amount of built-in damping. These factors make it unable to follow rapid variations in the coil current. Except for those relatively slow changes in signal that the movement can follow, the meter is deflected according to the *average value* of coil current. When the PMMC movement is used to measure a current or voltage that contains both a single AC component (at a frequency higher than the meter can follow) and a DC (static) component, it reads only the *DC component* since the average value of the AC component is generally zero.

Sensitivity and Coil Resistance

Although it is possible to make PMMC movements that have a **full-scale deflection (FSD)** current of only a few microamperes, a range of 0–20 μA is about the most sensitive range practical that is commercially available for general use. The construction of more sensitive meters would be too delicate for most purposes.

A typical, relatively sensitive, commercial microammeter of the pivot and jewel type has a range of 0–50 μA and a coil resistance of 2300 Ω. The same type of meter with a range of 0–100 μA has a coil resistance of 825 Ω because less sensitive movements generally have a lower DC resistance. A more sensitive movement in a meter, more turns of wire on the coil. To fit in the same physical space, the wire must have a smaller diameter. The smaller diameter causes an increase in the resistance of the wire because wire resistance is inversely proportional to its cross-sectional area and directly proportional to length.

Temperature Compensation

Both the magnetic field strength and spring tension tend to decrease with an increase in temperature. The resistance of the coil increases with temperature. The net result of these changes is that the meter tends to read slightly low at higher temperatures. For this reason, meters that must operate over a wide temperature range must have some form of compensation. This compensation may be implemented by connecting a resistance wire with a low temperature coefficient, such as manganin or similar metal, in series with the moving coil to swamp out the effect of any change in coil resistance. A better method, however, uses the circuit shown in Figure 5–7. In this circuit, R_M is a low temperature-coefficient resistance wire, and R_C is a copper resistance wire.

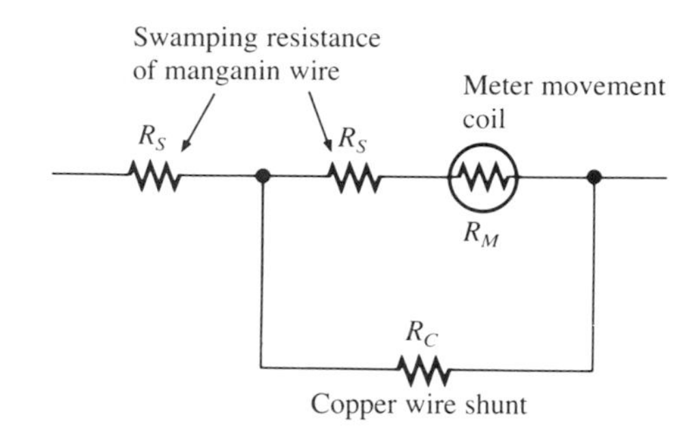

Figure 5–7 Use of swamping resistors to provide temperature compensation of the meter movement.

As the ambient temperature increases, the resistance of R_C increases. The increased resistance diverts more current through the coil. The increased current combined with the slight decrease in spring tension compensates for the decrease in magnetic field strength. Unfortunately, the added series resistance decreases the sensitivity of the movement. A typical PMMC movement with a full-scale accuracy of ±2% may have the following specification expressed as follows: "compensated for less than an additional ±2% change in full-scale accuracy over the range of −40 °C to +71 °C."

5–4 THE HOT-WIRE MOVEMENT

For some low-accuracy applications such as automobile instruments, a **hot-wire** meter movement is sometimes used. This device depends on the expansion of a heated wire to move a pointer. This type of meter is operated by passing the current to be sensed through a hot-wire. As a result, the meter is simple, rugged, and low cost, but it suffers from a nonlinear scale, lack of sensitivity, and error due to changes in ambient temperature.

5–5 THE GALVANOMETER

A **galvanometer** is essentially an application of the PMMC movement to detect the presence of extremely low current levels. Many true galvanometers have an *uncalibrated* zero-center scale. (Uncalibrated means that the scale divisions do not correspond to standard units.) This type of instrument indicates only the direction and relative magnitude of current. When direct current passes through the coil in one direction, the pointer is deflected to the left. When direct current passes through the coil in the opposite direction, the pointer is deflected to the right.

One of the earliest types of galvanometers, which is still used for laboratory measurements requiring very high sensitivity, is the *suspension galvanometer*. This is probably the earliest application of the taut-band principle. As shown in Figure 5–8, a coil of fine wire is suspended vertically by a fine metallic filament in the field of a permanent magnet. The metallic filament acts as a

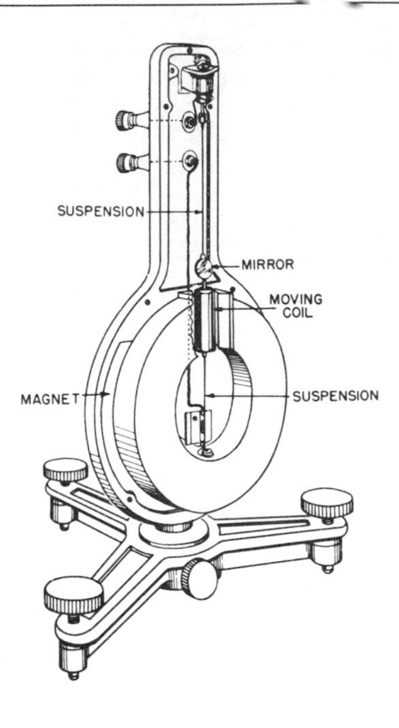

Figure 5–8 The suspension type galvanometer. Courtesy Solartron Electronics Inc., Weston USA Division.

torsion element analogous to the metal bands in the taut-band suspension system, but with a much lower restraining torque constant. The suspension filaments also serve as a conducting path between the galvanometer terminals and the moving coil.

Instead of a pointer, the suspension galvanometer has a small mirror attached to the moving coil. This mirror is used to deflect a focused light beam along a scale mounted at some distance from the galvanometer itself. This light beam is equivalent to a very long pointer with almost no mass.

5–6 DC AMMETERS

The DC **ammeter** is used to measure the steady current that flows from one point to another in a given circuit. The basic DC ammeter is capable of measuring currents from microamperes to several milliamperes. When used with external resistance shunts, its range can be extended to currents of several hundred amperes. Since the ammeter is the fundamental building block of all analog meters, it is then possible to create other types of meters that measure voltage or resistance.

The Basic Ammeter Circuit

The PMMC movement is inherently an ammeter. The amount of current required to deflect the meter to full scale is called the **full-scale deflection current,** I_{FSD}, and the resistance of the coil winding is designated R_M. Since it

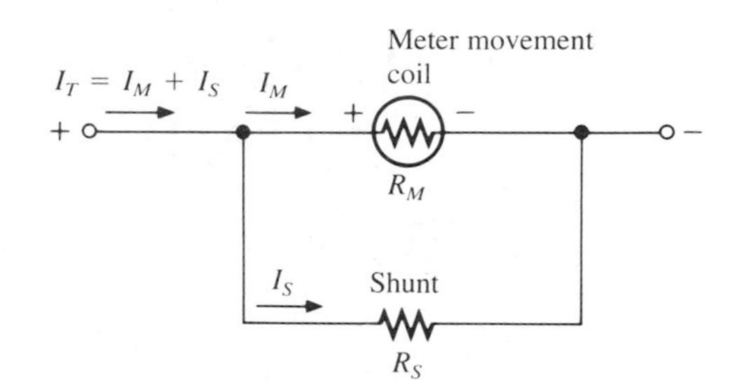

Figure 5–9 Basic DC ammeter circuit with parallel shunt resistance.

is desirable to keep the basic meter movement small and light, the full-scale deflection current for most movements is less than 100 μA. To measure larger currents, it is necessary to bypass or shunt excess current around the meter coil. Figure 5–9 shows the basic DC ammeter circuit, which consists of a meter movement with a **coil resistance**, R_M, and a parallel **shunt resistor**, R_S.

Since the meter movement and shunt resistor form a parallel circuit, the voltage across the meter movement and the shunt resistor must be equal, so that

$$V_{\text{SHUNT}} = V_{\text{MOVEMENT}}$$

$$I_S R_S = I_M R_M \tag{5–5}$$

From Equation 5–5, the shunt resistor, R_S, is then

$$R_S = \frac{I_M R_M}{I_S} \tag{5–6}$$

Using Kirchhoff's current law, the current through the ammeter shunt equals the total meter current minus the current through the meter movement, or $I_S = I_T - I_M$. If I_T equals the full-scale current of the ammeter circuit and I_{FSD} equals the full-scale current of the meter movement, the required shunt resistance is then found from

$$R_S = \frac{I_{\text{FSD}} R_M}{I_S} = \frac{I_{\text{FSD}} R_M}{I_T - I_{\text{FSD}}} \tag{5–7a}$$

An alternative calculation involves determining the factor, N, by which the meter scale is to be increased, so that

$$R_S = \frac{R_M}{N - 1} \tag{5–7b}$$

This parallel combination of a shunt resistor and the resistance of the meter movement results in an ammeter with a full-scale deflection of I_T, as illustrated by the following example.

EXAMPLE 5–1

Determine the size of the required shunt resistance as well as its power rating to use a 0–100 μA PMMC meter movement having an internal (coil) resistance of 1 kΩ as the basis for a 0–10 mA ammeter.

Solution:

(a) For a full-scale meter current of 10 mA, the shunt current is then 10 mA − 0.1 mA, or 9.9 mA. Using Equation 5–7*a*, the required shunt resistance is then

$$R_S = \frac{(0.1\text{ mA})\,(1\text{ k}\Omega)}{(10 - 0.1\text{ mA})}$$

$$= 10.1\ \Omega$$

(b) The maximum voltage across the shunt resistor at a full-scale meter current of 10 mA is 100 mV. Therefore, its power rating is

$$P = (100\text{ mV})\,(9.9\text{ mA})$$

$$= 0.99\text{ mW}$$

On the other hand, using Equation 5–7*b*, N is 10 mA/100 μA, or a ratio of 100:1. Therefore,

$$R_S = \frac{1\text{ k}\Omega}{100 - 1}$$

$$= 10.1\ \Omega$$

giving the same result as before.

The voltage drop across this ammeter circuit at the maximum current of 10 mA is equal to $I_S R_S$ or $I_M R_M$, and is equal to 100 mV. The total resistance presented by this ammeter is the parallel combination of the 1000-Ω meter movement and the 10.1-Ω shunt, equal to about 10 Ω. Thus, for a given PMMC movement, a higher current range requires a lower value of meter shunt resistance. This lower meter shunt resistance, in turn, lowers the value of total ammeter resistance.

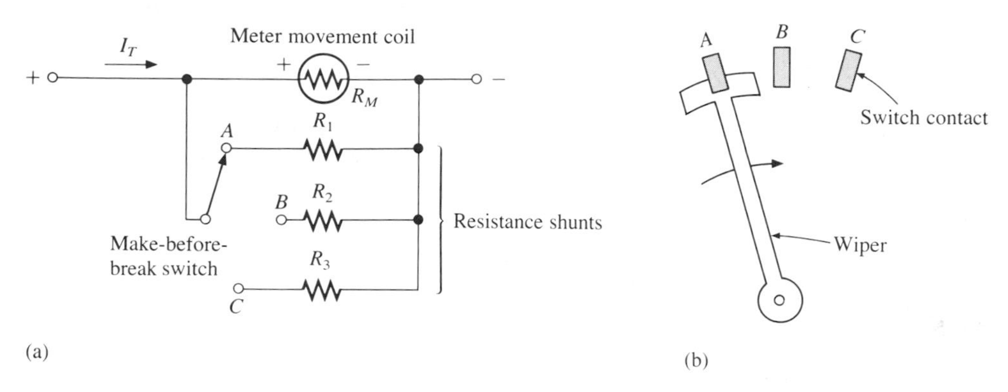

Figure 5–10 Multirange ammeter using switch-selectable shunts. (a) Schematic using switch-selectable resistance shunts. (b) Make-before-break switch detail.

The ideal DC ammeter would have no resistance. In this way, it would not add to the resistance of the circuit under test. Unfortunately, this is not possible. The effects of ammeter resistance must *always* be considered when making current measurements.

Multiple Range DC Ammeters

Multiple ranges may be added to the basic ammeter circuit by providing a series of switch-selectable shunts, as shown in Figure 5–10*a*.

With this circuit, it is very important that the switch is of the *make-before-break* type so that the meter movement is never without a shunt when the range switch moves from one position to another. Otherwise, the total current passes through the meter movement and destroys it. Figure 5–10*b* illustrates the details of the wiper and contacts of a make-before-break switch.

Figure 5–11 shows what is known as an **Ayrton shunt,** which eliminates

Figure 5–11 Ayrton shunt-type multirange ammeter.

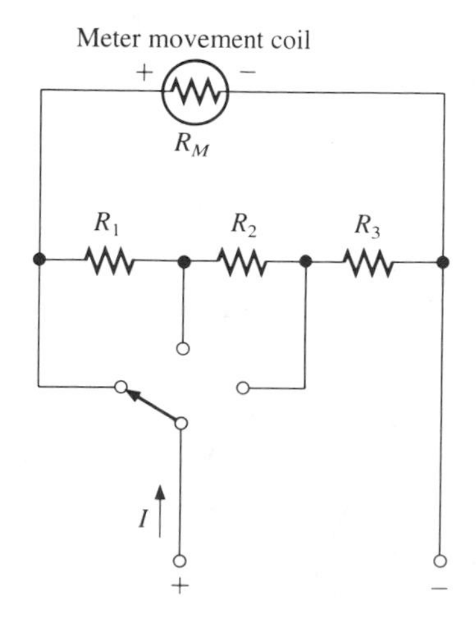

the need for a make-before-break type switch, but has the disadvantage of higher insertion resistance.

With the Ayrton shunt, the resistance between the switch arm and the common lead is the actual shunt. The resistance above the arm is in series with the movement and serves to increase the *effective* value of the movement resistance, R_M. Thus, each range has a different effective value of R_M as well as a different value of R_S.

Ammeter Loading Effects

Ideally, an ammeter inserted in a circuit should present no resistance. However, as has been mentioned previously, ammeters do have some internal resistance. The amount of internal resistance depends on the type of suspension employed in its construction as well as its full-scale range. As summarized in Table 5–1, the internal resistance of a DC ammeter can range from several ohms to several thousand ohms. As a result, the internal resistance presents a loading error. The internal resistance increases the resistance of the path of the current being measured.

Table 5–1 Typical Ammeter Internal Resistance Values

Full-scale current	Taut band	Pivot and jewel
50 μA	1–2 kΩ	2–5 kΩ
1 mA	20–100 Ω	50–150 Ω
10 mA	1–5 Ω	2–5 Ω

To demonstrate the loading effect error of an ammeter, consider first the circuit of Figure 5–12*a*. Current I_1 is the current without an ammeter in the circuit. The *expected* current through R_1 is then

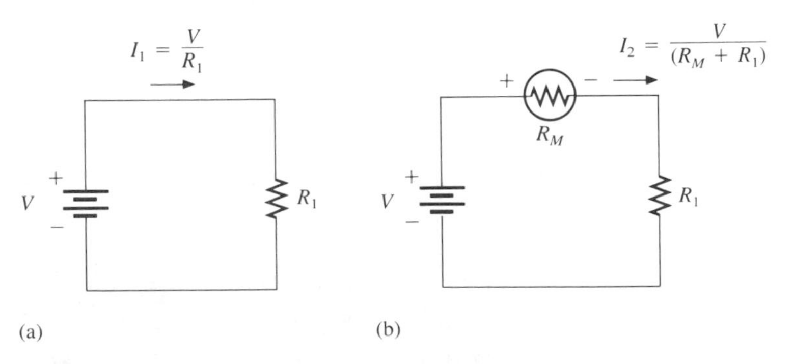

Figure 5–12 Ammeter loading effects. (a) Without ammeter. (b) With ammeter in circuit.

$$I_1 = \frac{V}{R_1} \tag{5–8}$$

When an ammeter having an internal resistance R_M is placed in series with R_1, as in Figure 5–12*b*, the *measured* current I_2 flowing through R_1 as indicated by the meter is now

$$I_2 = \frac{V}{R_1 + R_M} \tag{5–9}$$

Since the total resistance in the circuit is increased, the current through R_1 is decreased by the inclusion of the internal meter resistance. By dividing Equation 5–9 by Equation 5–8, we are then able to express the percent error that results from the inclusion of the ammeter in a circuit, so that

$$\begin{aligned} \%\ \text{error} &= \left(1 - \frac{I_2}{I_1}\right) \times 100\% \\ &= \left(1 - \frac{R_1}{R_1 + R_M}\right) \times 100\% \end{aligned} \tag{5–10}$$

EXAMPLE 5–2

If a DC ammeter having an internal resistance of 50 Ω is placed in a series with R_1 in the circuit of Figure 5–13, determine the percent error of the meter reading that results from ammeter loading.

Figure 5–13 Circuit for Example 5–2.

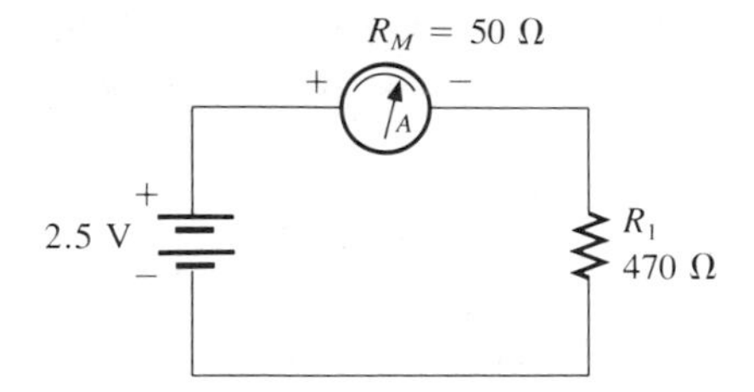

Solution:

Using Equation 5–10,

$$\begin{aligned} \%\ \text{error} &= \left(1 - \frac{470}{470 + 50}\right) \times 100\% \\ &= 9.6\% \end{aligned}$$

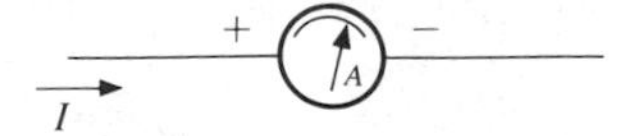

Figure 5–14 Direction of current flow through an ammeter versus meter polarity using conventional current flow.

Precautions in Using DC Ammeters

Since the ammeter is a delicate device, certain precautions should be followed to guard against damage through carelessness and misuse.

1. Never connect an ammeter directly across a source of voltage. An ammeter normally has a relatively low value of resistance. Ammeters should always be connected in *series* with a load of sufficient resistance to prevent excessive current from flowing through the meter. Excess meter current, which forces the pointer hard against the mechanical stop at the upper end of the scale, could permanently damage the pointer or destroy the coil or shunt.
2. Observe proper polarity when connecting an ammeter. A large current in the reverse direction causes deflection against the stop at the lower end of the scale and could go unnoticed despite the fact that the coil is overheated and damaged. If the pointer strikes the stop hard enough, it may be bent. Using conventional current flow, Figure 5–14 shows the proper polarity of the ammeter.
3. Initially, always set a multirange ammeter to its *highest* range setting *before* inserting it in a circuit. As with any changes made to a circuit, this should always be done with the power *off*. After power is applied to the circuit, then decrease the range setting to permit as high a scale deflection as possible without going off scale. This is done because the highest meter accuracy is near the *full-scale* reading. As an added precaution, a normally closed, momentary push-button switch is sometimes connected across an ammeter so that current passes through the meter only while the button is depressed.

5–7 DC VOLTMETERS

The DC **voltmeter** is used for measuring steady voltages from one point to another. In its simplest form, it is nothing more than a DC ammeter with a series resistor, and is capable of measuring voltages ranging from microvolts to kilovolts.

The Basic Voltmeter Circuit

The basic DC voltmeter circuit shown in Figure 5–15 consists of a resistor R_S, called the **multiplier resistor**, which is connected in series with a PMMC-movement DC ammeter. With full-scale voltage applied to the meter circuit,

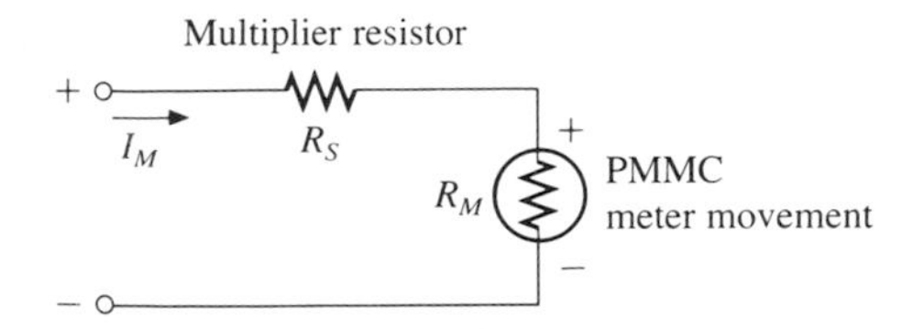

Figure 5–15 The basic DC voltmeter.

the total resistance of the circuit must limit the current through the meter to exactly the value needed for full-scale deflection, I_{FSD}.

The total meter circuit resistance is the sum of the external series resistance and the resistance of the moving coil

$$R_{TOTAL} = R_S + R_M \tag{5–11}$$

Since the total resistance must equal the applied voltage divided by the meter current, we may solve for the series multiplier resistor, R_S, as follows

$$R_{TOTAL} = R_S + R_M$$
$$= \frac{V_{FSD}}{I_{FSD}}$$

Rearranging, we obtain

$$R_S = \frac{V_{FSD}}{I_{FSD}} - R_M \tag{5–12}$$

EXAMPLE 5–3

It is desired to construct a voltmeter with a range of 0–10 V using a PMMC movement having a full-scale current of 100 μA and a coil resistance of 1 kΩ. Determine the required value and power rating of the multiplier resistor.

Solution:

From Equation 5–12,

$$R_S = \frac{10\text{ V}}{100\ \mu\text{A}} - 1\text{ k}\Omega$$
$$= 99{,}000\ \Omega$$

Such a resistance must then have a power rating of

$$P = (99\ \text{k}\Omega)\ (100\ \mu\text{A})^2$$
$$= 0.99\ \text{mW}$$

Multiple Range DC Voltmeters

Multiple ranges may be added to the basic DC voltmeter circuit of Figure 5–15 by providing a number of switch-selectable, series resistances. Figure 5–16 illustrates this arrangement for three voltage ranges.

Range 1 uses only R_1 as the external series resistance and is the lowest range. Range 2 adds R_2 in series with R_1; Range 3 uses the series resistance of the entire resistance string.

EXAMPLE 5–4

The required full-scale voltage ranges are to be 1 V, 5 V, and 10 V respectively. Determine the values of the three series resistances R_1, R_2, and R_3 to construct a voltmeter using the circuit of Figure 5–16, using a 100-μA meter movement having an internal resistance of 1 kΩ.

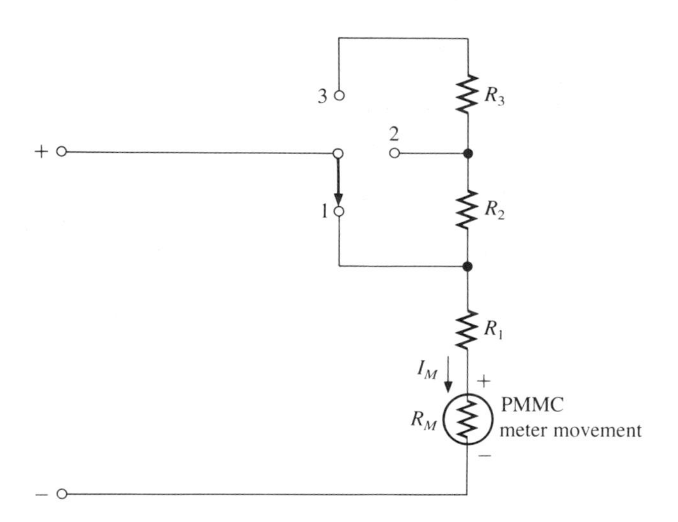

Figure 5–16 Multirange DC voltmeter with three voltage ranges.

Solution:

(a) For the 1-V range, using Equation 5–12

$$R_1 = \frac{1\ \text{V}}{100\ \mu\text{A}} - 1\ \text{k}\Omega$$
$$= 9{,}000\ \Omega$$

(b) For the 5-V range

$$R_1 + R_2 = \frac{5\text{ V}}{100\ \mu\text{A}} - 1\text{ k}\Omega$$
$$= 49{,}000\ \Omega$$

so that

$$R_2 = 40{,}000\ \Omega$$

(c) For the 10-V range

$$R_1 + R_2 + R_3 = \frac{10\text{ V}}{100\ \mu\text{A}} - 1\text{ k}\Omega$$
$$= 99{,}000\ \Omega$$

so that

$$R_3 = 50{,}000\ \Omega$$

Voltmeter Ohms-per-Volt Specifications

If the total resistance of a voltmeter is divided by the voltage required for full-scale deflection, we arrive at what is known as the voltmeter's *ohms-per-volt rating,* or **sensitivity factor** *(S)*. In the case of multirange voltmeters, the ohms-per-volt rating is a constant value regardless of the range selected. It is a function only of the current required for full-scale deflection of the basic meter movement, and is equal to $1/I_{FSD}$.

In a practical sense, the ohms-per-volt rating is a measure of the loading effect that a meter exhibits on a given range. If the ohms-per-volt rating of a voltmeter is known, it is then a simple matter to determine the resistance of the voltmeter by multiplying the ohms-per-volt rating by the full-scale voltage of the range of interest. For example, a voltmeter with a rating of 20 kΩ/V on its 100-V range would offer a resistance of

$$\text{Resistance} = (20\text{ k}\Omega/\text{V}) \times (100\text{ V})$$
$$= 2\text{ M}\Omega$$

For most applications, the higher the ohms-per-volt rating, the better.

EXAMPLE 5–5

Determine the ohms-per-volt rating of the voltmeter of Example 5–4.

Solution:

(a) For 1-V range

$$S = \frac{10\text{ k}\Omega}{1\text{ V}}$$
$$= 10\text{ k}\Omega/\text{V}$$

(b) For the 5-V range

$$S = \frac{50\ \text{k}\Omega}{5\ \text{V}}$$

$$= 10\ \text{k}\Omega/\text{V}$$

(c) For the 10-V range

$$S = \frac{100\ \text{k}\Omega}{10\ \text{V}}$$

$$= 10\ \text{k}\Omega/\text{V}$$

(d) Taking $1/I_{\text{FSD}}$

$$S = \frac{1}{100\ \mu\text{A}}$$

$$= 10\ \text{k}\Omega/\text{V}$$

The voltmeter's ohms-per-volt rating is a constant and independent of the voltage range, as shown by this example.

The ohms-per-volt rating of a voltmeter is also useful in calculating the size of multiplier resistor used with a given meter movement, so that

$$R_S = (S \times V_{\text{FSD}}) - R_M \tag{5–13}$$

where

R_S = multiplier resistor
S = ohms-per-volt rating of the meter
V_{FSD} = full-scale voltage
R_M = meter movement resistance

EXAMPLE 5–6

Determine the multiplier resistors of Example 5–4 using the ohms-per-volt rating, S.

Solution:

(a) From Example 5–5, $S = 10\ \text{k}\Omega/\text{V}$. Using Equation 5–10 for the 1-V range

$$R_1 = (10\ \text{k}\Omega/\text{V})(1\ \text{V}) - 1000\ \Omega$$
$$= 9000\ \Omega$$

(b) For the 5-V range

$$R_1 + R_2 = (10\ \text{k}\Omega/\text{V})(5\ \text{V}) - 1000\ \Omega$$
$$= 49{,}000\ \Omega$$

so that

$$R_2 = 40{,}000\ \Omega$$

(c) For the 10-V range

$$R_1 + R_2 + R_3 = (10\ \text{k}\Omega/\text{V})\ (10\ \text{V}) - 1000\ \Omega$$
$$= 99{,}000\ \Omega$$

so that

$$R_3 = 50{,}000\ \Omega$$

giving the same results as in Example 5–4.

Voltmeter Loading Effects

Since the PMMC is actuated by a current passing through it, an analog voltmeter will draw some amount of deflection current from the circuit under test. Thus, the meter acts as an electrical load on the portion of the circuit to which it is connected. In addition, this has the same effect as connecting a resistor equal to the resistance of the voltmeter across that portion of the circuit. If the current drawn by the meter is a significant load on the circuit under test, it can make a considerable difference in circuit operation. In this situation, like the DC ammeter, the DC voltmeter is said to *load down* the circuit.

EXAMPLE 5–7

As shown in Figure 5–17, the DC voltmeter of Example 5–4 is connected to the output of a signal source that has an internal equivalent series resistance (i.e., output impedance) of 5 kΩ and an open circuit terminal voltage of 0.9 volts.

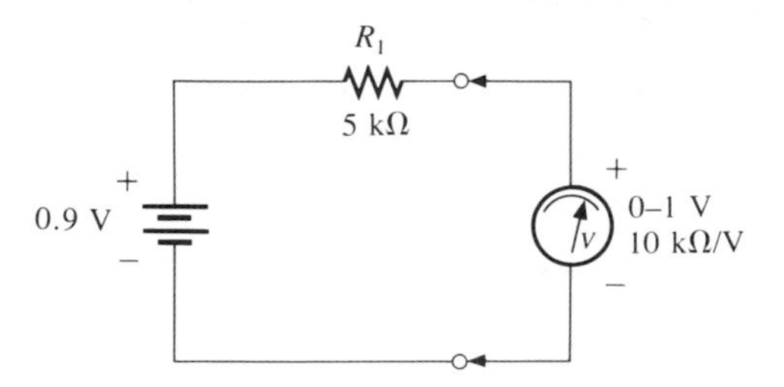

Figure 5–17 Circuit for Example 5–7.

(a) Determine the actual voltage as read by the meter.
(b) Determine the percent error introduced by voltmeter loading.
(c) If the voltmeter range is changed to the 5-V scale, determine the meter reading and the percent error.

Solution:

(a) From Example 5–4, the total meter resistance on the 1-V range is 10 kΩ. The meter and the resistance of the voltage source form a voltage divider, so that the actual voltage across the meter is

$$V_M = (0.9\text{ V})\left(\frac{10\text{ k}\Omega}{10\text{ k}\Omega + 5\text{ k}\Omega}\right)$$

$$= 0.6\text{ V}$$

(b) The percent error introduced by voltmeter loading is

$$\%\text{ error} = \left(\frac{9\text{ V} - 6\text{ V}}{9\text{ V}}\right) \times 100\%$$

$$= 33.3\%$$

(c) Using the 5-V range, the meter will read

$$V_M = (0.9\text{ V})\left(\frac{50\text{ k}\Omega}{50\text{ k}\Omega + 5\text{ k}\Omega}\right)$$

$$= 0.82\text{ V}$$

while the percent error is

$$\%\text{ error} = \left(\frac{0.9\text{ V} - 0.82\text{ V}}{0.9\text{ V}}\right) \times 100\%$$

$$= 8.9\%$$

Although the meter loading effect is reduced by using a higher meter range, both meter resolution and accuracy decrease. In general, however, use the lowest meter range possible; increased resolution and accuracy outweigh loss of performance from meter loading.

EXAMPLE 5–8

A voltmeter that uses a 50-μA PMMC meter movement is set to the 100-V range to measure the voltage across R_2 in the circuit of Figure 5–18.

(a) Determine the voltage read by the meter reading.
(b) Determine the voltage across R_2 without the meter loading effect.

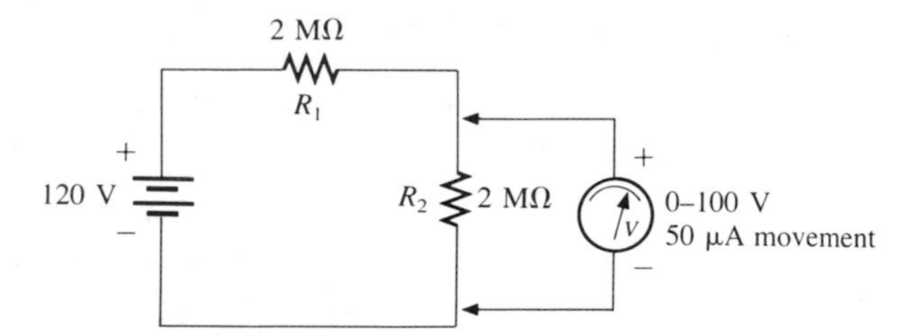

Figure 5–18 Circuit for Example 5–8.

Solution:

(a) The meter sensitivity is

$$S = \frac{1}{50\ \mu\text{A}}$$
$$= 20\ \text{k}\Omega/\text{V}$$

The voltmeter resistance on the 100-V scale is then

$$R_M = (100\ \text{V})(20\ \text{k}\Omega/\text{V})$$
$$= 2\ \text{M}\Omega$$

The parallel resistance of R_2 and the voltmeter is

$$R_P = \frac{(2\ \text{M}\Omega)(2\ \text{M}\Omega)}{2\ \text{M}\Omega + 2\ \text{M}\Omega}$$
$$= 1\ \text{M}\Omega$$

so that the voltage across R_2 with the meter connected is

$$V_{R_2} = (120\ \text{V})\left(\frac{R_P}{R_P + R_1}\right)$$
$$= (120\ \text{V})\left(\frac{1\ \text{M}\Omega}{1\ \text{M}\Omega + 2\ \text{M}\Omega}\right)$$
$$= 40\ \text{V}$$

(b) The voltage across R_2 without the meter connected is

$$V_{R_2} = (120\ \text{V})\left(\frac{R_1}{R_1 + R_M}\right)$$
$$= (120\ \text{V})\left(\frac{2\ \text{M}\Omega}{2\ \text{M}\Omega + 2\ \text{M}\Omega}\right)$$
$$= 60\ \text{V}$$

Precautions in Using DC Voltmeters

Like the ammeter, the DC voltmeter is a delicate instrument. Consequently, follow these precautions to guard against damage through carelessness and misuse.

1. Always be certain that the voltmeter range is adequate for the application. When using a multirange voltmeter, start with the *highest* range and work your way down to get the best scale deflection possible without going off scale.
2. Be certain that the *polarity* is correct. Incorrect polarity causes the meter to strike the mechanical stop at the low end of the scale and may cause damage.
3. Unlike ammeters, always connect voltmeters in *parallel* with circuit elements. They should never be connected in series.
4. Always be aware of the effects of meter loading and select the proper meter and range for the application. Select a higher range if necessary to decrease meter loading effects, as illustrated by Example 5–7.
5. Always be absolutely certain that both the meter range and insulation of test leads used are adequate for the application. Overlooking this precaution, particularly when measuring high voltages, could result in serious injury.

5–8 MEASUREMENT OF DC RESISTANCE USING VOLTMETERS AND AMMETERS

With a voltmeter and an ammeter, the measurement of DC resistance is a simple and straightforward task. The technique, known as the **voltmeter-ammeter method**, requires (1) applying a voltage across the device to be measured, (2) measuring both the voltage across and the current through the device and (3) applying Ohm's law to determine the resistance.

Figure 5–19 shows two possible circuits for using the voltmeter-ammeter method. In the circuit of Figure 5–19*a*, the ammeter reads the true load current, but the voltmeter reads the voltage across the ammeter plus the voltage across the load. In the circuit of Figure 5–19*b*, the voltmeter reads the true load voltage, but the ammeter reads the current through the voltmeter plus the current through the load. If the voltmeter ideally had infinite resistance and the ammeter had zero resistance, the circuits could be used interchangeably.

The circuit of Figure 5–19*a* is preferred either when the resistance of the ammeter is very small compared to the resistance of the device being measured, or the resistance of the voltmeter is not very large compared to the resistance of the device. In these cases, the voltage drop across the ammeter does not cause an appreciable error in the voltmeter reading, and the current drawn by the voltmeter does not affect the ammeter reading. The circuit of

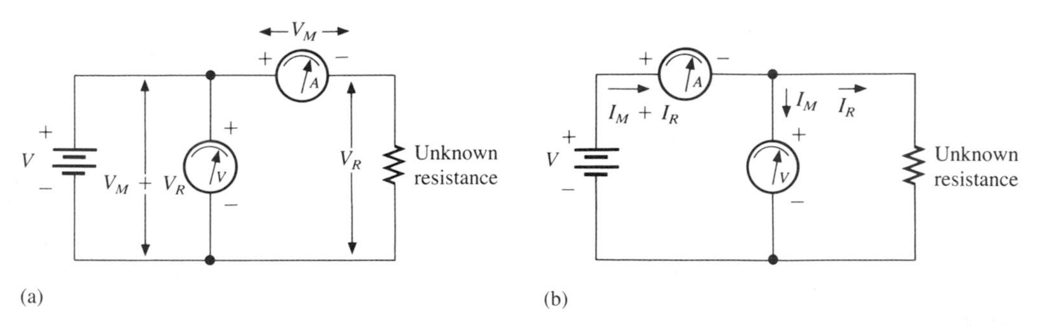

Figure 5–19 Voltmeter-ammeter method for measuring DC resistance. (a) Arrangement for when unknown resistance is relatively high. (b) Arrangement for when load resistance is relatively low.

Figure 5–19*b*, however, is preferred when the resistance of the ammeter is a significant fraction of the resistance of the device being measured, or the voltmeter resistance is substantially larger than that of the device. In this case, the voltage drop across the ammeter does not affect the voltmeter reading, and the current through the voltmeter is negligible compared to the current through the device under test. In summary, the circuit of Figure 5–19*a* is best when the load resistance is relatively high, and Figure 5–19*b* is best when the load resistance is relatively low.

5–9 DETERMINATION OF INTERNAL METER RESISTANCE

Many of the examples and problems given in this chapter are based upon knowing the meter's internal resistance (R_M). In general, this is given as part of the example or problem. In a practical situation, however, the internal resistance of a particular meter may not be known. To estimate errors introduced by DC ammeters and voltmeters in a given measurement situation, the internal meter resistance must be experimentally determined unless it is known from the manufacturer's data sheet or other source. To determine the internal resistance of a meter, one of two methods is generally used—either the *half-scale* method or the *voltage-current* method.

The Half-Scale Method

The **half-scale method** uses the arrangement shown in Figure 5–20. The value of R_1 is made much larger than the expected meter resistance. With the switch open, R_1 is adjusted so that the meter reads the *full-scale* current. The switch is then closed, and R_2 is adjusted so that the meter reads exactly *half scale*. Resistance R_2 is then disconnected from the circuit and its value measured with an ohmmeter. The measured resistance equals the meter's internal resistance. This is an application of Kirchhoff's current law in the form of a current divider.

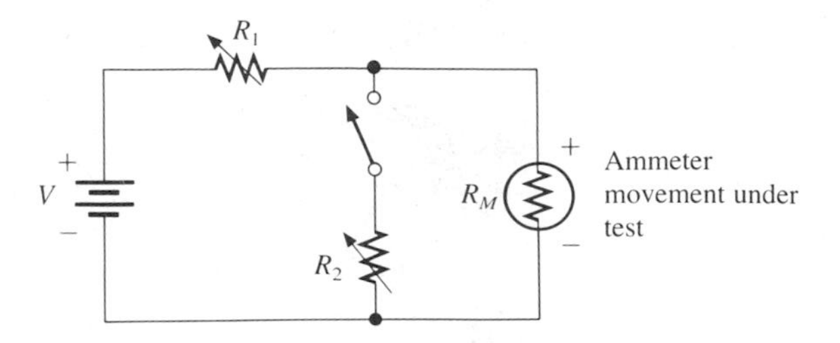

Figure 5–20 Circuit for the measurement of the internal meter resistance using the half-scale method.

The half-scale method presents only an approximation. The constant current source, formed by the voltage source and R_1, is not perfect, and the total current increases slightly when R_2 is connected. This increase in total current causes R_M to be slightly less than the true value of R_M when the meter reads half scale. The larger R_1 is with respect to R_M, the more constant the current and the more accurate the result. One method of getting around this problem is to add an ammeter in series with R_1 and adjust R_1 to keep the ammeter current constant whenever R_2 is adjusted.

The Voltage-Current Method

In the **voltage-current method,** using the arrangement shown in Figure 5–21, R_1 is adjusted so that the ammeter reads exactly *full scale*. The voltmeter must have a high ohms-per-volt rating, preferably greater than 50 kΩ/V. From the voltmeter reading, which is generally less than 1 V, the internal meter resistance is found by using Ohm's law.

Under no circumstances should an analog ohmmeter* be used to measure directly the internal resistance of a meter movement. The ohmmeter's internal battery would probably force an excessive current through the meter movement being tested, which would damage the movement.

*Ohmmeters are discussed fully in Chapter 7.

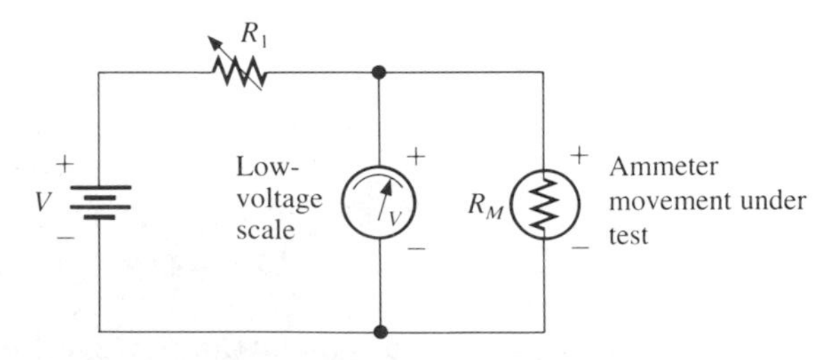

Figure 5–21 Voltage-current method of determining the internal resistance of a meter.

5–10 METER CALIBRATION

All measuring devices, whether simple meters or complex bridges and oscilloscopes should be checked periodically for accuracy. The accuracy of both DC ammeters and voltmeters is generally checked either by a **comparison** or **potentiometric** method.

Comparison Method

Checking ammeters or voltmeters using the comparison method involves the comparison of the reading of the meter to be calibrated against the reading of a meter whose accuracy is known. Figure 5–22 illustrates how a DC ammeter is calibrated by comparison. The circuit is nothing more than two meters in series, so that both should read the same amount of current flow. Meter M_1 is known to be accurate, but M_2 is being checked for accuracy. Resistors R_1 and R_2 are current limiting resistors. **Potentiometer** R_1 is adjusted for an exact reading for meter M_1, and the reading of M_2 can be compared with this value. This is done for several values of current. The results are then either summarized in a table or by a graph that presents the expected and observed values.

Besides constructing a calibration table or graph for a specific meter, the calibration data can also be expressed in the form of a least-squares straight line, as explained in Chapter 1. In this case, the independent variable is the *observed* value, and the dependent variable is the *expected* value. Using the resulting least-squares regression equation for a straight line, the observed value can then be substituted to obtain the correct value.

Figure 5–22 Comparison method of calibrating DC ammeters.

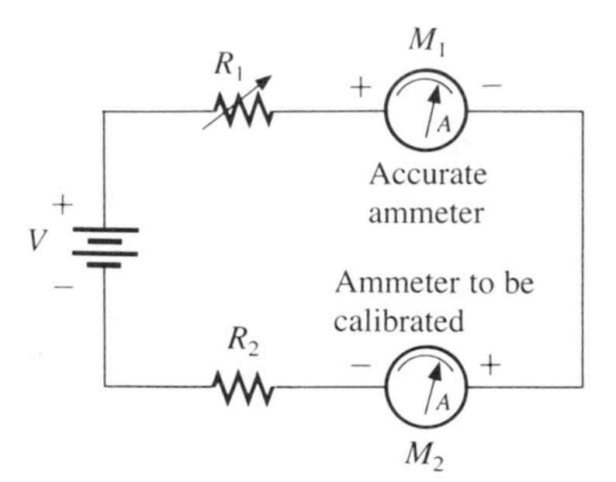

EXAMPLE 5–9

A DC ammeter having a full-scale reading of 1 mA was checked for calibration against an accurate 1-mA meter using the circuit of Figure 5–22. The expected and observed values obtained were

Expected value (M_1)	Observed value (M_2)
0 mA	0 mA
2	2.1
4	4.1
6	6.0
8	7.9
10	9.9

Using the least-squares technique, determine (a) the straight-line equation describing the calibration data, and (b) the actual meter current value when meter M_1 reads 1.5, 5.2, and 9.4 mA.

Solution:

(a) The slope and intercept are found from Equations 1–12 and 1–13 (or 1–14). From the data,

$$\begin{aligned} \Sigma X &= 30 \\ \Sigma Y &= 30 \\ \Sigma(XY) &= 218.80 \\ \Sigma(X^2) &= 217.6 \end{aligned}$$

then

$$m = \frac{(6)(218.80) - (30)(30)}{(6)(217.6) - (30)^2}$$

$$= 1.017$$

and

$$b = \frac{(30) - (1.017)(30)}{6}$$

$$= -0.085$$

so that the straight-line equation for the calibration data is

$$I_{\text{expected}} = -0.085 + (1.017)I_{\text{observed}}$$

(b) Based upon the above equation, the expected values are

For 1.5 mA

$$\begin{aligned} I &= -0.085 + (1.017)(1.5) \\ &= 1.44 \text{ mA} \end{aligned}$$

For 5.2 mA

$$\begin{aligned} I &= -0.085 + (1.017)(5.2) \\ &= 5.20 \text{ mA} \end{aligned}$$

For 9.4 mA

$$I = -0.085 + (1.017)(9.4)$$
$$= 9.47 \text{ mA}$$

DC voltmeters are calibrated by placing the voltmeter to be calibrated in parallel with the accurate voltmeter and a variable DC voltage source (as shown in Figure 5–23). The voltage source is adjusted for an exact reading for meter M_1, and the reading of M_2 is compared with this value. This is done for several values of voltage. As with ammeter calibration, the results are then either summarized in a table or a graph that presents the expected and observed values.

Potentiometric Method

The calibration of meters using the potentiometric method is highly accurate, since the potentiometer has an accuracy of 0.1% or better.

Using a potentiometer to calibrate a DC ammeter is accomplished using the circuit of Figure 5–24. The current value through the ammeter to be cali-

*Potentiometers are discussed in Chapter 8.

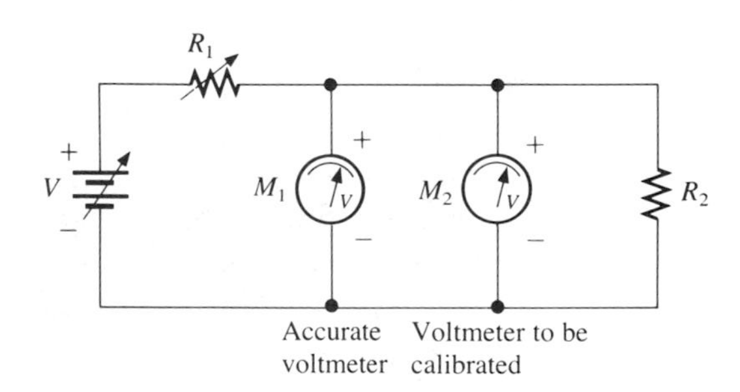

Figure 5–23 Comparison method of calibrating DC voltmeters.

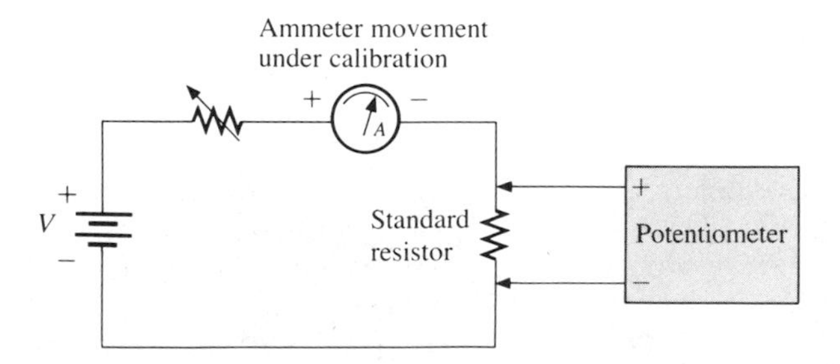

Figure 5–24 Potentiometric calibration of DC ammeter.

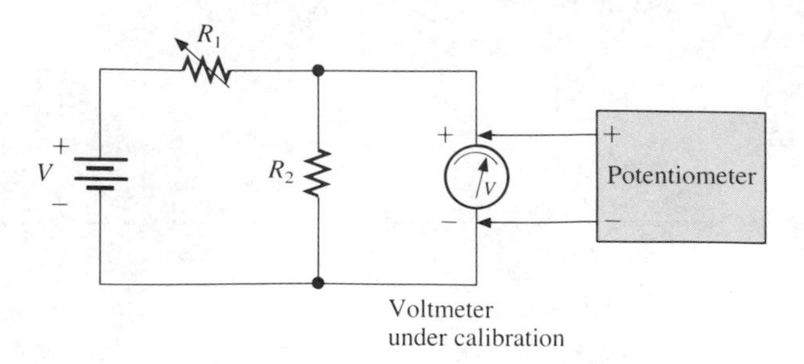

Figure 5–25 Potentiometric calibration of DC voltmeter.

brated is determined by using the potentiometer to measure the voltage across a highly accurate standard resistor, using Ohm's law. The rheostat is varied to obtain several readings over the meter's range. For each setting, the calculated current value is compared against the ammeter reading in the form of a table, graph, or regression line.

Figure 5–25 shows the arrangement used for calibrating DC voltmeters. This is similar to the circuit of Figure 5–22 except that a potentiometer is used in place of a standard meter that is known to be accurate. Since the potentiometer draws no current, it accurately measures the voltage drop across R_2 and the voltmeter under calibration.

5–11 SUMMARY

This chapter covers the construction and use of DC analog meters to measure current and voltage. The basic d'Arsonval movement is an integral part of most analog meters and is capable of measuring higher currents and voltages by addition of parallel or series multiplier resistors. Because all meters introduce a finite amount of loading error, this chapter discusses this error by considering either the meter's internal resistance or ohms-per-volt rating.

5–12 GLOSSARY

ammeter A low-resistance instrument that measures either AC or DC current flow.

Ayrton shunt A resistive voltage divider string connected in parallel with a galvanometer to provide variable meter sensitivity as well as meter damping.

coil resistance The DC resistance of the coil of a galvanometer.

comparison method A method whereby a given voltmeter or ammeter is calibrated by comparing its reading against a similar meter of known accuracy.

d-Arsonval movement A meter movement that uses a permanent magnet moving coil (PMMC).

full-scale deflection current (I_{FSD}) The current that gives the maximum current reading of a meter.

galvanometer An instrument used to measure the polarity and magnitude of very small electric currents.

half-scale method A method that determines the internal meter resistance by adjusting a resistance in parallel with the meter so that the meter then reads half scale. The parallel resistance then equals the meter's internal resistance.

hot-wire movement A low-accuracy, nonlinear meter movement based on the expansion and contraction of a heated wire as an electric current is passed through it.

multiplier resistance A series resistance used with an ammeter to create a voltmeter.

permanent magnet moving-coil (PMMC) movement: See **d'Arsonval movement.**

potentiometer An instrument that produces a known voltage that can be balanced against an unknown voltage (see Chapter 8).

potentiometric method A highly accurate method of calibrating DC ammeters and voltmeters by using a potentiometer to measure the DC voltage across the meter.

sensitivity factor The rating of a voltmeter as a measure of its resistive loading. It is expressed in terms of ohms/volt.

shorting wire A wire connected between the terminals of a PMMC meter when the meter is not in use or is being packaged. Its purpose is to protect the meter movement from external vibration by increasing its damping.

shunt resistor A precision resistance placed in parallel with an ammeter to increase its full-scale range.

taut band A suspension system used in the construction of galvanometers that eliminates friction caused by a jewel-pivot suspension

voltage-current method A method that determines the internal meter resistance of an ammeter by using a voltmeter to measure the voltage across the meter and the ammeter's reading of the current through the meter.

voltmeter A high-resistance instrument that measures either AC or DC voltage.

voltmeter-ammeter method A method that measures DC resistance by simultaneously measuring the voltage across and the current through a given circuit or element. The DC resistance is then found using Ohm's law.

5–13 PROBLEMS

1. For the PMMC ammeter circuit shown in Figure 5–26, determine the total current through the meter circuit when the meter movement reads
   (a) Full-scale
   (b) 50% full-scale
   (c) 10% full-scale

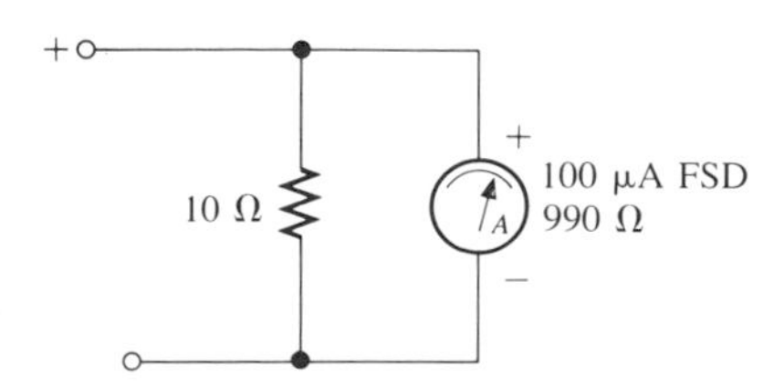

Figure 5–26 Circuit for Problem 1.

2. For the circuit of Figure 5–26 using a 50-Ω, 1-mA movement, determine the resistance and minimum power rating of the shunt required to increase the meter's full-scale range to
   (a) 10 mA
   (b) 50 mA
   (c) 150 mA
3. Repeat Problem 2 using Equation 5–7*b*.
4. Show that if the ammeter shunt resistance is equal to the meter's coil resistance, the ammeter's range is doubled.
5. If an ammeter having a 50-Ω, 1-mA movement and a 5.56-Ω shunt resistor is placed in series with the 2-kΩ resistor of Figure 5–27, determine the percent error of the meter reading that would result from ammeter loading.

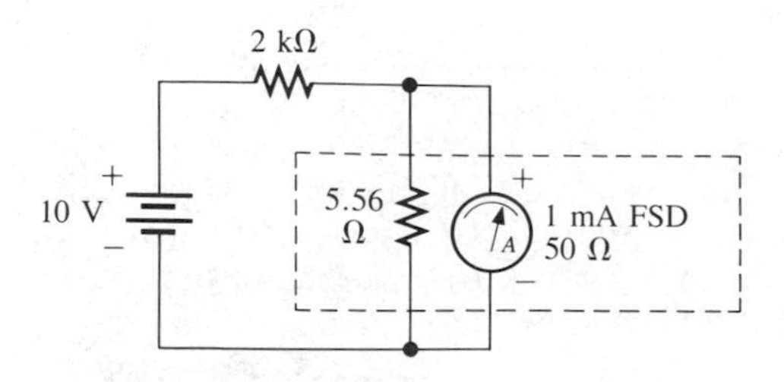

Figure 5–27 Circuit for Problem 5.

6. The three-range Ayrton shunt ammeter circuit of Figure 5–28 uses a 1-kΩ, 1-mA movement. Determine the three resistances required for full-scale readings of 10 mA, 50 mA, and 150 mA.

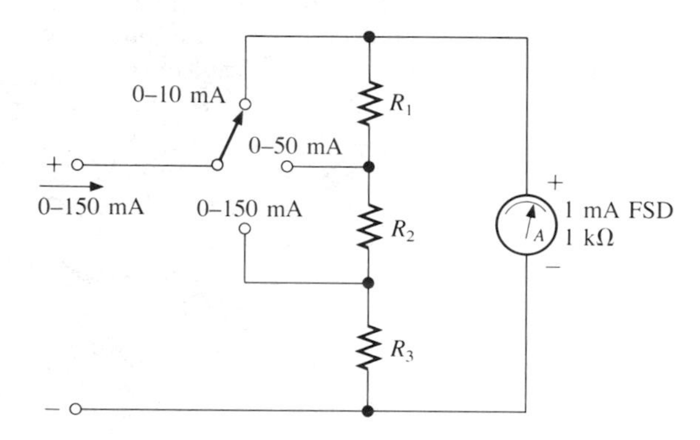

Figure 5–28 Circuit for Problem 6.

7. Determine the full-scale ranges of the ammeter circuit of Figure 5–29.

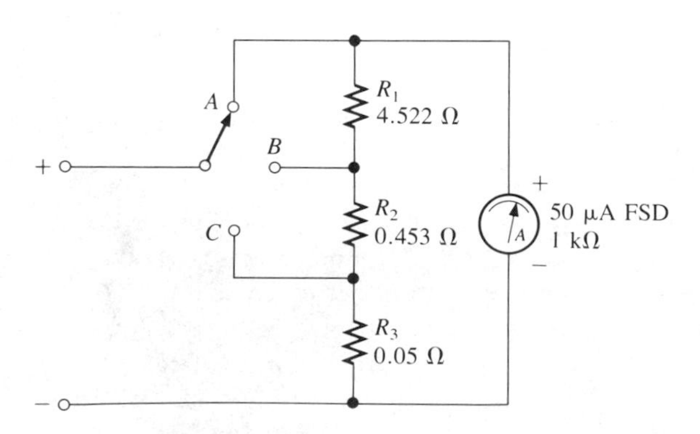

Figure 5–29 Circuit for Problem 7.

8. Given a 200-μA movement with a resistance of 1 kΩ, determine the value of the multiplier resistor required to convert it to a DC voltmeter having a full-scale reading of 150 V.

9. A PMMC movement with a sensitivity factor of 10 kΩ/V has a coil resistance of 50Ω.
   (a) Determine the multiplier resistance required for a 0–5-V voltmeter.
   (b) Using the scale shown in Figure 5–30, label the scale divisions in terms of the full-scale deflection current and the new voltage range.

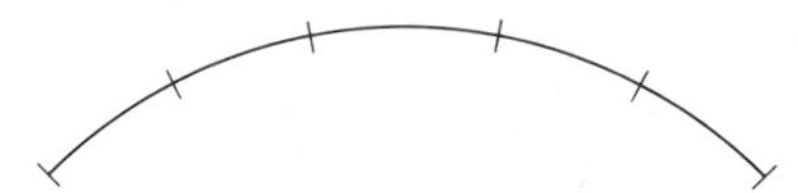

Figure 5–30 Meter scale for Problem 9.

10. A voltmeter, having a sensitivity of 2 kΩ/V on its 5-V range, is used to measure the voltage across R_2 in the circuit of Figure 5–31.
    (a) Determine the percent error introduced by the voltmeter.
    (b) If the voltmeter is switched to its 2.5-V range, determine the percent error.

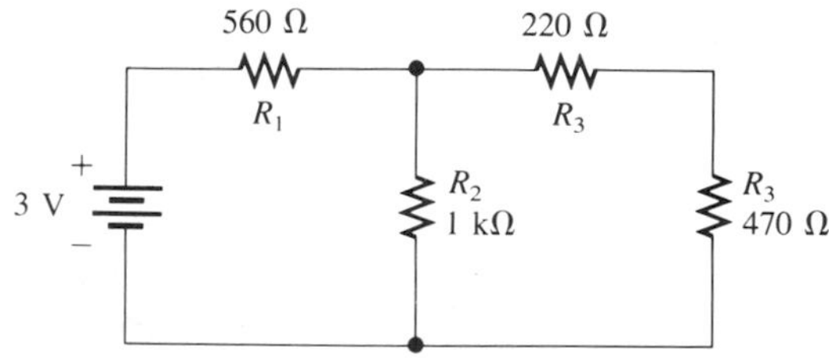

Figure 5–31 Circuit for Problem 10.

11. Which of the two voltmeter circuits shown in Figure 5–32 has the greater sensitivity factor?

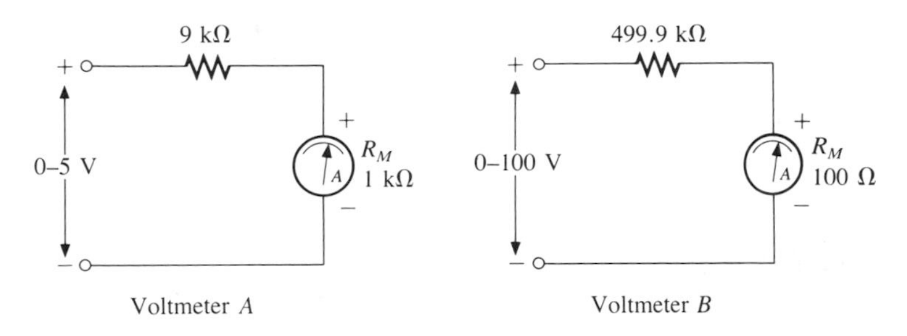

Figure 5–32 Circuit for Problem 11.

12. For the circuit of Figure 5–33, two different voltmeters are used to measure the voltage across the unknown resistance, R_2. Voltmeter 1, having a sensitivity factor of 5 kΩ/V, reads 1.49 V on its 10-V range. Voltmeter 2 reads 1.75 V on its 5-V range. Determine the sensitivity factor of voltmeter 2.

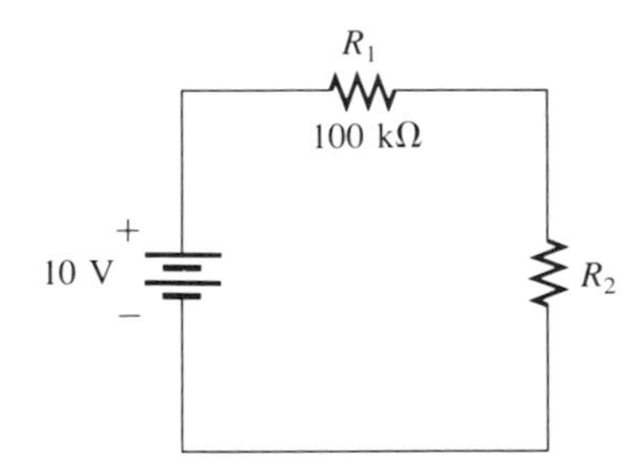

Figure 5–33 Circuit for Problem 12.

13. A laboratory DC voltmeter having a full-scale reading of 10 V was checked for calibration against an accurate meter. The expected and observed values obtained were

Expected value	Observed value
0 V	0V
2	2.05
4	4.10
6	5.95
8	7.89
10	9.91

(a) Determine and plot the straight-line equation along with the six data points describing the calibration data.

(b) Determine the actual meter voltage value when the laboratory meter reads 5.4 V.

6

AC INDICATING METERS

6–1 INSTRUCTIONAL OBJECTIVES

After completing this chapter, you will be able to

- Describe the operation of half-wave and full-wave rectifier AC meters.
- Explain how to correct meter readings for waveforms other than a sine wave.
- Describe the operation of the iron-vane, thermocouple, and electrostatic voltmeters.
- Describe the operation of the electrodynamometer for measuring AC currents and voltages.
- Describe the operation of single-phase and three-phase wattmeters.
- Describe the principle of the Hall-effect wattmeter.
- Explain how to measure the power factor of a load.
- Describe the operation of a varmeter.
- Describe how energy is measured using a watthour meter.
- Explain the purpose of instrument transformers.
- Explain the purpose of an AC-to-DC transducer.

6–2 INTRODUCTION

The measurement of AC electrical values involves a much broader range of techniques and equipment than is normally employed in the measurement of DC values. Because it is common practice regarding DC circuits for the average, RMS, and peak values to be equivalent, a given reading is referred to as simply the *DC value*. In an AC circuit, or a DC circuit with cyclic variations in amplitude, there are average, RMS, and peak values, all of which are a function of the waveform's shape. As a result, AC meter readings depend on the parameter that a particular instrument measures as well as the frequency of the waveform being measured.

6–3 RECTIFIER INSTRUMENTS

As discussed in Chapter 5, the deflection of a d'Arsonval movement is a function of the *average current* flowing through its coil as well as the direction of current flow. If a slowly varying (0.1 Hz or less) sine-wave current passes through a d'Arsonval meter movement, the pointer swings upscale during the positive half cycle and is driven against the stop at the lower end of the scale during the negative half cycle. This occurs because the direction of the deflecting torque changes with the direction of current flow.

If the frequency of the driving waveform were gradually increased, the pointer's swing would eventually begin to decrease in amplitude, and at some higher frequency would settle out to zero. At this frequency and above, the mechanical damping and inertia of the meter movement tend to average out the equal and opposite torques due to the reversals in the direction of current flow. At these higher frequencies, a deflection from zero occurs only if the AC waveform through the movement is not symmetrical, or if it has a DC component. In fact, at 60 Hz, a large AC voltage can exist across most d'Arsonval meter circuits, and the meter still reads zero even if current great enough to cause damage is flowing through the meter coil.

Half-Wave Rectifier

To use a d'Arsonval meter movement as an AC instrument, you must ensure that the meter current always passes through the meter in the one direction that produces an *upscale deflection*. One common method used is to rectify the meter current with a diode half-wave rectifier, shown in Figure 6–1. As always, the DC meter movement responds to the average value of the current passing through it. For a sine-wave input voltage, the average value of the resultant half-wave rectified voltage is

$$V_{AV} = \frac{V_P}{\pi} \tag{6–1}$$

Consequently, the AC voltmeter is not as sensitive as the DC voltmeter. To illustrate this decrease in sensitivity, consider that the voltmeter of Figure 6–1 is used to measure a 50-V RMS sine-wave input. The series resistor is chosen so that, ignoring the small voltage drop of the diode, the meter reads full scale when a 50-V DC voltage is applied. Suppose a 50-V RMS sine wave is applied to the input of the voltmeter circuit. The meter movement then responds to the average value of the half-wave rectified sine-wave current through the movement, so that

$$V_{AV} = \frac{1.414 \times 50 \text{ mV RMS}}{\pi}$$
$$= 22.5 \text{ V}$$

The meter now reads only 22.5 V for a sine wave, whereas the meter reads 50 V when a DC voltage is applied. Although this type of meter circuit reads both AC and DC voltages, it is now apparent that the circuit is not as sensitive to AC voltages as it is to DC voltages. In fact, when measuring AC sine-wave voltages, it is only 45% as sensitive as when measuring DC voltages.

As a practical matter, the AC meter circuit of Figure 6–1 would be designed to give a full-scale deflection in terms of a given *RMS voltage* instead of a DC voltage. If the meter were to have a full-scale deflection of 50 V RMS, then the total resistance of the multiplier and the meter coil would be 45% (i.e., $\sqrt{2}/\pi$) of the value chosen to give a full-scale deflection of 50 V DC.

EXAMPLE 6–1

Using the circuit of Figure 6–1, make an RMS-reading AC voltmeter having a full-scale deflection of 100 V. The meter movement has a full-scale deflection

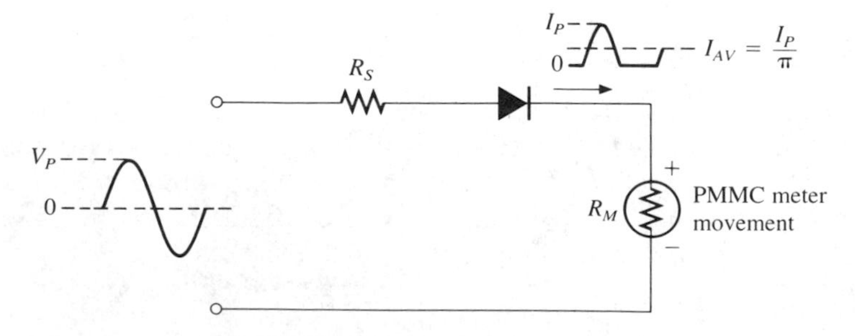

Figure 6–1 AC voltmeter using a diode half-wave rectifier.

current of 1 mA and an internal resistance of 100 Ω. Assuming a perfect sine wave and neglecting the voltage drop of the rectifier diode, determine the value of the series multiplier resistor.

Solution:

The average full-scale current through the PMMC meter movement with 100 V RMS applied to the circuit must equal 1 mA. The peak sine-wave voltage is then

$$V_p = \sqrt{2}\,\mathrm{V_{RMS}}$$
$$= 1.414 \times 100\ \mathrm{V}$$
$$= 141.4\ \mathrm{V}$$

Also, the peak sine-wave current is

$$I_P = \pi \times 1\ \mathrm{mA}$$
$$= 3.142\ \mathrm{mA}$$

The total circuit resistance is then

$$R_T = R_S + R_M$$
$$= \frac{V_P}{I_P}$$
$$= \frac{141.4\ \mathrm{V}}{3.142\ \mathrm{mA}}$$
$$= 45.0\ \mathrm{k\Omega}$$

Solving for R_S

$$R_S = 44.9\ \mathrm{k\Omega}$$

Although the half-wave rectifier circuit is commonly used, it has some serious limitations that are common to all rectifier AC meter circuits.

1. If the meter is to read RMS volts, the meter scale must be calibrated for the waveform that is being measured. This is because the ratio of the RMS value of the applied AC voltage to the average value of the resulting meter current, called the **form factor,** is a function of the waveform. A rectifier meter calibrated to measure the RMS value of a sine wave will not read correctly on a nonsinusoidal waveshape, such as a sawtooth. As is explained later in this section, the RMS reading must then be corrected for nonsinusoidal waveforms to give the correct reading.

2. The loading effect of the half-wave rectifier meter circuit is different for positive and negative half cycles. This is not a problem with full-wave rectification.
3. The rectifier-diode voltage drop causes the scale to be somewhat nonlinear if the meter is intended to measure low AC voltages.
4. Any DC component that is present causes an error in the AC reading.
5. Diode capacitance tends to bypass a portion of the AC signal, creating an error component that increases with frequency.
6. Because rectifier characteristics tend to be sensitive to temperature, significant errors may occur when ambient temperatures differ from the temperature at which the meter was calibrated.

An improved version of the half-wave rectifier voltmeter circuit is shown in Figure 6–2. Diode D_2 serves to keep the reverse voltage across D_1 to about 0.7 volts, minimizing any reverse leakage current through D_1 that might otherwise affect the meter reading. R_2 causes additional current flow through D_1 during the positive portion of the half cycle and serves to keep D_1 operating in a more linear portion of its characteristic curve. As a result, the circuit also provides a more constant meter loading effect for both half cycles of the applied AC voltage. While R_2 improves the linearity of the meter on the low end of the meter scale, it nevertheless reduces the meter's AC sensitivity.

Full-Wave Rectifier

As shown in Figure 6–3, an AC voltmeter can also be constructed using a full-wave bridge rectifier. The full-wave bridge rectifier produces twice as much average meter current as the half-wave rectifier does. Thus, it requires less meter damping and lower meter sensitivity. On the other hand, it has the disadvantage of having two diodes in series with the meter movement at all times, which results in increased nonlinearity at low levels of applied voltage.

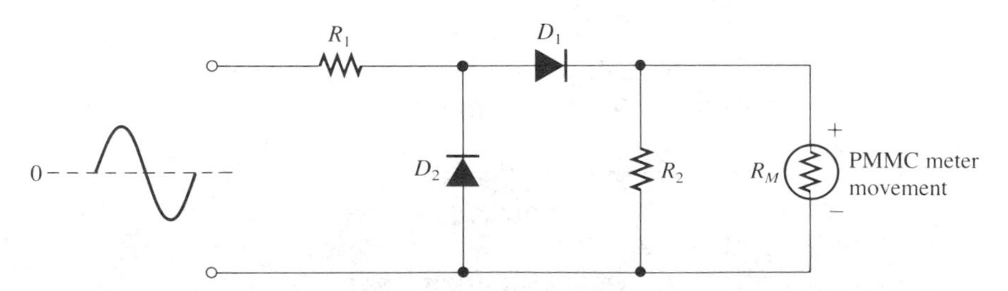

Figure 6–2 Improved diode rectifier circuit. Diode D_2 keeps the reverse voltage across D_1 to about 0.7 V and minimizes any reverse leakage through D_1 that would affect the meter reading.

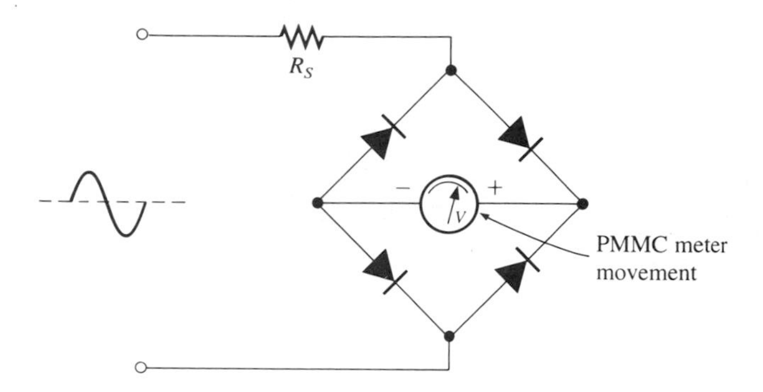

Figure 6–3 Full-wave bridge rectifier AC voltmeter.

EXAMPLE 6–2

An AC voltmeter, using the full-wave bridge rectifier circuit of Figure 6–3, is to measure the RMS value of a sine wave. The circuit uses a meter movement with a full-scale deflection current of 100 μA and an internal meter resistance of 500 Ω. Assuming ideal diodes, determine the value of the series multiplier resistor if the meter is to read 100 V RMS full-scale.

Solution:

The average meter current with 100 V RMS applied to the circuit must equal 100 μA. For a full-wave rectified sine wave, the peak meter current is

$$I_P = \frac{\pi}{2} I_{AV}$$

$$= 1.57 \times 100\ \mu\text{A}$$

$$= 157\ \mu\text{A}$$

whereas the peak applied sine-wave voltage is

$$V_P = \sqrt{2}\ V_{\text{RMS}}$$

$$= (1.414)(100\ \text{V})$$

$$= 141.4\ \text{V}$$

Consequently, the total circuit resistance must equal

$$R_T = R_S + R_M = V_P/I_P,$$

or

$$R_T = R_S + R_M$$

$$= \frac{V_P}{I_P}$$

$$= \frac{141.4 \text{ V}}{157 \ \mu\text{A}}$$

$$= 900.6 \text{ k}\Omega$$

Solving for R_S

$$R_S = 900.1 \text{ k}\Omega$$

Another type of full-wave rectifier circuit commonly found in the AC converter section of commercial multimeters is the **half-bridge full-wave rectifier**. Figure 6–4 shows a typical circuit of this type designed to have a full-scale reading of 2.5 V RMS. This circuit has the advantage of providing full-wave rectification, but with only one diode in the current path at any given time.

During the half cycle when terminal A is positive with respect to terminal B, diode D_1 conducts and diode D_2 is reverse-biased. During the opposite half cycle, D_2 conducts and D_1 is reverse-biased. The current flows through R_S, the conducting diode, and then is split between two parallel paths. With the values shown, ¾ of the current flows through R_2, while ¼ flows through the series connection of R_1, R_M, and R_3.

EXAMPLE 6–3

Assuming that the forward diode resistance is 250 Ω, (a) verify that the meter movement must have a full-scale rating of 50 μA using the resistance values shown in Figure 6–4, and (b) determine the ohms-per-volt rating of the meter.

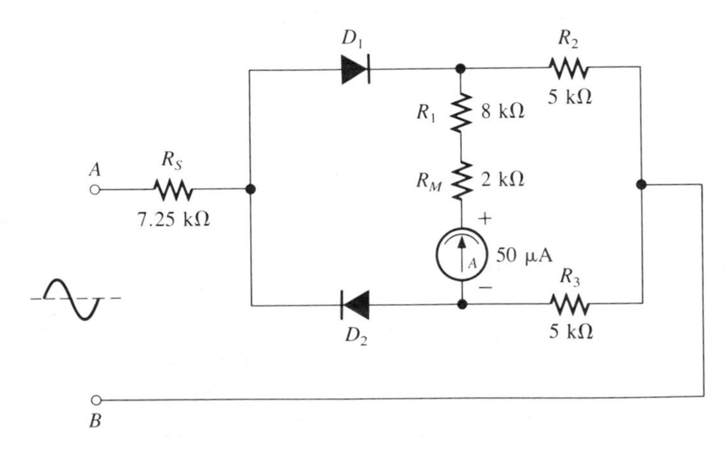

Figure 6–4 DC-to-AC converter using a half-bridge full-wave rectifier.

Solution:

As illustrated by the DC-equivalent circuit of Figure 6–5 with the forward-biased diode replaced by its equivalent resistance, the average (DC) equivalent source voltage is

$$V_{AV} = 0.9\ V_{RMS}$$
$$= 0.9 \times 2.5\ V$$
$$= 2.25\ V$$

The total resistance of the circuit is then

$$R_T = R_S + R_D + (R_1 + R_M + R_3) \parallel R_2$$
$$= 7.25\ k\Omega + 250\ \Omega + (15\ k\Omega \parallel 5\ k\Omega)$$
$$= 11.25\ k\Omega$$

The total average current is

$$I_T = \frac{2.25\ V}{11.25\ k\Omega}$$
$$= 200\ \mu A$$

so that, treating the circuit as a current divider, the average current through the meter movement is

$$I_M = \left(\frac{5\ k\Omega}{5\ k\Omega + 5\ k\Omega + 10\ k\Omega}\right)(200\ \mu A)$$
$$= 50\ \mu A$$

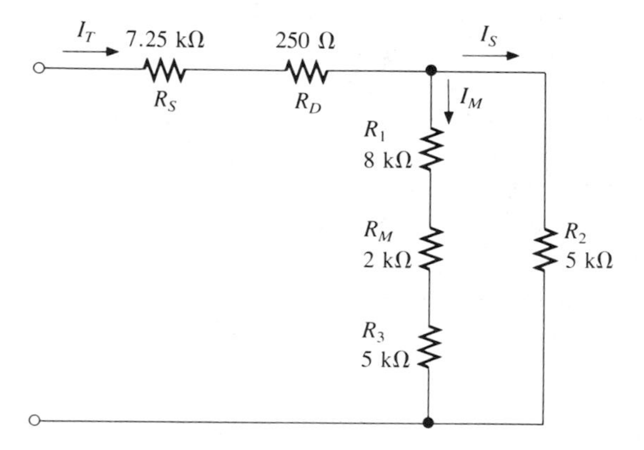

Figure 6–5 Half-bridge full-wave rectifier equivalent circuit for Example 6–3.

(b) The ohms-per-volt rating of the meter circuit is

$$S = \frac{R_T}{V_{\text{FSD}}}$$

$$= \frac{11.25\ \text{k}\Omega}{2.5\ \text{V}}$$

$$= 4.5\ \text{k}\Omega/\text{V}$$

Rectifier AC Voltmeters for Measuring Voltages below One Volt

One of the major drawbacks of the simple meter rectifier circuits discussed is their inability to measure small AC voltages accurately due to the diode voltage drop, which is typically 0.7 V. This problem may be overcome in two ways. The voltage being measured can first be amplified to a level where the diode drop is no longer a significant factor; then it can be rectified in the usual manner. Or the rectifier can be used as part of the feedback loop in an operational amplifier circuit forming a **precision rectifier.** The precision rectifier compensates for the diode drop and produces a nearly perfectly rectified output signal.

Meter Correction for Different Waveforms

As noted earlier, the meter movement following a rectifier circuit senses the average current flowing through it. For AC meters, the meter scale is generally calibrated to read the RMS value of the waveform rather than its average value. This calibration is usually based on the assumptions that the measured quantity is a sine wave and that the characteristics of the particular rectifier circuit used are applicable for that waveform.

To illustrate this point, assume that an AC voltmeter uses a full-wave rectifier circuit to measure the RMS voltage of a sine wave. With full-wave rectification, the average voltage is $2/\pi$, or 0.636 times its peak value, while the RMS value is 0.707 times its peak value. Consequently, the ratio of the peak to average value, known as the form factor, is 1.111 for a sine wave. Thus, the RMS meter scale indicates 1.111 times the sine wave's average value. Conversely, the average value of the sine wave is 1/1.111, or 0.900 times its RMS value. If the meter circuit uses half-wave rectification, the form factor is 2.222 because the meter scale reads 2.222 times the average value.

When rectifier AC meters are used to measure voltages or currents of signals having waveshapes other than that of a sine wave, the meter readings will be in error. Even though the meter reads the average value of the applied waveform, unless the waveform's form factor is the same as for a sine wave, the RMS value as indicated by the meter's scale will be misleading. A correction factor, based on the waveshape of the AC signal, must be applied to eliminate meter error.

Shown in Table 6–1 are the RMS values and the form factors for waveforms commonly encountered in electrical measurements. To determine the correct RMS value of a particular waveform based upon the reading taken from an AC meter that is of the averaging type calibrated for sine waves, simply multiply the RMS meter reading by the factor given in the last column of the table. These values are found by taking the form factor of the waveform being measured and dividing it by the form factor of a sine wave (1.111). For a triangle wave, since its average and RMS values are 0.5 and 0.577 times its peak value respectively, the form factor is 1.154. Dividing this by the sine wave's form factor gives 1.154/1.111, or 1.039. The RMS meter reading is then multiplied by 1.039 to obtain the correct RMS value.

Table 6–1 AC Meter Correction Factors for Different Waveforms

Waveform	RMS value	Form factor	Multiply AC meter reading by to obtain correct RMS value
 V_P, 0	$0.707V_P$	1.111	1.0
 V_P, 0	$0.5V_P$	1.571	1.414
 V_P, 0	$0.707V_P$	1.111	1.0
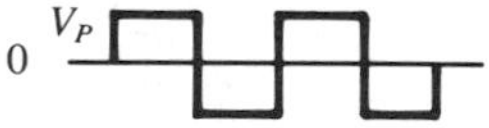 V_P, 0	$0.5V_P$	1.0	0.9
 V_P, 0	$0.707V_P$	1.414	1.273
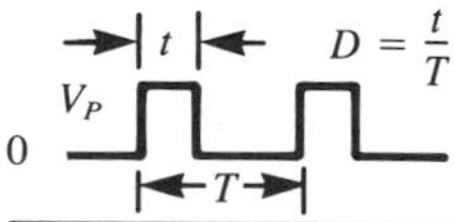 t, $D = \frac{t}{T}$, V_P, 0, T	$\sqrt{D}V_P$	$1/\sqrt{D}$	$0.9/\sqrt{D}$
 V_P, 0	$0.577V_P$	1.154	1.039
 V_P, 0	$0.577V_P$	1.154	1.039

EXAMPLE 6–4

If an AC voltmeter, of the averaging type calibrated for sine waves, reads 6.42 V RMS when measuring the voltage of the pulse train shown in Figure 6–6, determine (a) the correct RMS voltage, and (b) the peak value of the waveform.

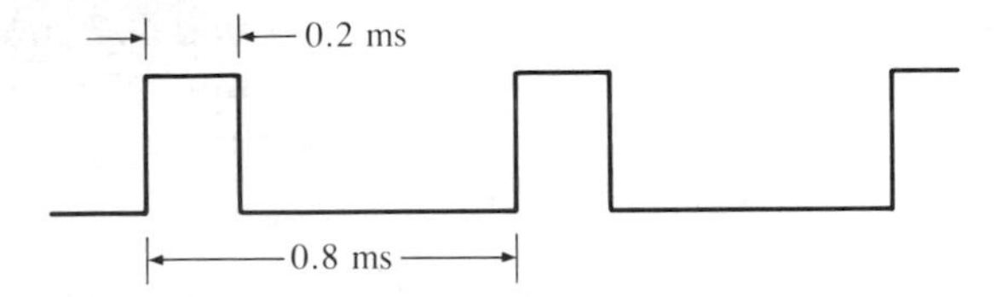

Figure 6–6 Pulse train for Example 6–4.

Solution:

(a) The duty cycle (D) of the pulse train is 0.2 ms/0.8 ms, or 0.25. The correct RMS voltage is then

$$V = \frac{0.9}{\sqrt{D}} V_{\text{Meter}}$$

$$= \frac{(0.9)(6.42 \text{ V RMS})}{\sqrt{0.25}}$$

$$= 11.56 \text{ V RMS}$$

(b) Since for this waveform the RMS value is $\sqrt{D}$ times the peak value, the peak voltage is then

$$V_P = \frac{V_{\text{RMS}}}{\sqrt{D}}$$

$$= \frac{6.42 \text{ V}}{\sqrt{0.25}}$$

$$= 12.84 \text{ V}$$

6–4 THE IRON-VANE METER

A movement commonly used for the measurement of both AC voltages and currents at low frequencies is the **iron-vane** type, which basically consists of a movable ironvane mounted inside of a single fixed coil. The soft-iron vane and a pointer are attached to a shaft mounted within jewel bearings. They are free to rotate over a limited range. The restoring torque is provided by a spiral spring.

The basic iron-vane movement is one which depends on magnetic repulsion, having both fixed and movable soft-iron plates, as illustrated in Figure 6–7. This construction produces corresponding north and south poles on opposite ends of the plates when current flows in the coil. Because the polarity of both plates is the same, they repel each other. When the coil passes an AC current, polarity reversal occurs simultaneously in both plates. The repelling action always produces torque in the same direction. Consequently, this type of movement may be used for either AC or DC measurements.

A variation of this same scheme has the fixed and movable plates in the shape of coaxially mounted cylinder segments with the shaft at the center. The movable plate, which is attached to the shaft through a radial member, is constrained to move along a circular path inside the coil, thus causing shaft rotation. Another form of iron-vane movement has one or more iron vanes that are inclined on the shaft. The fixed coil is also inclined to the axis of the shaft in such a way that, as the magnetic field becomes stronger, the shaft rotates to minimize the reluctance of the magnetic path; that is, to bring the vanes more in line with the coil's axis. When the iron-vane movement is used in a voltmeter, the coil is composed of several turns of fine wire. When used in an ammeter, the coil is usually wound with fewer turns of larger wire.

Iron-vane meters are relatively inexpensive and rugged. They are used extensively for *low-frequency* AC measurements where accuracies of 5 to 10% are satisfactory. Since coil reactance increases with frequency, causing changes in calibration, the iron-vane meter is usually calibrated for a specific

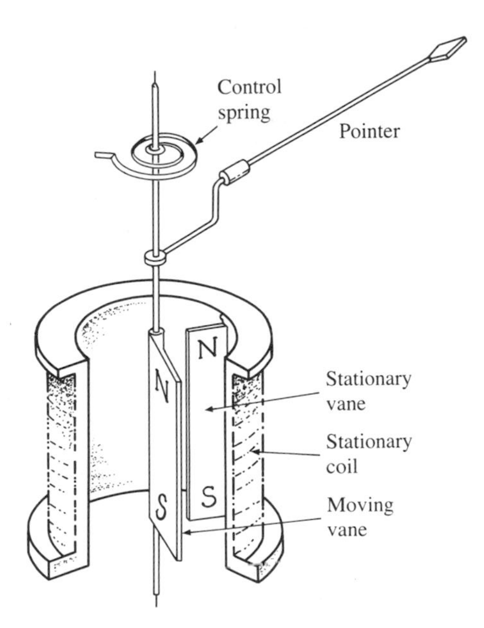

Figure 6–7 Iron-vane movement that depends on magnetic repulsion.

frequency. Voltmeter sensitivity tends to be lower than that of a PMMC meter, and ammeter resistance tends to be somewhat higher for the same ranges. Although it is capable of measuring DC quantities, hysteresis makes the iron-vane movement impractical for most DC measurements that require any degree of accuracy. Nevertheless, it does find use in indicators, such as automobile instruments, where accuracy is not important.

6–5 THE THERMOCOUPLE METER

Thomas Seebeck, a German physicist, found that when two dissimilar metals are joined, a voltage is generated at the junction. Moreover, that voltage is proportional to the junction temperature; as the junction temperature increases, so does the voltage across the junction. This **Seebeck effect** forms the basis for the **thermocouple meter**.

For the thermocouple meter, the meter current passes through a resistive element, whose power dissipation heats a small thermocouple. The change in voltage generated at the thermocouple junction forces a small DC current through a PMMC meter movement. As shown in Figure 6–8, the current flowing through the resistor from A to B causes the resistance wire to heat. The heat dissipated is transferred to the "hot junction" of the thermocouple at E. The junctions between the thermocouple leads and the copper meter leads (C and D) are known as **cold junctions** and form two more thermoelectric generators. Because these junctions are in series with the hot junction, they interfere to some extent with the voltage produced by the hot junction. However, most thermocouple instruments provide some means to compensate for the thermoelectric effect of the cold junctions.

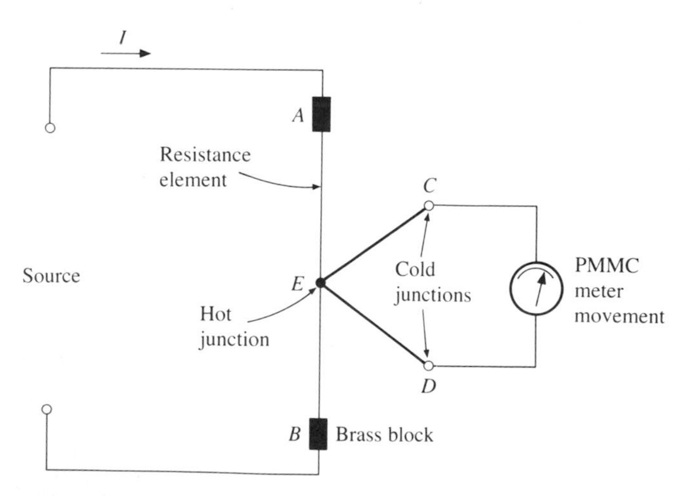

Figure 6–8 Thermocouple meter.

The heat produced at a thermoelectric junction is proportional to the square of the RMS current (I^2R) so that the meter has a nonlinear square-law response. Consequently, the scale is crowded at the lower end and spreads out at the upper end. The thermocouple meter finds its principal use in the measurement of radio-frequency currents up to about 50 MHz. At frequencies above 50 MHz, the skin effect becomes a significant factor so that its accuracy is reduced.

6–6 ELECTROSTATIC VOLTMETERS

When a capacitor has a charge on it, there exists a force between the plates that is proportional to the square of voltage between the plates. Using this basic principle, **electrostatic voltmeters** are constructed in one of two forms: the attracted-disk and rotating-plate types.

Attracted-Disk Electrometer

A very simple form of electrostatic voltmeter is the **attracted-disk electrometer.** Simply, a horizontal circular plate is suspended a small distance above a fixed plate that has a larger area. Concentric with, almost touching, and in the same plane as the moving plate, is an annular "guard ring" that causes the electrostatic lines of force between the two plates to be perpendicular to the plates and uniformly distributed over their areas. The force of attraction, F (in newtons), for a given difference of DC potential, V (in volts), moving plate area, A (in m^2), and plate spacing, d (m), is given by

$$F = \frac{8.85 \times 10^{-12} V^2 A}{2d^2} \tag{6–2}$$

The force is usually measured by suspending the movable plate from one arm of a laboratory balance. Balance is achieved when the moving plate is restored to the plane of the guard ring, i.e., to the proper spacing (d).

With this method, voltage can be measured *directly,* rather than indirectly, by the current that it produces. This principle is particularly suitable for the *absolute* measurement of very large potential differences. Since it consumes no power except during the brief time when the capacitor plates are in motion, it represents an infinite impedance to DC. When an AC voltage is applied, the average force is given by

$$F_{AV} = KV^2_{RMS} \tag{6–3}$$

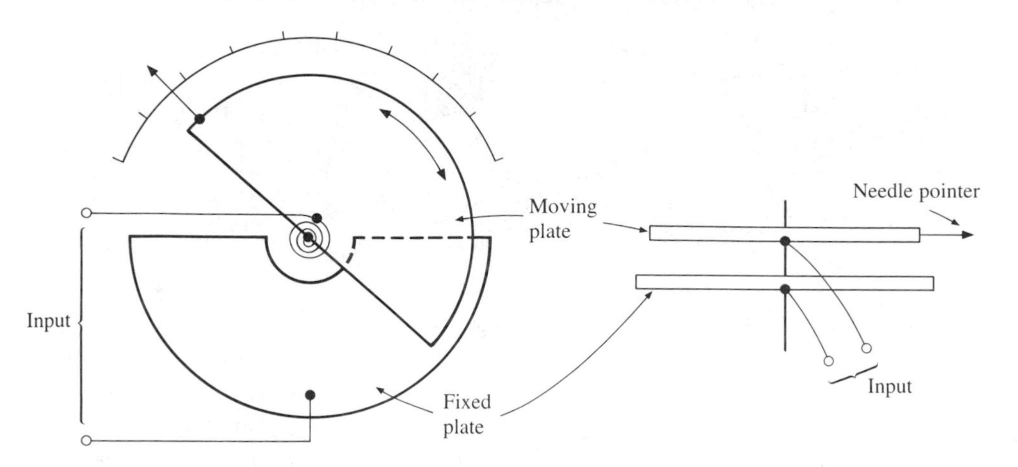

Figure 6–9 The rotating plate electrostatic voltmeter.

where K is a proportionality constant. As a consequence, the average force is proportional to the square of the applied RMS voltage, thereby making the meter independent of the shape of the applied-voltage waveform.

Rotating-Plate Electrostatic Voltmeter

A more modern and compact version of the electrostatic voltmeter consists of an assembly of two or more semicircular plates similar to a variable-tuning capacitor as illustrated in Figure 6–9. Alternate plates are fixed to the frame, while the intervening plates are mounted on a shaft that supports a pointer and is restrained by a spiral spring. Opposite charges on the plates attract and produce a torque in the direction which brings the plates closer. The pointer rotates until the developed torque is equal to the restraining torque of the spiral springs. Here the torque is proportional to the square of the RMS voltage.

Unlike the iron-vane meter, the electrostatic voltmeter can be used with either AC or DC over a wide range of frequencies. An instrument intended for AC use may be calibrated on DC. This type of instrument is used almost solely for measurements at high AC voltages where minimum meter loading is important.

6–7 THE ELECTRODYNAMOMETER

The **electrodynamometer** movement works on the same principle as the PMMC movement, except that the permanent magnet is replaced with an electromagnet (Figure 6–10). Since the polarities of both the electromagnet and

Figure 6–10 The electrodynamometer.

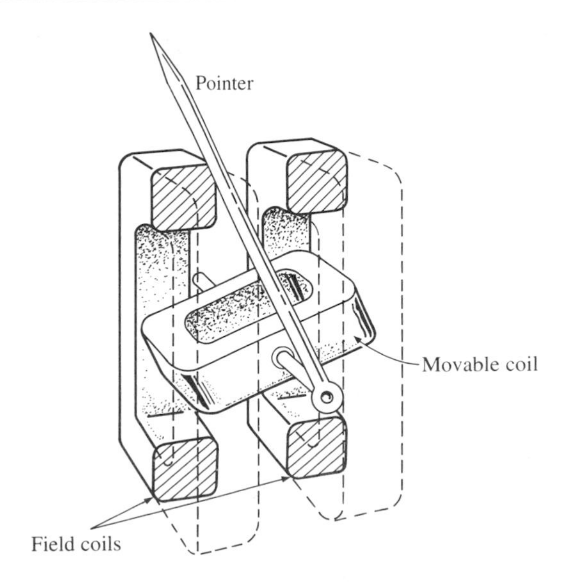

the moving coil change with the direction of the current flow, the resultant torque is always unidirectional (i.e., an upscale meter deflection) regardless of the polarity of the applied signal. Consequently, the electrodynamometer requires no rectification when used in an AC meter circuit. Furthermore, since it responds to the true RMS value of a waveform, it is equally suited for both AC *and* DC measurements.

To minimize the effects of eddy currents, it is a normal practice not to use either a metallic coil form or iron core in the electrodynamometer intended for AC measurements. In this case, the magnetic flux produced by the field coils is much lower than that generated by a PMMC device. In order to produce a deflecting torque equivalent to that of a PMMC movement, however, the current in the electrodynamometer's moving coil must be larger than that of the PMMC movement. As a consequence, electrodynamometers tend to be less sensitive than PMMC movements.

Air-Damping System for the Electrodynamic Movement

Since the electrodynamic instrument has no permanent magnetic field, and often no metallic coil form, eddy current damping is not practical. The electrodynamometer commonly uses a form of air damping that consists of a lightweight vane pivoted in an enclosed chamber. For the meter to move, the vane must force air through a restricted path around the periphery of the vane. This tends to damp out any rapid movements and greatly reduces any tendency for overshoot and oscillation.

Electrodynamometer Response

Like the PMMC movement, the torque developed by the moving coil in the electrodynamometer is proportional to the product of the magnetic field strength and the current through the moving coil. The magnetic field strength is proportional to the field current. If the field winding and moving coil are connected in series, the resulting torque is proportional to the square of the current. Like the PMMC movement, the deflection is proportional to the torque, so that the deflection is directly proportional to the square of the current.

If the electrodynamometer is used for DC or AC (RMS) measurements, its scale is nonlinear with crowded scale markings at the low end, progressively spreading out toward the high end. When the movement is used to measure AC, the instantaneous torque is proportional to the square of the instantaneous current. This torque is always up-scale and pulses in step with the applied waveform. Since the movement cannot follow these rapid variations in torque, it takes up a position in which the average torque is equal to the restraining torque of the control springs. As a consequence, the meter deflection is proportional to the average value of the current squared (i.e., mean-of-the-square).

Because the RMS value of the meter current is equal to the square root of the average value of the current squared (i.e., the root of the mean of the square), it is necessary only to calibrate the meter as a square-root function to obtain an instrument that reads the *true RMS value* of a waveform, independent of wave shape. Conveniently, the RMS value of a DC current is equal to the DC, or average, value of the current. This means that the AC electrodynamometer may be calibrated using a DC source. A 1-A DC current produces the same deflection as a 1-A (RMS) AC current. Unfortunately, the reactance of the coils in an electrodynamic instrument increases rapidly with frequency. This increase limits the use of the electrodynamometer movement to frequencies below approximately 200 Hz. The electrodynamometer is used extensively at power line frequencies, and forms the basis for a variety of highly accurate instruments.

Measurement of Current and Voltage

Like the PMMC movement, the electrodynamometer is basically a current-actuated device. It may also be used as a voltmeter, as shown in Figure 6–11*a*, by connecting the moving (potential) coil in series with the fixed field current coil and supplying an external multiplier resistor in a manner like the PMMC voltmeter. When the electrodynamometer is used to measure current (Figure 6–11*b*), it is common to wind the field coils with conductors of adequate size to carry the current to be measured (up to 5 A) but to shunt the moving coil to keep its winding relatively small and light. This shunt must have an impedance that has the same ratio of inductance to resistance as the moving coil, so that the current in the moving coil will be in phase with the current in the field coils.

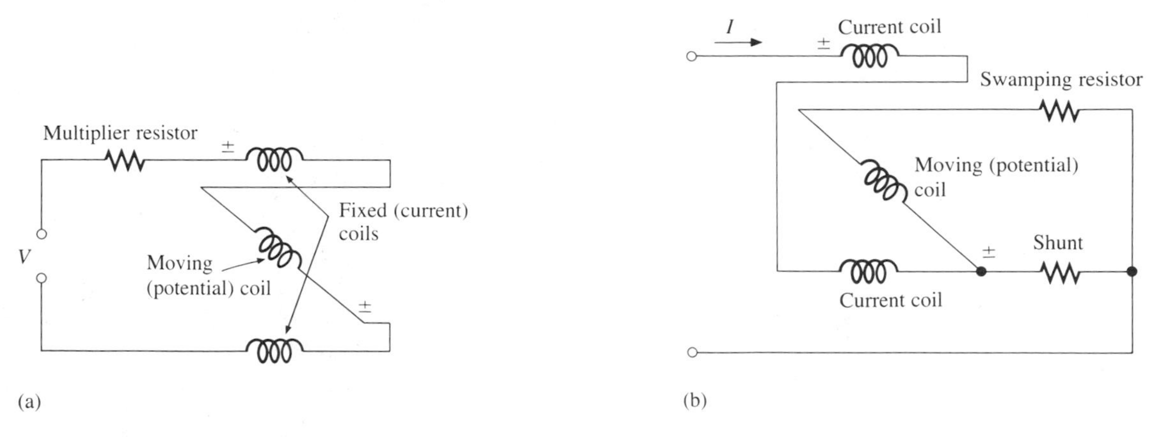

Figure 6–11 Electrodynamometer connections. (a) Voltmeter. (b) Ammeter.

The electrodynamometer has a relatively high power consumption compared to a rectifier AC meter, since it needs to supply its own field flux. When used to construct a voltmeter, the sensitivity may be as low as 10 Ω/V, as compared to 5 kΩ/V and higher for a typical rectifier instrument. Due to its low sensitivity, meter loading must always be a consideration. Dynamometer instruments are relatively expensive and are usually used only in standards and other high-accuracy applications.

6–8 SINGLE-PHASE WATTMETERS

Basic Dynamometer Wattmeter

Single-phase, low-frequency (below 400 Hz) **wattmeters** for the measurement of power are almost universally constructed with an electrodynamometer movement. Shown in Figure 6–12*a*, the fixed (current) coils are connected in series with the load and carry a current approximately equal to the load current. The moving (potential) coil in series with a fixed resistor is connected across the power line and carries a small current proportional to the line voltage.

In the wattmeter, the moving coil current, i_P, is approximately v_L/R, since R is much greater than the resistance of the moving coil. The current in this voltage branch is usually in the range of 10 to 50 mA maximum. The total current, i_T, is equal to the sum of the load current, i_L, and the coil current, i_P. If the load current is much larger than the coil current, the total current is approximately equal to the load current.

The resultant instantaneous torque is a function of the product of the instantaneous coil currents, i_T and i_P. Assuming that i_T is approximately equal to the load current, and i_P is a linear function of load voltage, the instantaneous torque is then proportional to the instantaneous power, so that

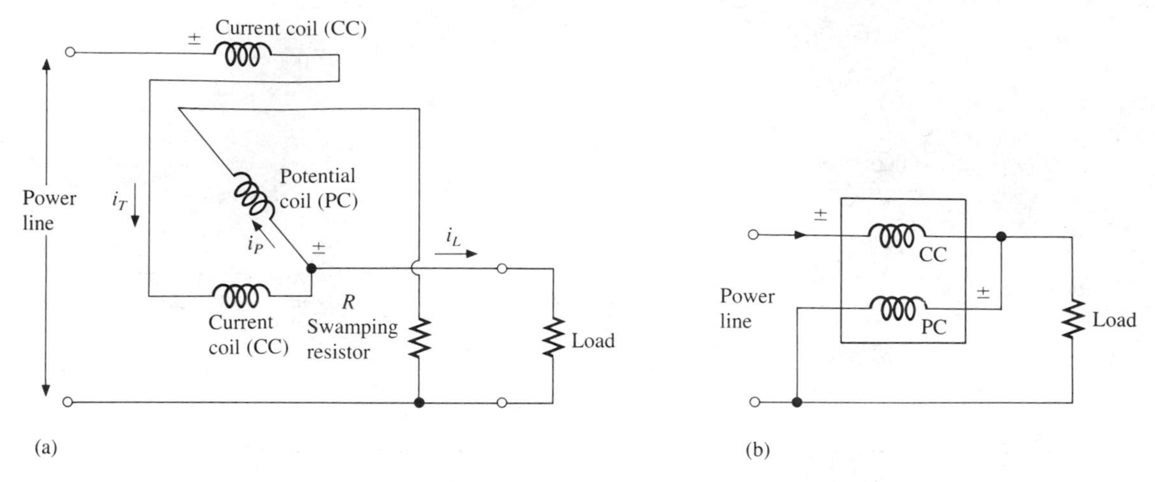

Figure 6–12 Dynamometer-type single phase wattmeter. (a) Internal diagram. (b) Simplified circuit.

$$\text{Instantaneous torque} = (i_T \times i_P)k \propto i_L \times v_L \tag{6–4}$$

Because the inertia of the moving element is relatively large at frequencies above a few hertz, the actual deflection of the moving coil for AC power is proportional to the average torque, and hence the average real power.

$$\tau_{AV} = \frac{K}{T}\int_0^T (v_L i_L)dt \tag{6–5}$$

where

K = a constant
T = period of one cycle

Thus, meter deflection is a function of the average real power, and meter accuracy is independent of waveshape and power factor. Figure 6–12*b* shows a commonly-used schematic representation of the connections to a single-phase wattmeter.

Wattmeter Connections

The dynamometer wattmeter has four external connections: two terminals for voltage and two terminals for current. The voltage terminals provide connec-

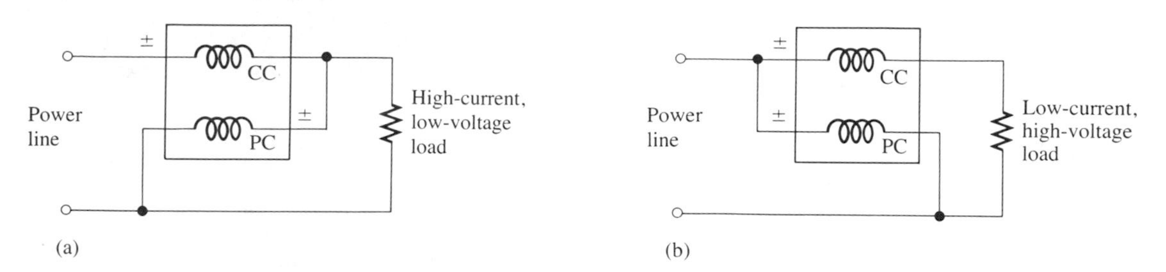

Figure 6–13 Connecting the wattmeter to measure power. (a) For high-current, low-voltage loads. (b) For high-voltage, low-current loads.

tions to the moving coil-resistor branch, and the current terminals provide connections to the stationary coils. One terminal of each pair is marked "±." Connect the ± terminal of each branch to *the same wire of the incoming power line* to ensure that the meter reads up scale. A wattmeter can be connected so that the voltage coil is placed either before or after the current coil. When the wattmeter is connected as shown in Figure 6–13*a*, the voltage branch senses the correct load voltage, but the current coil senses the voltage coil current in addition to the load current. On the other hand, when the wattmeter is connected as shown in Figure 6–13*b*, the current coil senses the true load current, but the voltage coil senses the load voltage plus the voltage drop across the current coil. In both cases, the meter reads high. The connection of Figure 6–13*a* is preferable for those situations with high-current, low-voltage loads, and the connection of Figure 6–13*b* is preferable for low-current, high-voltage loads.

A slightly more sophisticated wattmeter construction eliminates the problem of the wattmeter reading higher than the true value. In the "compensated wattmeter" circuit (Figure 6–14) which requires only three connections, the current coil consists of two bifilar windings, each with the same number of turns. One winding uses relatively heavy wire and carries the current for the load and the current coil; the other winding is of finer wire and carries only the

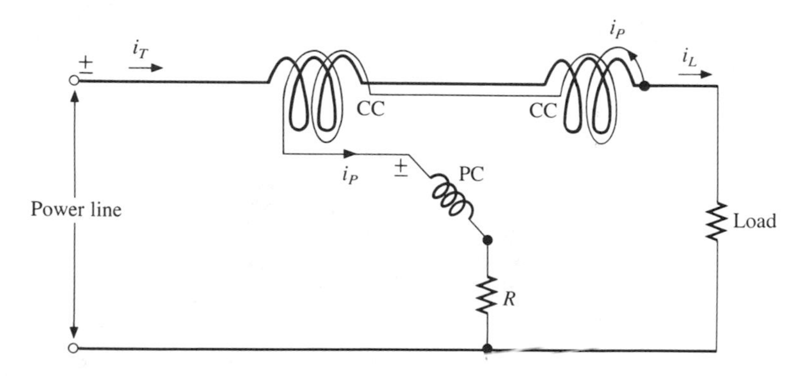

Figure 6–14 Compensated wattmeter circuit.

potential coil current. The direction of the currents is such that the component of magnetic flux due to the potential coil current is completely cancelled out.

6–9 THE HALL-EFFECT WATTMETER

Edwin H. Hall, while studying for his Ph.D. at Johns Hopkins University, found that a strip of conducting material that carries current in the presence of a transverse magnetic field (as shown in Figure 6–15) produces a difference of potential between the two planes of the conductor that are perpendicular to the field and the current. Although the discovery was made in 1879, its practical application was delayed until semiconductor materials with sufficient output were available. From this phenomenon, known as the **Hall effect,** it is now possible to construct a wattmeter with no moving parts, other than the actual meter movement.

The moving charges in the semiconductor experience a force that tends to push them to one side of the conducting strip. This creates an imbalance of charge. This imbalance creates a potential difference between the two sides of the material. The direction of the force acting on the moving charges may be determined by the usual rule for the force acting on a current-carrying conductor in a magnetic field. Equilibrium is established when the force due to the resulting electric field equals the deflecting force. Based on the current (I), the magnetic field strength (B), the thickness of the material (d), and a property of the conducting material known as the *Hall coefficient* (R_H), the magnitude of the generated voltage is expressed by

$$V_H = \frac{R_H I B}{d} \tag{6–6}$$

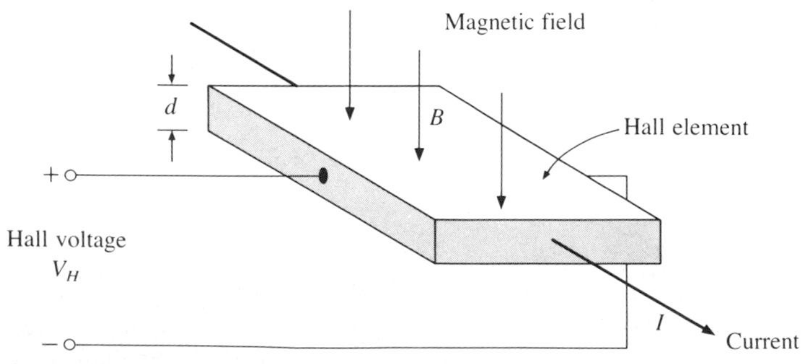

Figure 6–15 The Hall effect.

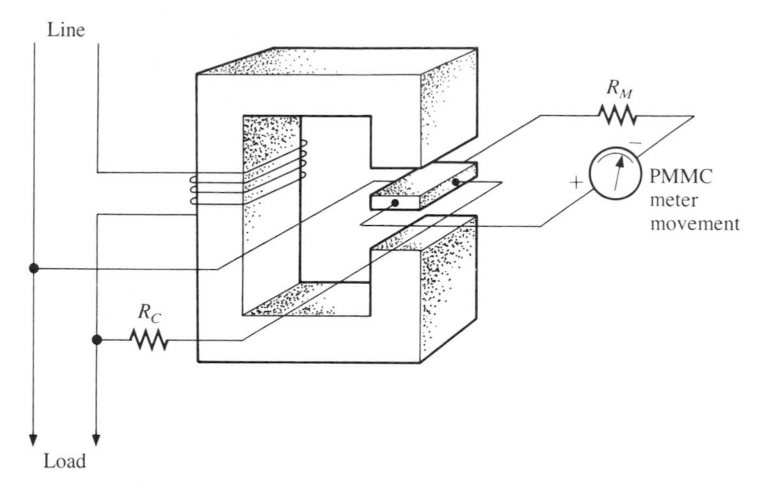

Figure 6–16 Hall watt transducer.

Since R_H and d are constants, the Hall element is essentially a device that produces an output that is proportional to the product of current and flux density (i.e., a multiplier). Figure 6–16 shows a Hall watt transducer* in which the magnetic field is a function of load current, and the current through the Hall element is a function of load voltage. The Hall output voltage is then a function of the product of load voltage and current and is proportional to the instantaneous power dissipated by the load.

A d'Arsonval meter movement or a suitable electronic circuit may be used to average the Hall output voltage and produces a reading that represents average power.

6–10 POLYPHASE POWER MEASUREMENTS

Polyphase Measurements Using Single-Phase Wattmeters

In a perfectly balanced polyphase system, it is necessary to measure only the power consumed in one leg of the load, such as one of the connections shown in Figure 6–13, and multiply it by the number of phases to obtain the total power.

In a three-phase system it is also possible to measure the total power

*The term transducer refers to any device that transfers energy (electrical, mechanical, kinetic) from one system to another. A microphone is a transducer because it converts acoustical energy, sound, into electrical energy, voltage. Other instrumentation transducers are discussed at the end of this chapter and in Chapter 10.

using only two single-phase wattmeters even if the load is unbalanced, provided a "neutral" line is not present. Figure 6–17 shows the connection for the "two wattmeter method" of three-phase power measurement.

If the power factor of the load is greater than 0.5 (a leading or lagging phase angle of $\cos^{-1}(0.5) = 60°$ between voltage and current), both wattmeters read positive, and the total power (P_T) is equal to the sum of the two meter readings: $P_T = P_1 + P_2$. If the power factor is less than 0.5, one of the meters gives a negative reading, and the current leads of the negative reading meter should be reversed. In this case, the total power is equal to the algebraic difference between the normally "positive reading" meter and the one that initially gave a negative reading: $P_T = P_1 - P_2$.

When the power factor is 1 (no phase difference), this then corresponds to a purely resistive load, so that both wattmeters read the same ($P_1 = P_2$). On the other hand, when a power factor of 0 (90° phase angle) which corresponds to a purely reactive load occurs, both wattmeters have the same reading, but one has a polarity opposite of the other ($P_1 = -P_2$). Finally, when the power factor of the load is 0.5, one wattmeter reads zero, while the other reads the total power delivered to the load.

Three-Phase Wattmeters

Three-phase, or **polyphase**, wattmeters are constructed by mounting two electrodynamometer movements on a common shaft. This is equivalent to the "two wattmeter method." It is connected in a similar fashion (Figure 6–17). The torque acting on the shaft is the sum of the two torques produced by the individual moving coils. When properly connected, the total power is summed and indicated on a single scale.

In a three-phase, four-wire system, three wattmeters are connected as

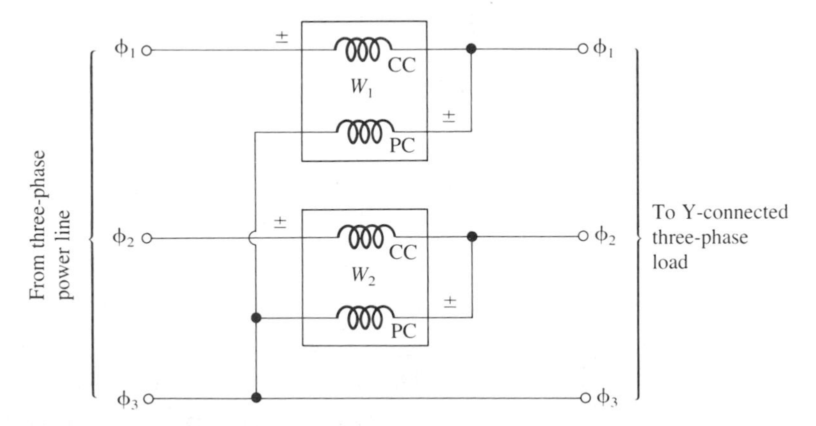

Figure 6–17 Two-wattmeter method for measuring power in a three-phase system.

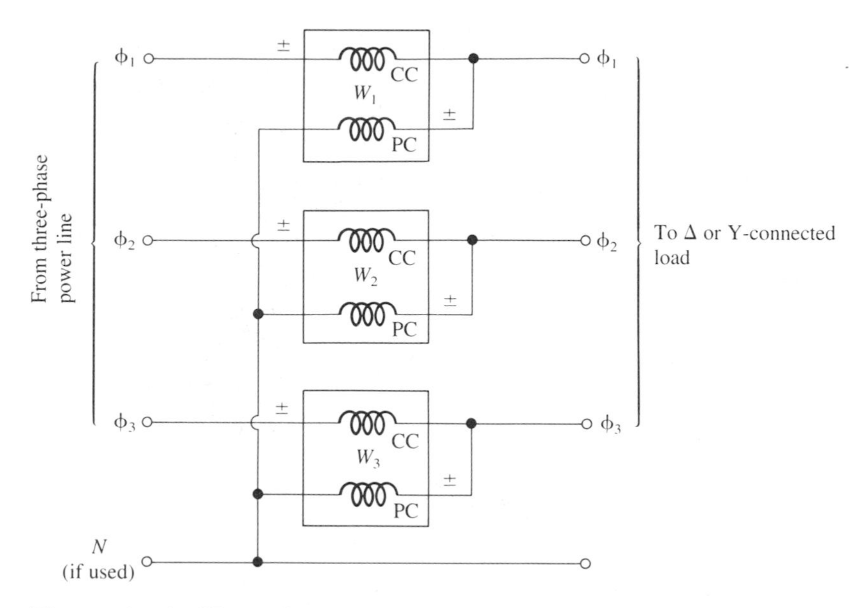

Figure 6–18 Using three wattmeters to measure power in a three-phase, four-wire system.

shown in Figure 6–18. Total three-phase power is the sum of the three meter readings. This follows from what is sometimes referred to as **Blondel's theorem,** which states: in a polyphase system with N wires, the total power can be measured with $N - 1$ wattmeters. Consequently, in a three-phase, four-wire system, three wattmeters are needed. However, if the load is balanced, one wire (i.e., the neutral wire) can be eliminated. In this case, only two wattmeters are necessary to measure total power. In all cases, however, the total power is the algebraic sum of the wattmeter readings.

6–11 POWER FACTOR METERS

The power factor is the ratio of the true power, as measured in watts, to the apparent power (volt-amperes) of a system. To determine the power factor, we could use a voltmeter, ammeter, and a wattmeter to make simultaneous measurements. However, it is more convenient to use a power factor meter in situations where the power factor must be continuously monitored. For example, installations that use a lot of inductive motors can create a situation where the power factor is significantly less than the local power company will allow. Since wattmeters measure real power, a low-power factor lowers the value of the real power consumed and the power company is "cheated" out of its rightful charges for electricity supplied.

Quadrature Coil Meter

The power factor in single-phase circuits is measured using a modified version of the electrodynamometer movement. In the power factor meter, the moving element consists of two coils mounted in quadrature as shown in Figure 6–19. One of these coils is in series with a resistor and has a current that is in phase with the load voltage. The other coil, because it is in series with an inductive load, has a current that lags the load voltage by 90°. The stationary windings produce a magnetic field that is supplied by the load current.

If the load current and voltage are in phase, the axis of coil *A* lines up with the axis of the current coil. If the load current lags the load voltage by 90°, the axis of coil *B* lines up with the axis of the current coil. For intermediate phase differences, the crossed coils assume a corresponding intermediate position depending upon the relative magnitudes of the torques produced by each crossed coil. This type of meter does not require a restoring torque, and consequently has no springs. Variations of this principle are also used for polyphase power factor meters.

Induction or Polarized Vane Meter

A second type of power factor meter, used almost exclusively for balanced three-phase systems, is the **induction**, or **polarized vane meter**. In Figure 6–20, it consists of a current coil with a single winding and is connected in series with one phase line. This coil is wound concentrically with the shaft. It is used to polarize moving iron vanes fixed to each end of the shaft. A balanced three-phase potential winding similar to the stator winding of a three-phase induction motor is mounted over the current coil. This winding produces a rotating magnetic field.

The vanes are either attracted or repelled by the rotating field. They

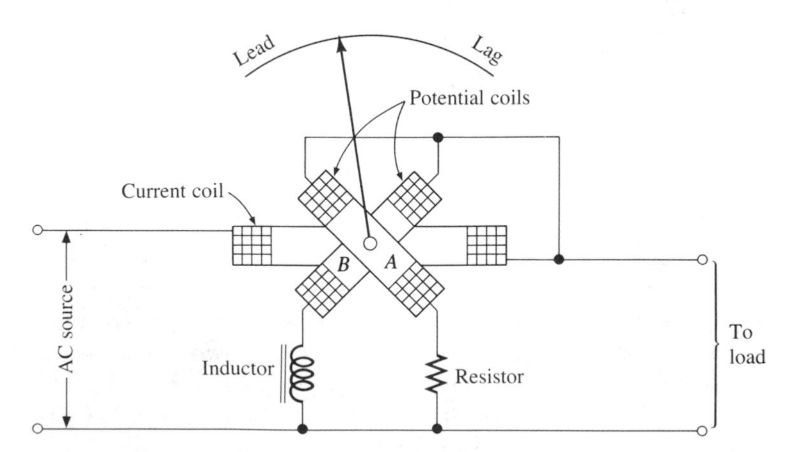

Figure 6–19 Construction of a single-phase quadrature-coil power factor meter.

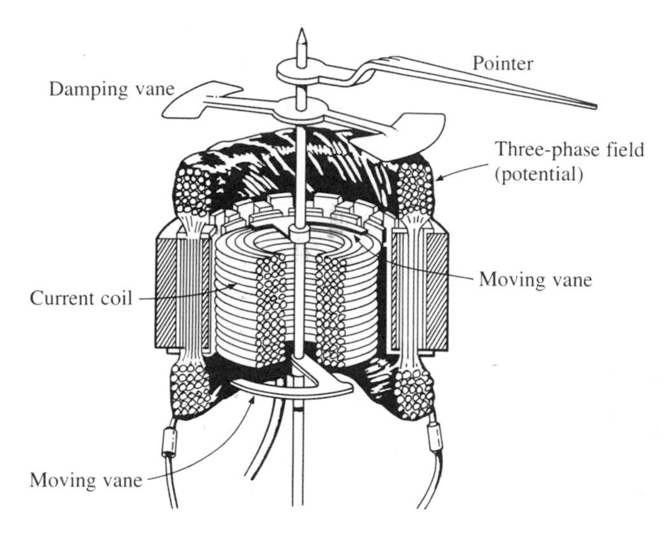

Figure 6–20 Polarized-vane (induction-type) power factor meter.

eventually take up a position where this field is zero at the same time that the field produced by the current coil is zero. Consequently, the instantaneous position corresponds to the difference in phase between the line current and the line voltage.

Power Factor Measurement Using Two Wattmeters

In three-phase, three-wire circuits, the power factor can be calculated from the readings of two wattmeters connected in the standard configuration for measuring real power (Figure 6–17). If P_L is the larger reading (which is always positive), and P_H is the smaller reading (which may be either positive or negative) the power factor is given by

$$\text{Power factor} = \frac{1 + \alpha}{2(1 - \alpha + \alpha^2)^{1/2}} \qquad (6\text{–}7)$$

where α is the ratio P_L/P_H. This relationship is summarized in the graph of Figure 6–21.

EXAMPLE 6–5

Determine the total power and the power factor of the three-wire system of Figure 6–17 if one wattmeter reads 8 W, and the other reads 2 W after initially giving a negative reading.

Solution:

(a) In this case, the total power is equal to the algebraic difference between the normally "positive reading" meter and the one that initially gave a negative reading,

$$P_T = P_H - P_L$$
$$= 8 \text{ W} - 2 \text{ W}$$
$$= 6 \text{ W}$$

(b) Since one wattmeter initially gave a negative reading, the power factor must be less than 0.5. Using Equation 6–7,

$$\alpha = -\frac{2 \text{ W}}{8 \text{ W}}$$
$$= -0.25$$

$$\text{PF} = \frac{1 - 0.25}{(2)(1 + 0.25 + 0.25^2)^{1/2}}$$
$$= 0.33$$

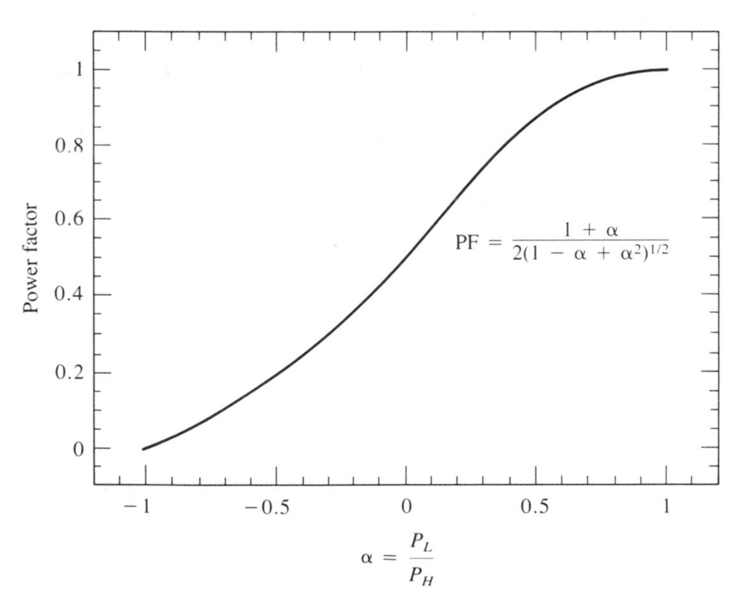

Figure 6–21 Graph of power factor versus the ratio P_L/P_H.

6–12 VARMETERS

It is possible to measure reactive voltamperes by shifting the voltage applied to an ordinary wattmeter mechanism by 90°. The resulting torque is the product of the quadrature components of current and voltage rather than the in-phase components. The torque is a function of *reactive power,* or **vars**. In a single-phase circuit, the 90°-phase shift may be obtained by the use of resistors and capacitors in series with the potential coil of the meter. In a three-phase circuit, two wattmeters may be used with the quadrature voltages obtained by using a pair of special autotransformers connected in an open-delta connection. The arithmetic sum ($P_1 + P_2$) of the two wattmeter readings is equal to the total reactive power delivered to the load. This is analogous to the two wattmeter method of measuring real power. Figure 6–22 shows how a two-element varmeter with its auxiliary phase-shifting autotransformers are connected to measure reactive power.

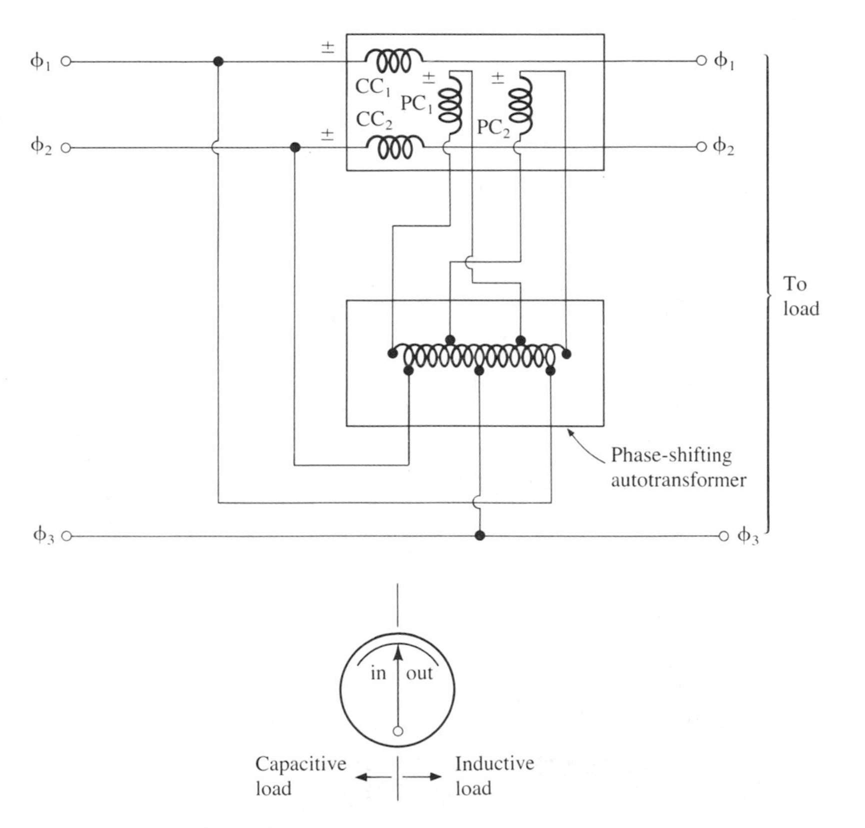

Figure 6–22 Two-element varmeter to measure reactive power.

Varmeter scales are usually the "zero-center" type. They are marked "IN" and "OUT" at their left and right-hand ends, respectively. By convention, when the meter reads "vars OUT," it indicates that reactive power flows from the supply to the load. Such is the case when a generator is supplying an inductive load (i.e., an induction motor). On the other hand, a "vars IN" reading indicates that the load is capacitive.

6–13 ENERGY MEASUREMENT—THE INDUCTION WATT-HOUR METER

Energy is the product of the power and the time during which the power is used. Dimensionally, energy has the units of watt-seconds (Ws). Because of the large quantities frequently involved, however, instruments are generally arranged to indicate energy in terms of either watt-hours (Wh) or kilowatt-hours (kWh). Although many techniques have been used to measure energy in the past, the induction watt-hour meter is now used almost exclusively for the measurement of AC energy at power-line frequencies. The induction watt-hour meter integrates the product of time and power over the period during which power flows and continuously displays the result.

As illustrated in Figure 6–23, the moving element of an induction watt-hour meter is an aluminum disk mounted on a vertical shaft. It is free to rotate

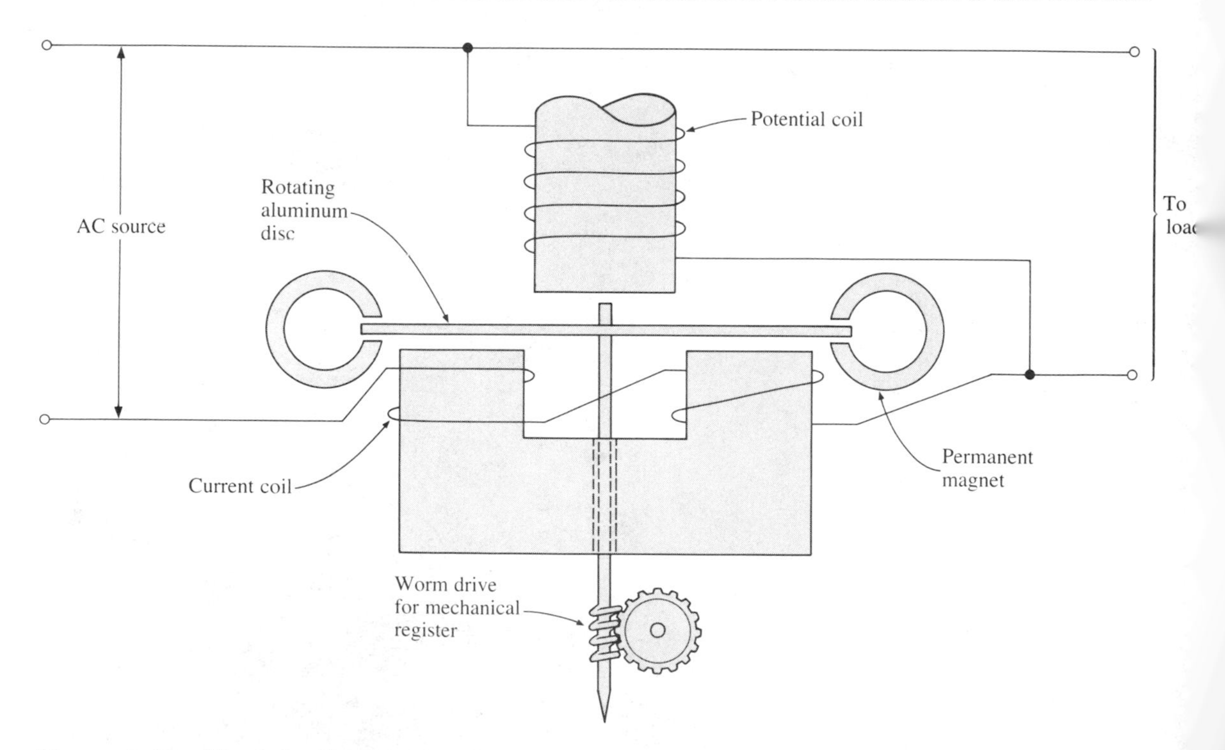

Figure 6–23 The induction-type watthour meter with a rotating aluminum disc.

between the poles of potential and current coils. These poles produce a resultant flux that sweeps across the disk and produces eddy currents in the disk. These eddy currents react with the flux and cause the disk to rotate. The torque produced is proportional to the component of current that is in phase with the voltage. The torque is, thus also proportional to power.

A retarding torque is produced by a permanent magnet assembly. The disk cuts the flux of permanent magnets as it rotates. This creates additional eddy currents that develop a torque that opposes rotation. The magnitude of this torque is proportional to the speed of the disk. The disk turns at a speed where the driving torque and the retarding torque are equal. The driving torque is proportional to power, but the retarding torque is proportional to speed. The speed of the disk is proportional to power, and the number of revolutions is a measure of the power integrated over a period of time, i.e., watt-hours. To display the instantaneous energy value, the shaft is coupled through a gear train to a mechanical counter with a register calibrated in watt-hours or kilowatt-hours, as shown in Figure 6–24.

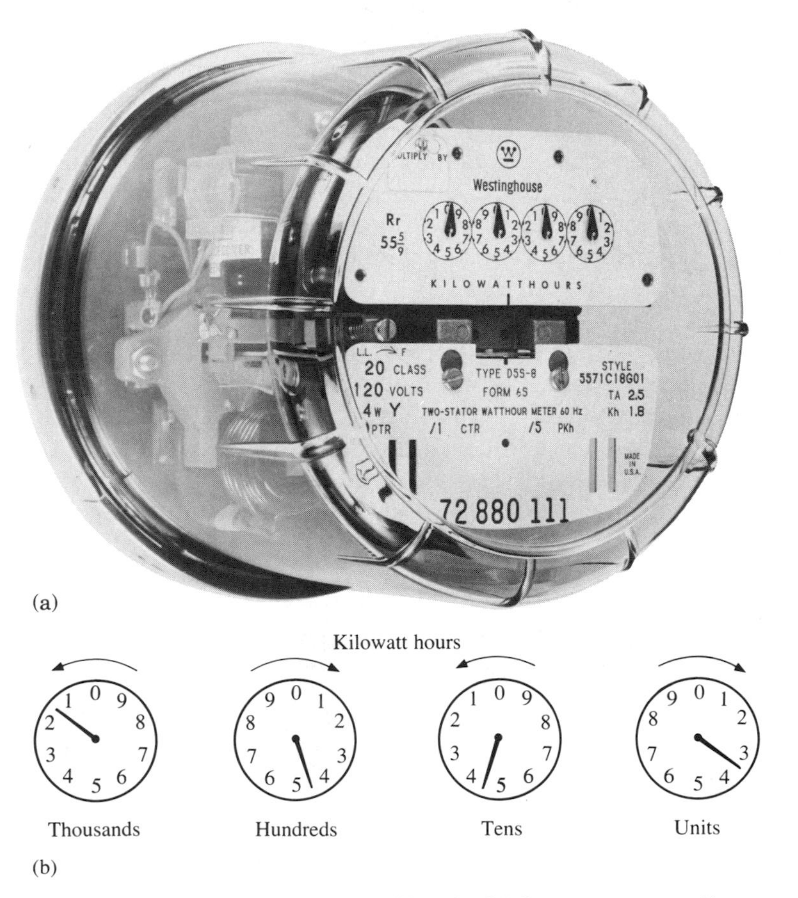

Figure 6–24 Watt-hour meter. (a) Assembly. (b) Dial arrangement. Courtesy Westinghouse Electric Corporation, Measurements and Control Division.

6–14 INSTRUMENT TRANSFORMERS

In the areas of power transmission and heavy industry, voltage levels may reach hundreds of thousands of volts while currents may be in the thousands of amperes. When the magnitudes of AC voltage and current to be measured are too large to be handled practically or safely by available instrumentation, it is a standard practice to step the levels down using what are known as **instrument transformers.** Basically, an instrument transformer supplies an instrument with a precise fraction of the quantity being measured, such as voltage or current. The scale of the measuring instrument is calibrated to account for the step-down ratio of the instrument transformer, so the resulting display indicates the actual quantity.

Potential Transformers

Shown in Figure 6–25, the **potential transformer**, or *PT*, is a step-down transformer having a precisely defined turns ratio. The PT usually has a fused primary. It is normally designed to step the maximum measured voltage down to a level of 120 V, which has become a standard for panel-board instruments and meters. Among the parameters that must be tightly controlled in a PT are the voltage ratio, the insulation level, and the phase-angle error. Ratio and phase-angle errors may change as the power system voltage is changed. It is important that these variations do not exceed the required tolerances. In addition, ratio and phase-shift errors are also influenced by the load imposed on the transformer by instruments and meters. The load on a potential transformer is called the **burden.**

The potential transformer is basically a power transformer that has low magnetizing current, low core-loss current, low leakage reactance, and very small ratio and phase-angle errors. The PT is designed for a very small secondary load. A typical PT burden might be 200 VA or less. One side of the secondary should be grounded as a precaution with the large primary voltages encountered. PTs are available in a wide range of shapes, sizes, and ratings for both indoor and outdoor applications (Table 6–2).

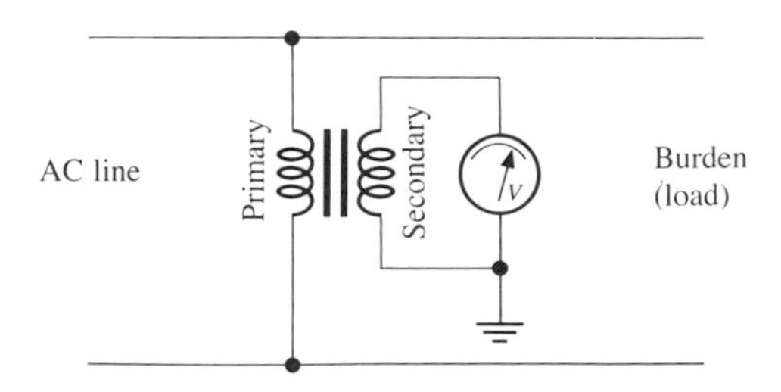

Figure 6–25 A potential transformer connected in parallel with the AC line.

Table 6–2 Typical standard potential transformer ratios

Primary	Secondary
2400/4160 V (wye)	120 V
2400 V	120 V
4200 V	120 V
4800 V	120 V
7200 V	120 V
8400 V	120 V
12000 V	120 V
14400 V	120 V

Current Transformers

The **current transformer**, or *CT*, is designed to supply a precise fraction of the measured current to instruments, meters, and relays (Figure 6–26). Current transformers are iron-core devices with a primary winding that consists of a few turns of large current-carrying capacity. This winding must be sufficiently insulated from the core and the secondary winding that it can safely withstand full power-circuit voltage. The secondary winding consists of several turns of relatively small wire. The turns ratio is normally designed to supply 5 A to the secondary circuit when the rated full-load current flows in the primary (Table 6–3).

The current flow in the primary of a current transformer is determined, for all practical purposes, by the power circuit in which the transformer is installed. This current is almost completely independent of the characteristics of the transformer itself or of the secondary burden imposed upon it. Ideally, the secondary of a current transformer should be short-circuited, corresponding to zero burden. In actual practice, however, the secondary load impedance is kept very low, consisting of only one or more instrument coils. If the secondary

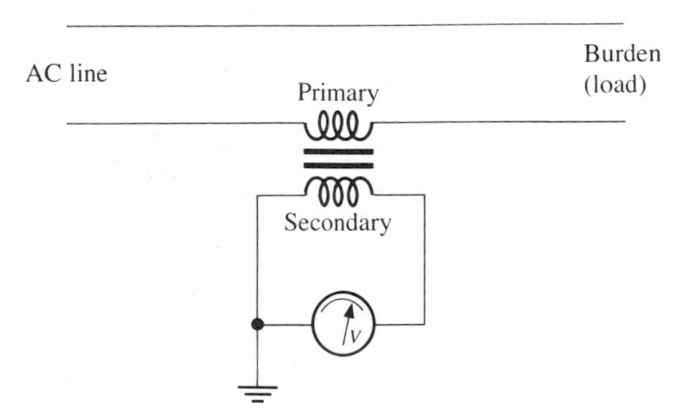

Figure 6–26 A current transformer connected in series with one AC line.

Table 6–3 Typical standard current transformer ratings.

Primary	Secondary	Primary	Secondary
75 A	5 A	800 A	5 A
100 A	5 A	1200 A	5 A
150 A	5 A	1500 A	5 A
200 A	5 A	2000 A	5 A
300 A	5 A	3000 A	5 A
400 A	5 A	4000 A	5 A
600 A	5 A		

load impedance becomes large enough to be significant, then the burden is said to be too high. Current transformers must never be operated with the secondary open-circuited; the resulting high voltages would cause damage. As a standard practice, the secondary windings of a CT are short-circuited *before* disconnecting any of the devices that it supplies.

Many CTs have no primary winding of their own. They use the power-circuit conductor passing through a window in the CT to act as a single-turn primary. The popular hand-held clamp-on ammeters, shown in Figure 6–27, use a variation of this principle. The sensitivity of the clamp-on ammeters may be increased by looping the current-carrying conductor through the window a

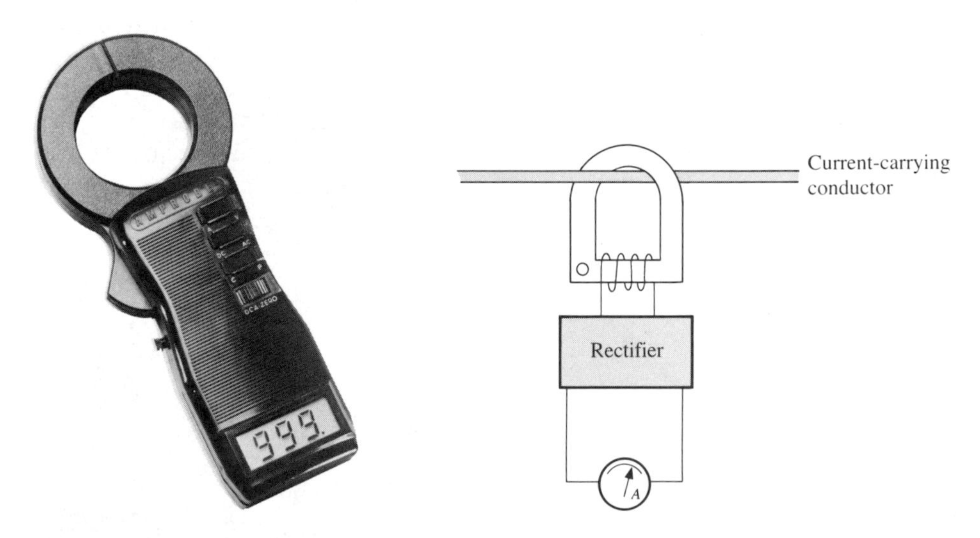

Figure 6–27 Hand-held clamp-on ammeter. Photograph courtesy Amprobe Instruments.

given number of times. For example, a conductor that carries 1 A and passes through the window twice causes the meter to read 2 A. This works because the magnetic fields produced by each conductor add.

6–15 AC-TO-DC TRANSDUCERS

Transducers are available that convert AC signals to a DC voltage or provide a standard current-loop output. These are particularly convenient for such uses as driving DC instruments, recorders, and interfacing with computers. Figure 6–28 illustrates a situation in which a transducer converts an AC current to be read by a DC meter. The current-loop signal is suitable for transmission over considerable distances. It is also compatible with standard process control equipment.

Transducers are available that convert the following AC signals into a standard current-loop signal of either 0 to 1 mA DC or 4 to 20 mA DC

AC volts
AC amperes
AC watts
AC vars
Power factor
Phase-angle
Frequency

Maximum input voltage are 150 V; maximum current is about 5 A. These transducers operate at 50/60 Hz except for the frequency transducers, which are available in several common power frequency ranges.

6–16 SUMMARY

This chapter discussed meters that measure AC quantities. By using a rectifier, a DC meter can be used to measure AC voltage, and the reading can be

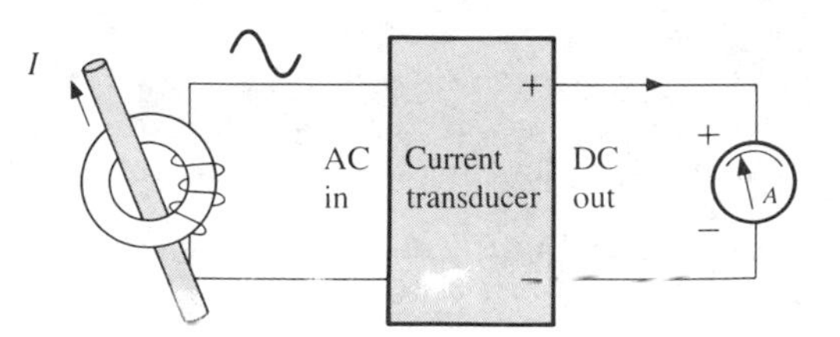

Figure 6–28 AC-to-DC current transducer application.

corrected for RMS voltages of waveforms other than sinewaves. In addition to voltmeters and ammeters, power is measured using wattmeters and varmeters. Other meters are capable of measuring power factor and energy. AC and DC meter movements are listed in Table 6–4.

Table 6–4 Classification of AC and DC meter movements

Meter type	Suitability	Major use/feature
d'Arsonval	DC only	DC current and voltage measurements in low- and medium-impedance circuits
moving iron	DC or AC	rough indication of large currents or voltages at power-line frequencies
electrodynamometer	DC or AC	precise AC voltage and current measurements at power-line frequencies
electrostatic	DC or AC	high-voltage measurements
electronic	DC or AC	Up to high audio frequencies without RF probe. High input impedance
rectifier	AC only	with a d'Arsonval movement, used for medium-sensitivity applications at audio frequencies
thermocouple	DC or AC	RF currents, low input impedance

6–17 GLOSSARY

attracted-disk electrometer An electrostatic AC voltmeter suitable for the direct measurement of very large voltages. The RMS voltage meter reading is independent of the shape of the applied voltage waveform.

Blondel's theorem The total power of an N-wire polyphase system can be measured using $N - 1$ wattmeters.

burden The load on a potential transformer.

cold junction The junction of thermocouple wires with a conductor leading from a measuring instrument. The reference temperature is either room temperature or that of the freezing point of water.

current transformer An instrument transformer designed to supply a precise fraction of the measured current to an instrument, meter, or relay.

electrodynamometer An RMS meter movement whose principle is similar to that of the PMMC movement, except that the permanent magnet is replaced with an electromagnet. It is equally suited for both AC and DC measurements.

electrostatic voltmeter A voltmeter based on a capacitor created by the force between two parallel plates. The force is proportional to the square of the voltage between the plates.

form factor The ratio of the applied RMS value to the average value of a given waveform.

half-bridge full-wave rectifier A circuit that converts an AC voltage to a DC voltage with a full-wave rectifier that has only one diode in the current path at any given time instead of the usual two diodes.

Hall effect The effect created by passing a current through a semiconductor bar situated in a magnetic field that is perpendicular to the direction of the current. A voltage is developed that is perpendicular to both the magnetic field and current.

induction meter An instrument, also called a *polarized-vane* meter, that measures the power factor of balanced three-phase systems.

instrument transformer A transformer that supplies an instrument with a precise fraction of the quantity being measured, such as voltage or current.

iron-vane meter A low-frequency AC meter used to measure either voltage or current. It consists of movable iron vanes mounted within a fixed coil.

polarized-vane meter See **induction meter.**

polyphase wattmeter An instrument for measuring three-phase power using two electrodynameter movements on a common shaft.

potential transformer An instrument transformer that steps down the measured primary voltage to a secondary voltage of usually 120 V RMS. It is basically a power transformer designed for a very small secondary burden.

precision rectifier A rectifier circuit using an operational amplifier to minimize the voltage drop that exists across a conventional diode rectifier.

quadrature coil meter An instrument that measures power factor using a pair of coils mounted so that their axes are 90° apart.

rotating-plate electrostatic voltmeter An electrostatic voltmeter consisting of two or more semicircular plates. Alternate plates are either fixed or rotating on a shaft which supports a pointer and is restrained by a spring. It can be used to measure both AC and DC voltages.

Seebeck effect The effect whereby a voltage is produced at a junction of two dissimilar metals when heated. This effect is the basis of the thermocouple.

thermocouple meter An instrument that measures RF current by using a thermocouple to detect the heat given off at a thermoelectric junction. Its meter scale is nonlinear.

three-phase wattmeter See **polyphase wattmeter.**

transducer Any device by which energy is transferred from one system to another, particularly when the input and output energy differs in form.

varmeter A "zero-center" meter that measures the reactive power of a circuit by shifting the voltage applied to an ordinary wattmeter by 90°.

Var Acronym for volt-ampere reactive.

wattmeter A meter that measures real, or active, power.

6–18 PROBLEMS

1. The peak current through an AC ammeter using half-wave rectification is 78 μA. Determine the average value of current through the meter movement.
2. Repeat Problem 1 for an AC ammeter using full-wave rectification.
3. A PMMC meter movement having a full-scale deflection of 500 μA and a coil resistance of 150 Ω is to be used as part of an AC voltmeter with half-wave rectification. If the voltmeter's full-scale reading is 15 V RMS, determine the meter's (a) DC sensi-

tivity, (b) AC sensitivity, and (c) the value of the multiplier resistor required.

4. Repeat Problem 3 using full-wave rectification.

5. An AC voltmeter has a 100-μA movement and a half-wave rectifier. If it is set on its 10-V range to measure the voltage across the 15-kΩ resistor of Figure 6–29, determine (a) the meter reading and (b) the percent error due to meter loading.

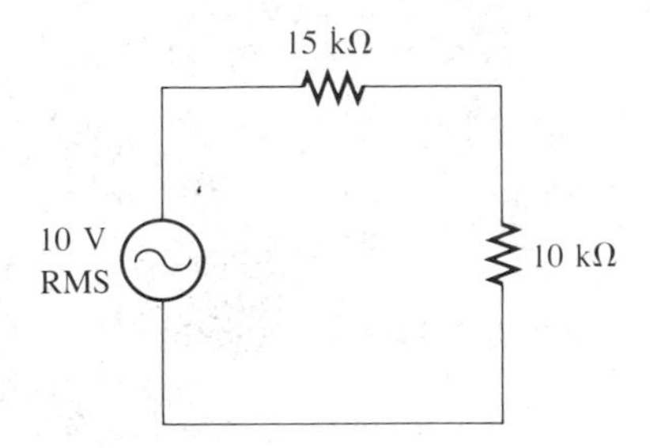

Figure 6–29 Circuit for Problem 5.

6. Repeat Problem 5 using full-wave rectification.

7. An AC voltmeter has the following specifications:

   rectification: half-wave
   meter movement: 100 μA
   full-scale voltage: 10 V
   limiting error: ±2% at full-scale

   When used to measure the voltage across the 10-kΩ resistor of Figure 6–29, determine the minimum voltage read by the meter.

8. Repeat Problem 7 using full-wave rectification.

9. If an AC voltmeter using a PMMC movement calibrated for sinewaves reads 5.32 V RMS when measuring the voltage of a triangle wave, determine (a) the correct RMS voltage and (b) the peak value of the triangle waveform.

10. Repeat Problem 9 for a half-wave rectified sine wave reading 10.94 V.

11. If a thermocouple AC voltmeter has a full-scale reading of 5 A, what is the meter reading when the pointer is deflected 25% of full-scale?

12. What is the full-scale reading of a thermocouple meter if, at mid-scale, it reads 4 A?

13. Two wattmeters are connected to a balanced three-wire system as described in Figure 6–17. The meters initially read positive values of 4.5 W and 9.2 W. Determine (a) the total three-phase power and (b) the power factor of the load.

7

SPECIALIZED METERS: ANALOG AND DIGITAL

7–1 INSTRUCTIONAL OBJECTIVES

At the completion of this chapter, you will be able to

- Describe the function and operation of a direct-coupled electronic voltmeter.
- Explain the purpose and function of a chopper-stabilized DC voltmeter.
- Describe how a precision rectifier is used to permit a PMMC meter to measure AC voltages.
- Describe the differences between a series and a shunt ohmmeter.
- Describe how resistance measurements are made using an ohmmeter.
- Explain the purpose of a volt-ohm-milliammeter (VOM).
- Describe how a digital meter works in terms of a ramp generator, voltage comparator, D/A converter, V/F converter, and A/D converter.
- Compare the advantages and disadvantages of digital and analog meters.

7–2 INTRODUCTION

The basic PMMC meter discussed in Chapter 5 has disadvantages when measuring either AC or DC voltages and currents. This chapter discusses the addition of a high-impedance amplifier with the PMMC meter movement to create an analog meter system with a higher sensitivity. A PMMC meter can also be used to measure DC resistance despite the fact that its scale is nonlinear. For most service applications, the functions of a voltmeter, ammeter, and ohmmeter are generally combined in a single instrument, called a volt-ohm-milliammeter (VOM).

There are now digital meter systems that reduce measurement errors and are easier to use than their analog counterparts. Despite the apparent advantages of digital meters, however, many still prefer the analog meter for certain applications.

7–3 THE ELECTRONIC ANALOG VOLTMETER

The electronic analog voltmeter combines a d'Arsonval meter movement with an amplifier to create a meter system with both a higher input impedance and sensitivity than could be obtained without the amplifier. Before the advent of solid-state devices, the vacuum tube was the active device, and such a meter was referred to as a *vacuum tube voltmeter* or *VTVM*. Today's equipment has replaced the vacuum tube with bipolar transistors and field effect transistors for most electronic voltmeter applications. A typical service-grade unit normally has an input impedance greater than 10 MΩ on all DC voltage ranges, and greater than 1 MΩ on all AC voltage ranges.

Electronic DC Voltmeter

Two different types of amplifiers are used in analog electronic DC voltmeters: direct-coupled and chopper-stabilized. The direct-coupled (DC) amplifier is the least expensive of the two and performs satisfactorily when the most sensitive scale is in the 0.1 to 1.0-V range. For more sensitive ranges, output level shifts and gain changes due to thermal effects and sensitivity to variations in power supply voltage make the DC amplifier impractical. The DC amplifier's DC stability (that is, its ability to maintain a constant output for a given input) is not sufficient for input signals of less than 100 mV.

Figure 7–1 shows the circuit of a simple electronic DC meter having both voltage and current ranges. The amplifier has a gain of ten and produces an output voltage of 10 V for a full-scale meter deflection. The input impedance of the operational amplifier, which is connected as a noninverting amplifier, is greater than 10 MΩ. The overall meter loading on the voltage ranges is approximately 1 MΩ: the resistance divider string (800 kΩ + 100 kΩ + 100 kΩ) in parallel with the input impedance of the op-amp circuit. The 1-Ω current shunt simply produces a voltage drop equal to 1 V per A. The meter has two scales: one calibrated from 0 to 1.0 for the 1V/1A and 10V/10A ranges; the other is calibrated from 0 to 5.0 for the 5V/5A range.

Expensive laboratory electronic DC voltmeters often overcome the DC stability problem by using a **chopper stabilized amplifier.** A *chopper* circuit converts a DC input voltage to an AC signal that is then amplified by an AC

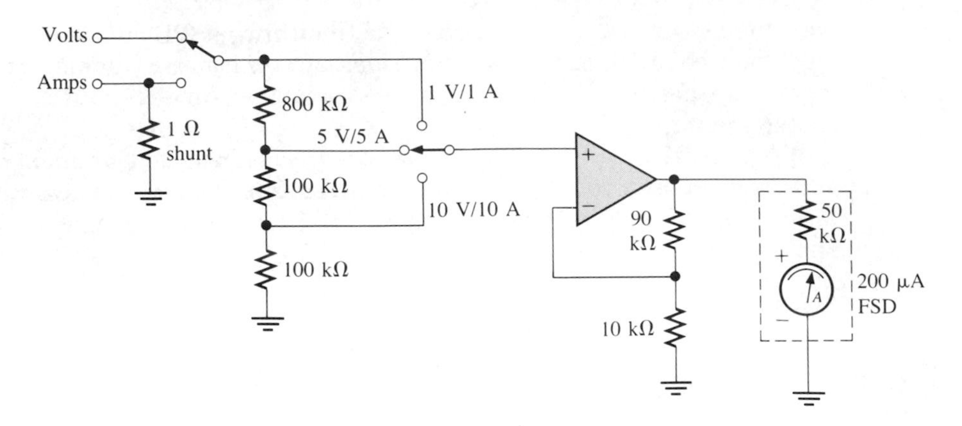

Figure 7–1 Direct-coupled electronic meter that measures DC voltage and current.

amplifier. An AC amplifier eliminates the DC drift problem since any shift in DC level within the amplifier is blocked from reaching the output by AC coupling circuits. At the output of the amplifier, a rectifying circuit synchronized with the chopper converts the amplified AC signal back to an equivalent DC level for application to a meter movement.

The chopper converts the DC signal to a proportional AC signal by interrupting the DC signal at a constant rate. Older meter circuits use an electromechanical device to accomplish this, but current instruments now use electronic circuits, like that of Figure 7–2. In this circuit, two pairs of photo-

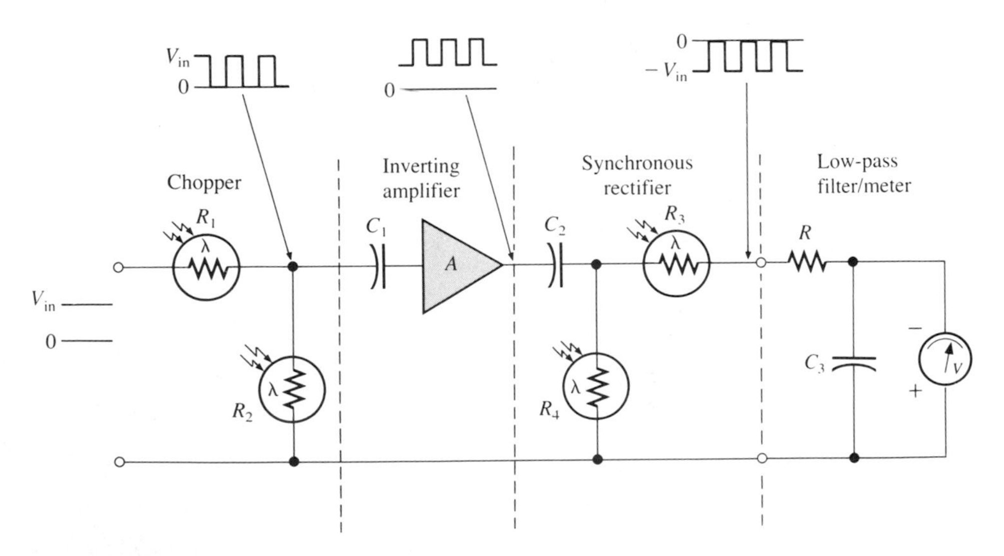

Figure 7–2 Chopper-stabilized DC voltmeter.

resistors* connected as single-pole, double-throw (SPDT) switches provide a low-noise, high-impedance chopper action. Mounted inside the meter case, the photoresistors are illuminated in alternate pairs by an oscillator operating at a frequency of about 90 to 100 Hz. Photoresistors R_1 and R_3 are illuminated for the first half cycle, while R_2 and R_4 are illuminated for the other half cycle. When illuminated, the photoresistors have a low ON resistance and act essentially like short circuits. When the photoresistors are dark, they exhibit a high OFF resistance, which effectively makes it an open circuit. This action is analogous to a diode that is forward biased and then reverse biased.

When the photoresistor pair, R_2 and R_4, is illuminated, both the input to the amplifier and the output side of capacitor C_2 are grounded, allowing C_2 to charge to the output voltage of the amplifier. During the second half of the chopper cycle, photoresistors R_1 and R_3 are illuminated, and the DC signal is applied to the input of the amplifier. This causes the amplifier output voltage to drop by an amount proportional to the size of the input signal. Capacitor C_2 then discharges through R_4 and the meter circuit by a corresponding amount. Since only the discharge current flows through the meter circuit, the meter is driven upscale by an amount proportional to the average discharge current. The discharge current, in turn, is proportional to the amplitude of the DC input signal. The lowpass filter (R-C_3) helps to remove any remaining AC component from the output before it reaches the meter movement.

Electronic AC Voltmeter

Electronic AC voltmeters differ from their DC counterparts only in that the AC voltage must be converted to DC *before* being applied to the meter movement. For relatively large AC signals where diode voltage drop and amplifier drift are not a problem, the measured voltage is first rectified and then applied to a DC electronic meter circuit. For relatively small AC voltage levels, however, the signal is passed through an AC amplifier before rectification. Both of these methods are sufficient where an average responding meter (PMMC) type is satisfactory. As is pointed out in Chapter 6, nonsinusoidal waveforms cause the meter to read either high or low depending on the form factor of the waveform.

A simple electronic AC voltmeter circuit using an operational amplifier is shown in Figure 7–3. It is a combination of a voltage-to-current converter and a precision diode rectifier. By putting the diode inside the feedback loop of the operational amplifier, the effect of diode voltage drop is minimized. The current waveform through the meter is a half-rectified sine wave, even for input voltages less than 0.7 V peak. During the half cycle when the diode is conducting, the instantaneous current through the meter (i_M) is given by

*A photoresistor is a light-sensitive device whose resistance is inversely proportional to light intensity. For higher-speed operation, photodiodes are used. Both are discussed further in Chapter 10.

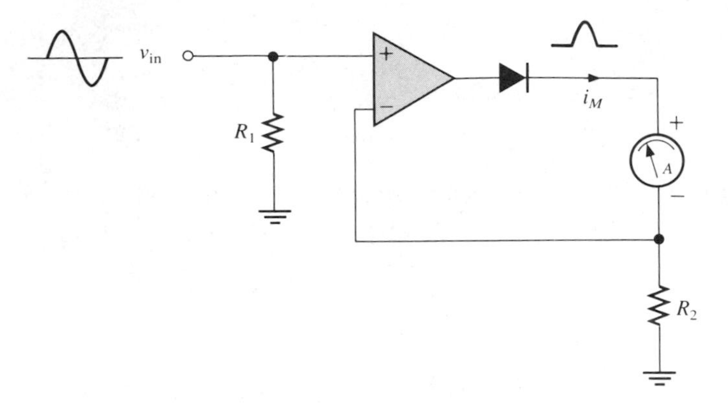

Figure 7–3 AC electronic voltmeter with precision half-wave rectifier.

$$i_M = \frac{V_{in}}{R_2} \tag{7–1}$$

For a sine wave, the average current through the meter is

$$I_M(\text{average}) = 0.318\ I_M(\text{peak}) \tag{7–2a}$$

$$= 0.45\ I_{RMS} \tag{7–2b}$$

On the other hand, a full-wave rectifier bridge can be used in the op-amp's feedback loop as shown in Figure 7–4. Diodes D_1 and D_4 are forward biased on the positive half cycles of the input signal. Consequently, current flows from the + to − terminals of the PMMC meter. During the negative half cycle of the input signal, diodes D_2 and D_3 are forward biased, and current flows through the meter as during the positive half cycle. In either case, the instantaneous current through the meter is

$$i_M = \frac{v_{in}}{R_2} \tag{7–3}$$

For a sine wave, the average current through the meter is

$$I_M(\text{average}) = 0.636\ I_M(\text{peak}) \tag{7–4a}$$

$$= 0.9\ I_{RMS} \tag{7–4b}$$

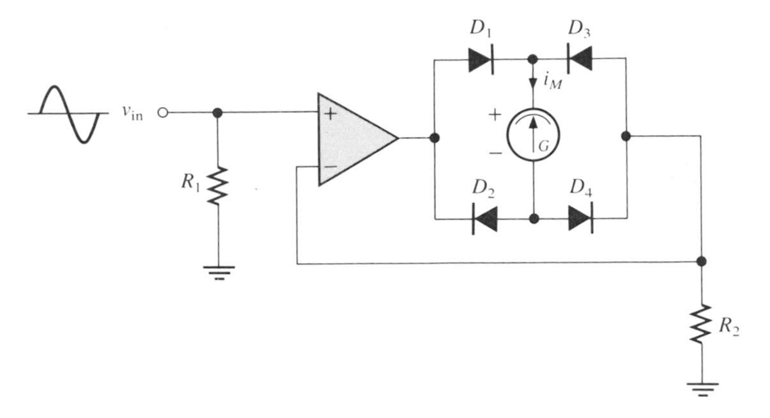

Figure 7–4 AC electronic voltmeter with precision full-wave bridge rectifier.

Probes for Electronic Voltmeters

Accessory probes are available for many electronic voltmeters which increase their range of applications. The RF (radio frequency) probe permits voltage measurements to about 250 MHz. It is particularly useful for measurements on radio and TV amplifier stages. The simple RF probe shown in Figure 7–5 is designed to operate into an 11-mΩ load and produces a negative DC voltage equal to the RMS value of the RF voltage being measured. Most of the simple RF probes of this type have a maximum applied input voltage rating of about 30 V RMS and/or 600 VDC.

The peak-to-peak probe measures the peak-to-peak voltage of complex waveforms, typically over the frequency range from 5 kHz to 5 MHz; this makes it particularly useful for TV and video servicing. The probe shown in Figure 7–6 produces a negative DC voltage equal to the peak-to-peak voltage of any waveform, complex or sinusoidal. The maximum applied input voltage is about 80 V peak-to-peak and/or 600 VDC.

Though both the RF and peak-to-peak probes tend to become inaccurate at low voltage levels due to the nonlinearity of the diodes, they are nevertheless still useful for relative readings down to several volts.

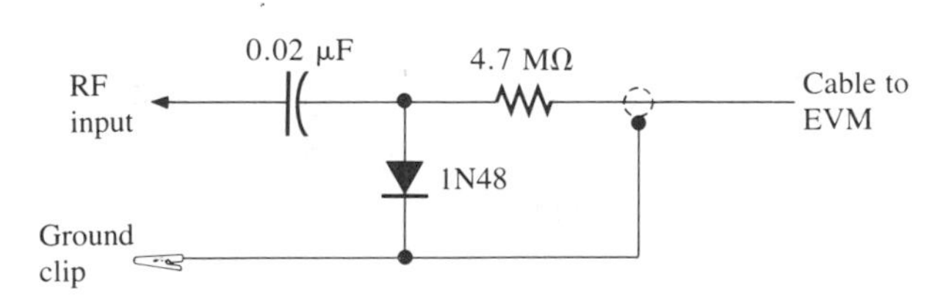

Figure 7–5 RF probe circuit.

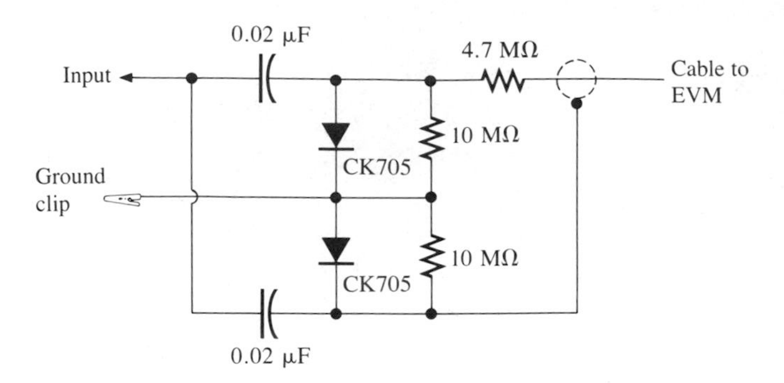

Figure 7–6 An electronic voltmeter probe whose negative output voltage equals the peak-to-peak input voltage.

7–4 ANALOG OHMMETERS

An **ohmmeter** is an instrument using a PMMC movement, battery, and precision resistors to measure DC resistance. The current through the indicating meter is inversely proportional to the unknown resistance. There are two major types of analog ohmmeters in use: the series and the shunt.

Series

The circuit of a simple **series ohmmeter** is shown in Figure 7–7. It consists of a battery, an adjustable resistor (R_{ADJ}), a PMMC movement with its associated internal meter resistance (R_M), and the unknown resistance to be measured (R_X). When the test leads are shorted together ($R_X = 0$), R_{ADJ} is initially adjusted to give a full-scale deflection. When the leads are open circuited ($R_X = \infty$), no current flows, so the meter has no deflection. Consequently, the meter current is inversely proportional to the measured resistance. The meter scale reads resistances between 0 and ∞; 0 Ω is at the right end of the scale, and ∞ is at the left. The current through the meter movement is given by

$$I_M = \frac{V_{BAT}}{R_X + R_{ADJ} + R_M} \tag{7–5}$$

Figure 7–7 A simple series ohmmeter.

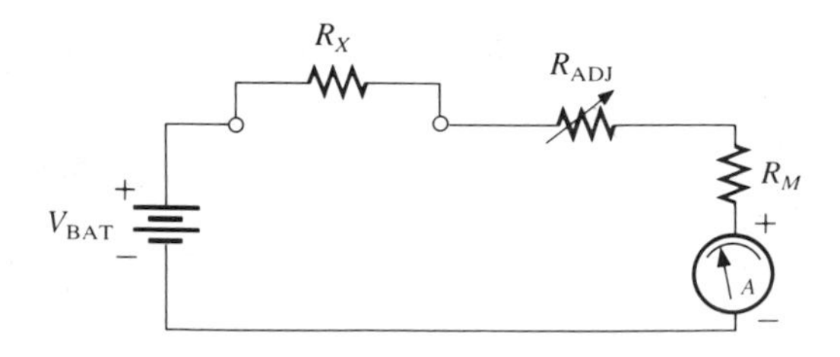

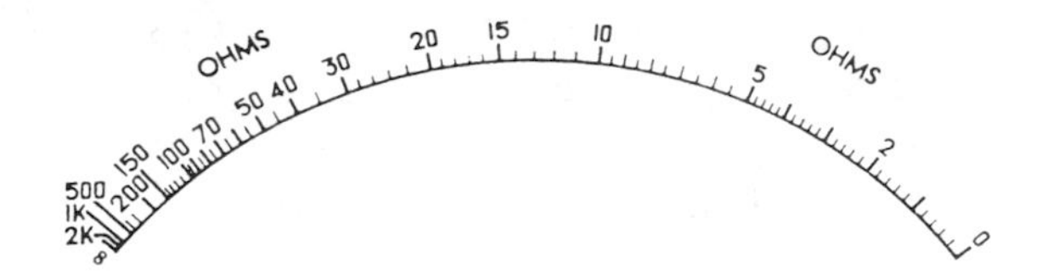

Figure 7–8 Nonlinear meter scale for a series ohmmeter.

Since the meter current (I_M) is a *nonlinear* function of R_X, the meter scale is also nonlinear. The graduations become more crowded as the scale approaches infinity (as shown in Figure 7–8).

The only measure of meter sensitivity is the *mid-scale resistance,* which is equal to the total resistance of the meter circuit ($R_{ADJ} + R_M$). For most applications, the resistance readings on the high-resistance end of the scale (i.e., the lower left-hand third of the meter scale) are too crowded for accurate readings.

EXAMPLE 7–1

It is desired to construct a series ohmmeter using a 50-μA PMMC meter movement having a coil resistance of 1.5 kΩ. For the circuit of Figure 7–7, the battery voltage is 1.5 V. Determine (a) the value of R_{ADJ}, and (b) the unknown resistance R_X required for a mid-scale reading.

Solution:

(a) With $R_X = 0$, the full-scale current is 50 μA. From Equation 7–5

$$R_{ADJ} + R_M = \frac{1.5 \text{ V}}{50 \ \mu\text{A}}$$

$$= 30 \text{ k}\Omega$$

so that $R_{ADJ} = 28.5 \text{ k}\Omega$.

(b) For the meter to read half scale, $I_M = 25 \ \mu\text{A}$. Applying Equation 7–5 again and solving for R_X,

$$R_X = \frac{1.5 \text{ V}}{25 \ \mu\text{A}} - (28.5 \text{ k}\Omega - 1.5 \text{ k}\Omega)$$

$$= 30 \text{ k}\Omega$$

Of course, the same result could have been obtained directly from part (a) since the mid-scale resistance equals the total meter resistance, $R_{ADJ} + R_M$.

The simple series ohmmeter of Figure 7–7 functions correctly as long as the battery voltage remains constant. However, the terminal voltages of all batteries decrease with use, making the ohmmeter scale incorrect. Changing R_{ADJ} to compensate for the change in battery voltage still causes an error because the ohmmeter's mid-scale reading becomes equal to the new value of $R_{ADJ} + R_M$.

Figure 7–9 shows the addition of a meter shunt resistor to minimize the problem associated with decreasing battery voltage. With the ohmmeter leads shorted ($R_X = 0$), R_S is adjusted so that the meter reads full scale. The total circuit resistance is

$$R_{total} = R_{ADJ} + (R_S \parallel R_M) \tag{7–6}$$

As R_{ADJ} is generally made much larger than the parallel combination of R_S and R_M, Equation 7–6 reduces to

$$R_{total} \simeq R_1 \tag{7–7}$$

If the unknown resistance to be measured equals R_{ADJ}, the meter then reads half scale because the total resistance is now doubled. Resistances R_S and R_M form a current divider, so the current through the battery (I_{BAT}) is divided into the current through the meter movement and the shunt

$$I_{BAT} = I_S + I_M \tag{7–8}$$

Since the meter deflection is now half its full-scale deflection when $R_X \simeq R_{ADJ}$, currents I_S and I_{BAT} are also halved. Consequently, the measured mid-scale resistance is still equal to the ohmmeter's internal circuit resistance.

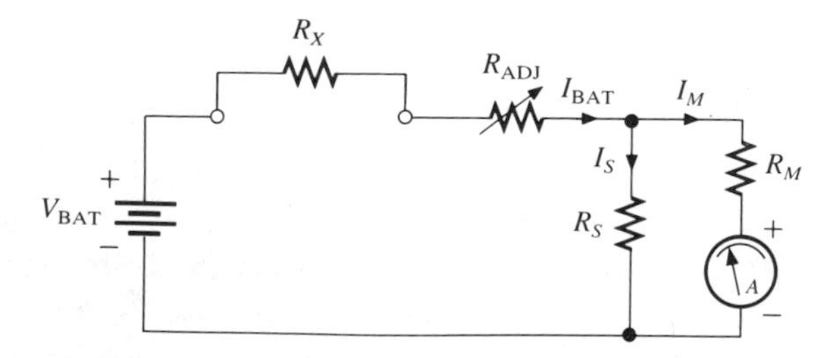

Figure 7–9 Addition of a meter shunt for minimizing effects of decreasing battery voltage on meter calibration.

The current through the battery is

$$I_{BAT} = \frac{V_{BAT}}{R_X + (R_S \parallel R_M) + R_{ADJ}} \tag{7–9}$$

However, as the parallel combination of R_S and R_M is made smaller than R_{ADJ}, the battery current can be approximated as

$$I_{BAT} \simeq \frac{V_{BAT}}{R_X + R_{ADJ}} \tag{7–10}$$

Since the voltage across the meter movement is

$$V_M = I_{BAT}\,(R_S \parallel R_M) \tag{7–11}$$

the meter current is

$$I_M = I_{BAT}\,\frac{(R_S \parallel R_M)}{R_M} \tag{7–12}$$

EXAMPLE 7–2

If a 50-μA movement having a resistance of 50 Ω is used in the ohmmeter circuit of Figure 7–9 with a 1.5-V battery and $R_{ADJ} = 15$ kΩ, determine the value of R_S required for a mid-scale reading of 15 kΩ.

Solution:

For a half-scale reading, the meter current is 25 μA. The total meter resistance is then approximately $2R_{ADJ}$, or 30 kΩ. The battery current is

$$I_{BAT} = \frac{1.5 \text{ V}}{30 \text{ k}\Omega} = 50\ \mu\text{A}$$

Consequently, 25 μA flows through R_S as well as the meter, which implies that the resistance of the shunt must equal that of the meter, or 50 Ω.

Shunt

An ohmmeter circuit that is well-suited to the measurement of low-resistance values is the **shunt ohmmeter**, the basic circuit of which is shown in Figure 7–10. When the leads are open-circuited, R_{ADJ} is adjusted to give a full-scale deflection. This corresponds to $R_X = \infty$. When the leads are short-circuited, no current flows through the meter, corresponding to $R_X = 0$. The lower the unknown resistance that is being measured, the more current that is shunted around the meter and, consequently, the lower the meter deflection. Thus, the meter scale reads zero at the left end of the scale and ∞ at the right. This is the opposite of the series ohmmeter.

Since test leads always have some small amount of resistance, the meter must always be calibrated for the particular leads being used. The value of external resistance that gives half-scale deflection is equal to the parallel combination of R_{ADJ} and R_M,

$$R_X(\text{midscale}) = \frac{(R_{ADJ})(R_M)}{R_{ADJ} + R_M} \tag{7–13}$$

(Since current always flows in the shunt ohmmeter circuit, provide an on/off switch to extend battery life.)

Figure 7–10 A shunt ohmmeter.

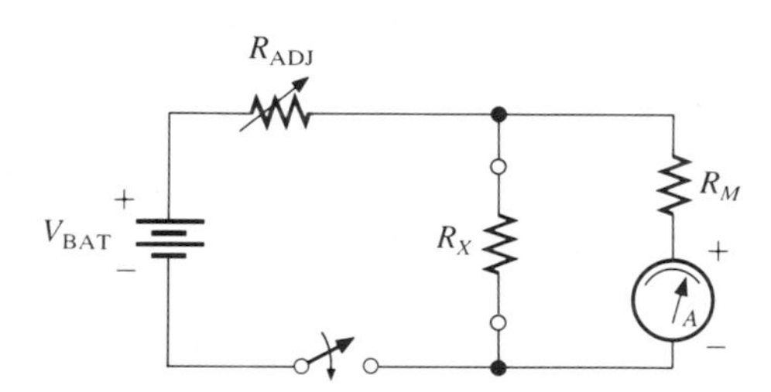

EXAMPLE 7–3

A 10-mA, 2-Ω meter movement and a 1.5-V battery are to be used to construct a shunt ohmmeter using the circuit of Figure 7–10. Determine the value of R_{ADJ} and the mid-scale reading.

Solution:

(a) For the meter to read full scale, R_X must be infinite. Therefore,

$$R_{ADJ} + R_M = \frac{1.5\ \text{V}}{10\ \text{mA}}$$

$$= 150\ \Omega$$

so that

$$R_{ADJ} = 148\ \Omega$$

(b) From Equation 7–13, the external resistance required for half-scale ($I_M = 5$ mA) deflection is

$$R_X = \frac{(148)(2)}{148 + 2}$$

$$= 1.97\ \Omega$$

7–5 THE VOLT-OHM-MILLIAMMETER

The **volt-ohm-milliammeter (VOM),** or **multimeter,** as it is often called, is nothing more than a multirange combination of three basic meters; it is a voltmeter, ohmmeter, and ammeter packaged in a single case. The actual circuit operation for each function is identical to that of the individual meter circuits described in Section 7–4 and Chapters 5 and 6.

The Analog Multimeter

For reasons of economy and size, a single PMMC movement with multiple scales is switched between the three circuits; this is done either by separate function and range switches, or by a single combination function/range switch. Consequently, in a single relatively compact unit, there is the equivalent of a number of individual meters, although only one function and range can be used at a time. Most analog VOMs, such as the Simpson Model 260 (Figure 7–11) measure both AC and DC voltages, but only DC currents flow through the meter.

A typical analog VOM such as the Simpson 260 has the following specifications:

meter movement: 50 μA full scale

DC sensitivity: 20 kΩ/V

AC sensitivity: 5 kΩ/V

frequency response: 20 Hz to 200 kHz

accuracy: ±2% of full-scale on the DC ranges
±3% of full-scale on the AC ranges

As shown in Figure 7–12, the schematic for the Simpson 260 VOM contains circuits for both DC and AC voltages and for current ranges that are typical of those previously discussed in Chapters 5 and 6. In addition, it contains other features, such as output and decibel measurements, that add to the versatility of the analog VOM.

Figure 7–11 Simpson Model 260 VOM. Courtesy Simpson Electric Company.

Output Measurements

A jack marked "output" is connected to the jack for the + lead through a DC-blocking capacitor. By connecting a test lead to the "output" jack rather than the + jack, it is possible to measure only the AC component of a signal that has both an AC and a DC component.

Decibel Measurements

Most analog VOMs include a scale calibrated in decibels. To measure a signal level in decibels, select the appropriate AC range, read the decibel scale, and add to the "dB offset" indicated for the particular range on the meter's scale faceplate. The decibel readings obtained will be correct only if a zero dB power level of 1 mW across a 600-Ω load is used, and if the voltage is measured across 600 Ω. A correction factor must be applied for other reference levels and impedances. It is important to remember that the dB scale, like the AC scale, is accurate only for a sinusoidal waveform and that either any distortion or DC component in the above causes an erroneous reading.

Ohmmeter Precautions

Two batteries are used in the ohmmeter section of the circuit of Figure 7–12. A 1.5-V battery (B_1) is used for the $R \times 1$ and the $R \times 100$ ranges, while a 7.5-V battery combination (B_1 and B_2) is used for the $R \times 10{,}000$ range. This is

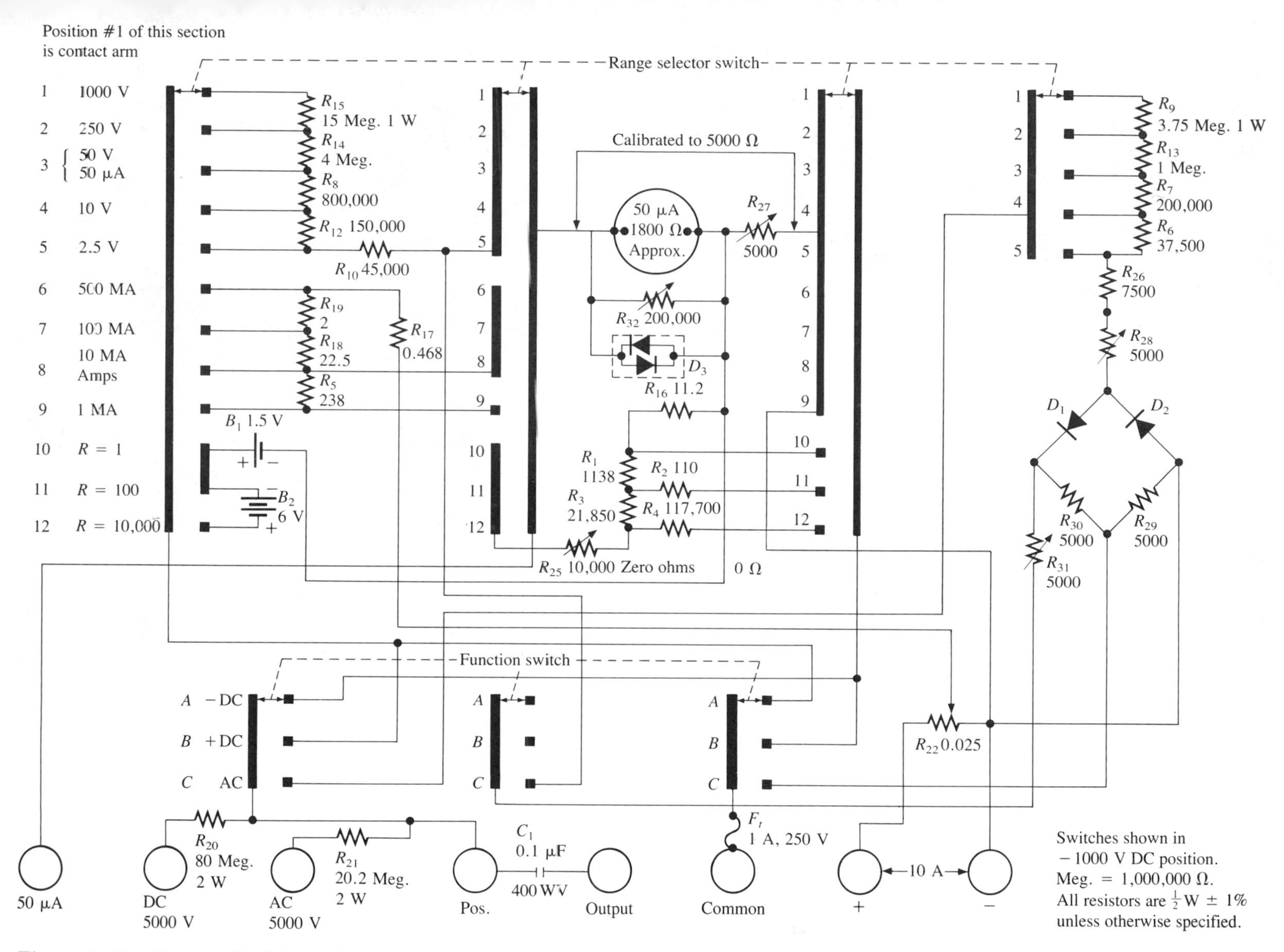

Figure 7–12 Simpson Model 260 VOM schematic.

important to know, particularly when using the ohmmeter to test transistor junctions or making in-circuit measurements in solid-state equipment. The 7.5 V level is high enough to break down a reverse biased base-emitter junction, while 1.5 V is comparatively safe for such tests. In the forward-biased direction, since they present nonlinear resistances, semiconductor junctions read different values on different resistance ranges.

While the positive terminals of both batteries are connected to the "common" jack in the Simpson 260, this is not true for all VOMs. An easy way to check lead polarity is to use a separate voltmeter.

Advantages and Disadvantages

The performance of the VOM is poorer in terms of both its sensitivity and input impedance when compared to the electronic analog voltmeter. The sensitivity of the VOM and, consequently, its input impedance, is based on the voltage level being measured and is typically 20 kΩ/V for a VOM having a 50-μA movement. On the other hand, electronic analog voltmeters frequently have input impedances greater than 10 MΩ, so meter loading is not usually a problem. However, many electronic voltmeters may not work properly in the presence of strong electromagnetic fields, such as those produced by transformers or RF transmitters. The VOM, on the other hand, is essentially unaffected by these factors.

7–6 DIGITAL VOLTMETERS

In contrast to analog meters that use a pointer deflection on a continuous scale, a digital meter displays measurements as *discrete* numbers. Digital meters have become increasingly popular. They frequently reduce human error, eliminate mistakes due to parallax, and often have greatly increased resolution on a given meter range. Some models even include automatic polarity and range-altering features that reduce both operator skill level requirements and the possibility of damage due to overload.

Despite the many advantages offered by digital meters, a significant number of users still prefer the analog meter display when making certain measurements, especially those measurements made during the adjustment of equipment. This is understandable in that sometimes both the direction and relative magnitude of a signal are more readily apparent with the analog meter, which provides the user with better visual feedback as an adjustment is made.

Digital meters are available to measure AC and DC voltages, AC and DC currents, resistance, and ratios, which they can measure either separately or in combination. Perhaps the currently most popular instrument is the digital multimeter, or digital VOM, like the Fluke Model 73 shown in Figure 7–13. Such a meter typically provides several ranges apiece for the measurement of resistance and AC and DC voltage and current. Many of the more expensive

Figure 7–13 Fluke Model 73 digital multimeter. Courtesy of John Fluke Mfg. Co., Inc.

digital instruments are also provided with interface circuitry to output data to computers, printers, and data-logging equipment, used for the purpose of either creating a permanent record of a measured parameter or controlling a process variable.

Most digital instruments are built around a **digital voltmeter** (DVM). Once a basic DVM circuit is equipped with the proper transducer, a wide range of physical parameters such as current and resistance can be measured and displayed.

Internally, all DVMs use some sort of analog-to-digital conversion scheme to convert the analog input voltage to an appropriate digital output. There are a number of techniques used for making this conversion, some of which are discussed in later sections of this chapter. However, two circuits are common to these techniques; these are the ramp generator and voltage comparator, which are discussed in the following sections.

Ramp Generators

Most practical ramp generators are based on the idealized constant current source-charging capacitor circuit of Figure 7–14. An ideal constant current source charges a capacitor of capacitance, C, producing a linear voltage ramp across the capacitor. Since the slope of the ramp voltage is I/C, the capacitor charges linearly. Moreover, the rate at which the capacitor charges is proportional to the current through it and inversely proportional to its size.

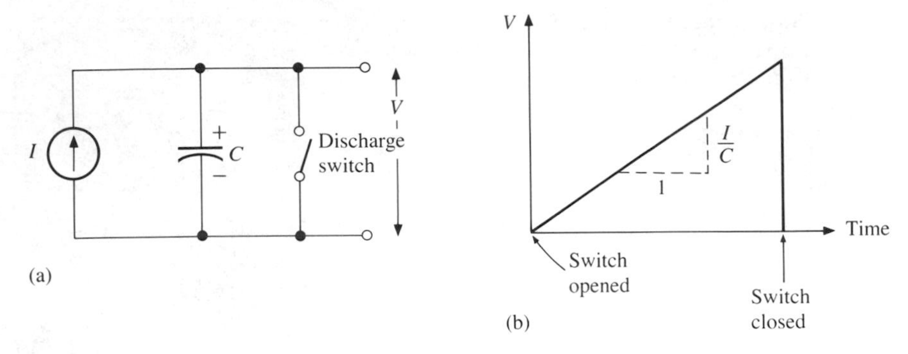

Figure 7–14 Linear ramp voltage generation. (a) Idealized circuit. (b) Output voltage versus time.

In a practical ramp generator circuit, however, an electronic circuit simulates the constant current source, which is represented by the equivalent circuit of Figure 7–15*a*. The initial part of the ramp is linear, but, due to resistance and the decreasing difference between capacitor voltage and the available source voltage, the ramp eventually begins to fall off from the ideal straight line (as shown in Figure 7–15*b*). A good approximation of an ideal ramp voltage is obtained by using only the lower portion of the curve. The ramp voltage is directly proportional to time over the linear portion of the curve, and time is easy to measure with digital logic.

Figure 7–16 illustrates two circuits for charging a capacitor from a constant current source. In the zener diode-PNP transistor constant current source of Figure 7–16*a*, if the base-emitter voltage drop of the transistor is neglected, the voltage across resistor R_E is equal to the voltage across the zener diode. Thus, the emitter current is constant and is approximately equal to V_Z/R_E. Because the collector and emitter currents are approximately equal in high-beta transistors, the capacitor-charging current is also a constant.

In the circuit of Figure 7–16*b*, the operational amplifier circuit acts as a

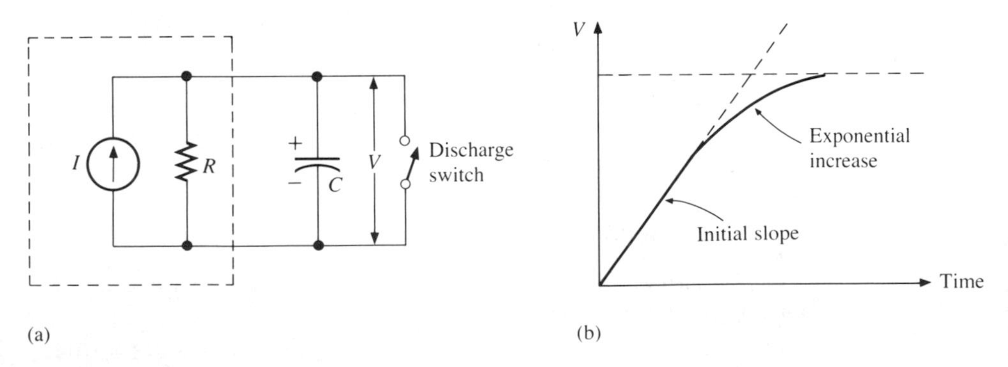

Figure 7–15 Practical constant current source. (a) Equivalent circuit. (b) Output voltage showing that the ramp tends to be exponential rather than linear due to source shunt resistance.

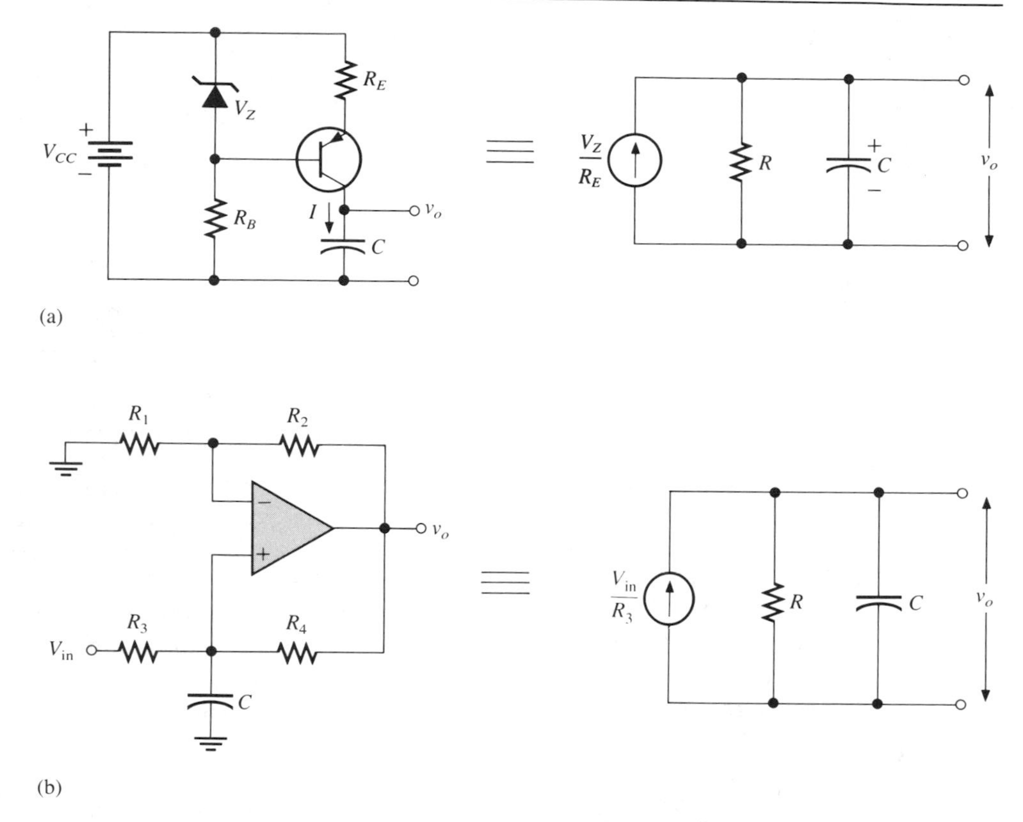

Figure 7–16 Charging a capacitor from a constant current source. (a) Transistor-zener diode circuit. (b) Op-amp circuit.

constant current source, provided that $R_1/R_2 = R_3/R_4$. In this case, the magnitude of the capacitor-charging current is controlled by V_{in}. The value of this charging current is V_{in}/R_3. The ramp voltage v_o is given by

$$v_o = V_{in}t\,\frac{R_1 + R_2}{R_2R_3C} \tag{7–14}$$

Voltage Comparator Circuit

Figure 7–17 shows the connections to a basic operational amplifier comparator. The comparator has two input terminals, V_1 and V_2, and a single output terminal, V_o. The voltage comparator compares two analog input voltages and produces a digital output signal. The output is high if V_1 is more positive than V_2; otherwise, the output is low (if V_1 is less than V_2). This circuit is used in **analog-to-digital (A/D) converters** to determine if an analog input signal (V_1) is greater than or less than a reference voltage (V_2).

Figure 7–17 Voltage comparator.

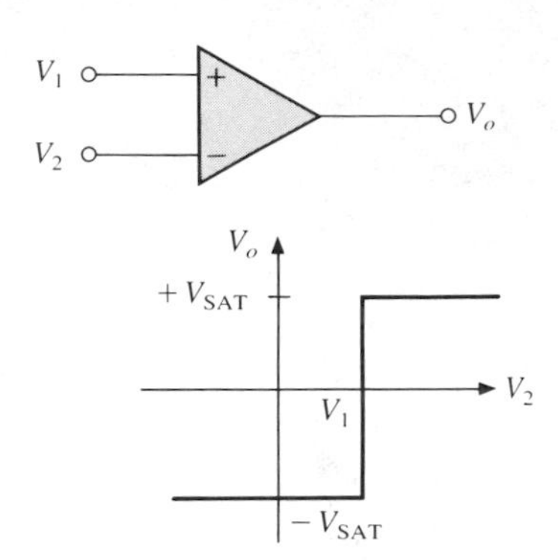

Digital-to-Analog and Analog-to-Digital Converters

A **digital-to-analog (D/A) converter** is a circuit that produces an analog output voltage that is a function of a weighted parallel binary input. The output voltage varies in steps, with the smallest possible step being controlled by the least significant bit (LSB) of the input. For example, a four-bit D/A converter would have 4^2, or 16, possible output levels. The output might range from 0 to 15 mV, the difference between any two adjacent levels being 1 mV. If the input were stepped sequentially through all 16 possible binary combinations (i.e., 0 through 15), the output voltage would look like a staircase increasing in 1-mV steps from 0 to 15 mV.

Most D/A converters use a weighted resistive ladder network in the form of an integrated circuit combined with an operational amplifier, like that of Figure 7–18, to convert binary inputs to an equivalent analog output voltage. The output voltage is incremented in steps of $-1/2^N$ times the voltage corresponding to a logic 1 of any of the binary inputs, where N is the number of bits. For example, if a 4-bit weighted resistor network is used, and the 4-bit inputs are set at $b_0 = 1$, $b_1 = 0$, $b_2 = 1$, and $b_3 = 0$ ($0101_2 = 5_{10}$), then the output voltage of the D/A converter of Figure 7–18 equals $-5/16$ times the logic 1 voltage of the logic family used. If TTL devices are used, then the logic 1 voltage is nominally 5 V. Thus this 4-bit D/A converter has voltage increments of $(-1/16)(5\text{ V}) = -312.5\text{ mV}$ over the output range from 0 (all inputs at logic 0) to -5 V (all inputs at logic 1). If the binary inputs are 0101_2, as above, the output voltage is -1.5625 V.

There are currently two A/D converter schemes in use: the ramp and the dual-slope method.

The ramp A/D converter, or **single-slope converter**, as shown by the block diagram of Figure 7–19, has four basic sections: a ramp generator, voltage comparator, clock, and digital counter.

At some point, a signal resets the binary counter to zero, simultaneously starting the ramp generator. As long as the ramp voltage is less than the voltage being measured (V_{in}), the output of the comparator is high, enabling the "AND gate," which passes the clock pulses along to the counter. When the

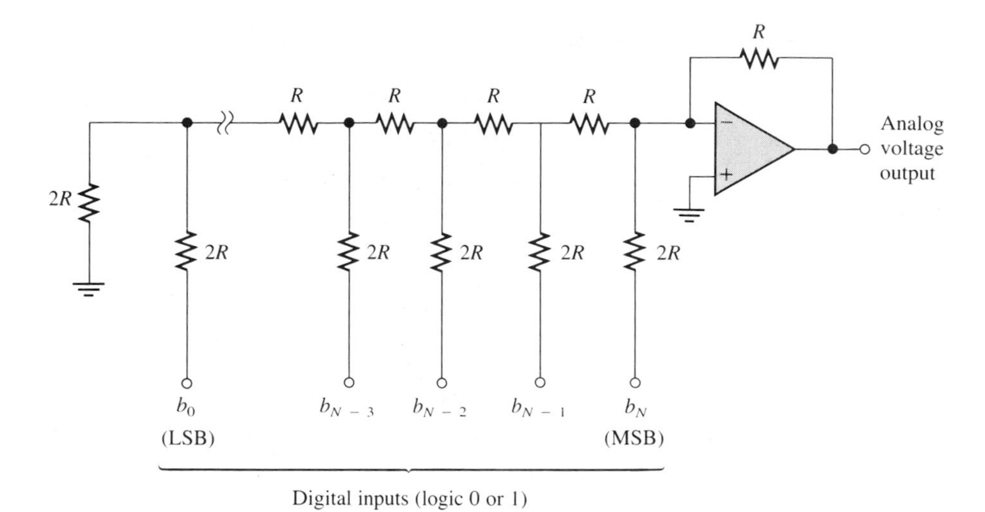

Figure 7–18 Operational amplifier R-$2R$ resistive ladder D/A converter.

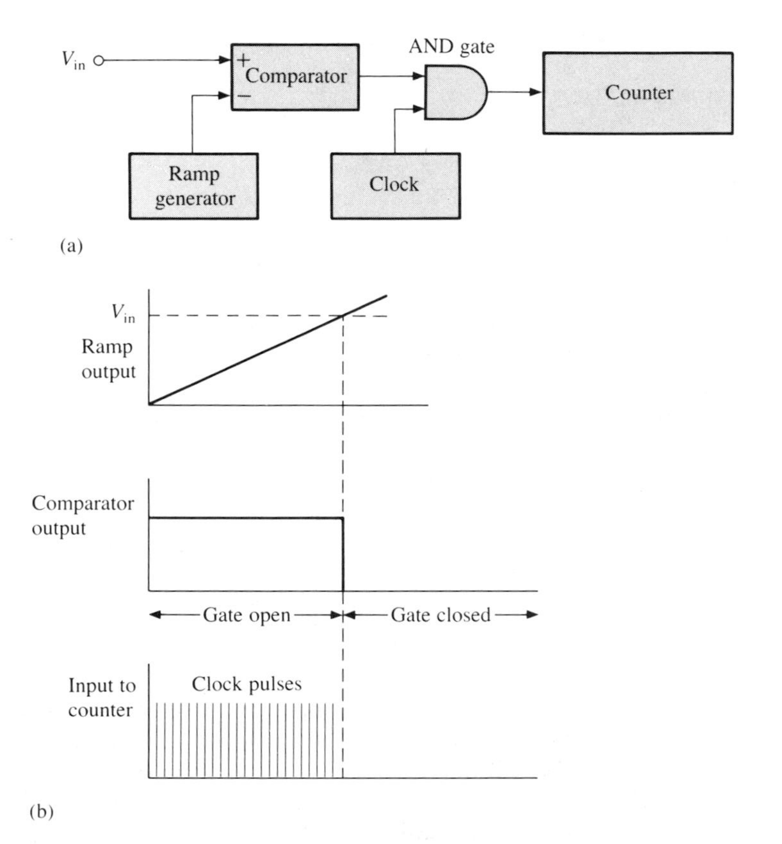

Figure 7–19 Ramp-type A/D converter. (a) Block diagram. (b) Waveforms.

output of the ramp generator exceeds the instantaneous value of V_{in}, the output of the comparator goes low, the clock pulses are prevented from reaching the counter, and the conversion cycle is ended. The total count is proportional to the value of V_{in}.

EXAMPLE 7–4

A ramp A/D converter has a ramp generator with a slope of 1 V/ms and a clock frequency of 100 kHz. If the analog input voltage is 2 V, what is the count at the end of a conversion cycle?

Solution:

With a slope of 1 V/ms, the ramp voltage takes 2 ms to reach a level of 2 V. In the 2-ms period that the "AND gate" is enabled, (2 ms) $\times$ (100 kHz), or 200, clock pulses reach the counter. At the end of the conversion cycle, the counter holds a multibit binary count (i.e., 11001000_2) that is equal to the decimal equivalent of 200. This is then converted to decimal form and displayed as 2.00 volts.

The ramp A/D converter is simple in concept. Its accuracy is limited by both the linearity and slope of the ramp and the frequency stability of the clock frequency.

Of the two A/D conversion types, the dual-slope A/D conversion is probably the more common in current use. Figure 7–20*a* shows a typical dual-slope conversion circuit. The system has a ramp generator whose input is switched between the analog input voltage (V_{in}) and a negative reference voltage (V_{REF}). The analog switch is controlled by the most significant bit (MSB) of the counter. When the MSB is a logic 0, the voltage being measured is connected to the ramp generator input. When the MSB is a logic 1, the negative reference voltage is connected to the ramp generator.

The conversion cycle can be broken down into two distinct steps. First, the counter is reset and V_{in} is connected to the input of the ramp generator. This causes a positive-going ramp voltage, so, assuming an 8-bit counter, the counter will count up from 00000000_2 to 01111111_2. The slope of the ramp during this period is given by V_{in}/RC. When the binary count reaches 10000000, $-V_{REF}$ is connected to the input of the ramp generator. This causes the ramp to go negative with a slope equal to $-V_{REF}/RC$, and the counter will continue to count until the ramp voltage again reaches zero. When the ramp voltage goes below zero, the comparator output goes low, disabling the AND gate to the clock signal and stopping the counter. The count is converted to a decimal form and displayed.

As shown in Figure 7–20*b*, a large input signal produces a large slope

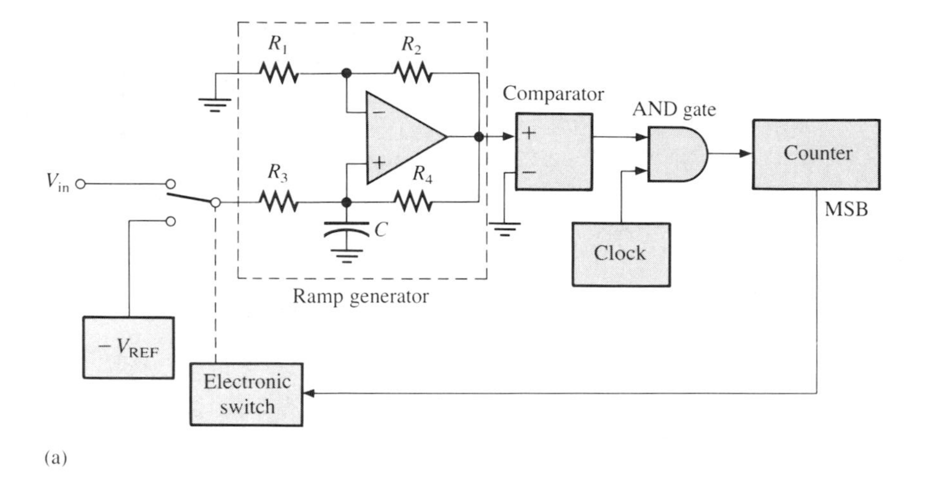

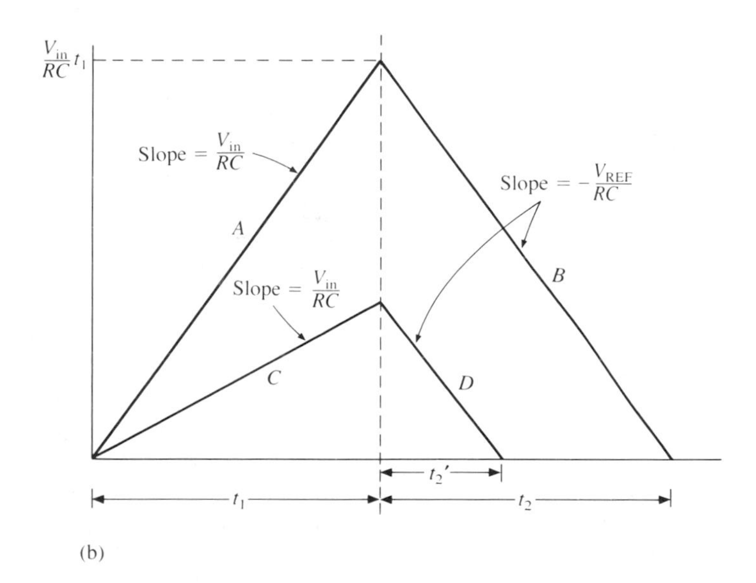

Figure 7–20 Dual-slope A/D converter. (a) Block diagram. (b) Charging and discharging slopes.

(like that of line A), and a smaller input signal produces a smaller slope (like that of line C). Since t_1, the time for the MSB of the counter to reach a logic 1, is constant, the peak value of the ramp is proportional to the value of V_{in}. The slope of the negative-going ramp is a constant, meaning that large values of the peak ramp voltage take proportionately longer than smaller values to return to zero. Thus, time t_2 is a linear function of the value of V_{in}, and the number of clock pulses counted during time period t_2 is proportional to the analog input voltage V_{in}. The most significant bit of the count is not included in the final displayed value, as it is used only to control the selection of V_{in} or $-V_{REF}$. Time period t_1 produces a full count; time period t_2 must not exceed t_1, or the most significant bit of the counter will again change to a logic 0 on the way down, preventing the ramp from reaching zero. This then establishes an upper range limit for V_{in}, that is equal to the absolute value of $-V_{REF}$. The products of the slope and time must be equal during time periods t_1 and t_2 because the changes in ramp value are equal

$$\frac{V_{in}}{RC}t_1 = \frac{V_{REF}}{RC}t_2 \tag{7–15}$$

Time periods t_1 and t_2 in Equation 7–15 can be replaced with the corresponding values of the decimal count (N_1 and N_2, respectively) to represent the changes in the count during these two periods, since time and count are proportional. As a result,

$$\frac{N_2}{N_1} = \frac{V_{in}}{V_{REF}} \tag{7–16}$$

Component tolerances and long-term drift in component values and clock frequency do not significantly affect the accuracy of the dual-slope converter. The clock frequency and component values must not change *significantly* only from period t_1 to period t_2. Consequently, only good *short term* stability is required.

EXAMPLE 7–5

A digital voltmeter using a dual-slope A/D converter has a reference voltage of -10.00 V and has the ability to display a full-scale reading of 1000. Find the reading at the end of period t_2 if the value of analog input voltage V_{in} is 2.5 V.

Solution:

Since the full-scale reading is proportional to N_1 and the actual reading is proportional to N_2, from Equation 7–16,

$$N_2 = (1000)\,\frac{2.5\text{ V}}{10\text{ V}}$$
$$= 250$$

Dual-slope A/D converters provide high accuracy, economy, and good immunity to noise. Whole numbers of cycles average out to zero; thus if period t_1 is equal to some integer multiple of the noise frequency, noise rejection can approach infinity.

Due to their nature, dual-slope converters tend to be slower than many other types; nevertheless, their advantages make them well-suited for use in the laboratory and as portable digital voltmeters where extreme speed is not required.

Voltage-to-Frequency Converter

The **voltage-to-frequency (V/F) converter** changes a DC input voltage into a string of pulses whose repetition rate is proportional to the magnitude of the input voltage. The pulses are counted for a fixed period of time, and the final count is proportional to the measured voltage. Since random noise tends to have an average value of zero, the V/F converter usually has an excellent immunity to noise.

Figure 7–21 shows the elements of a basic V/F conversion system. V_{in} generates a ramp the slope of which is proportional to the amplitude of V_{in}. Each time the amplitude of the ramp exceeds the absolute value of V_{REF}, the comparator output goes high and triggers the monostable multivibrator, which in turn discharges the capacitor and produces an output pulse to the counter. The cycle repeats at a frequency that is a linear function of the amplitude of V_{in}. For example, an input voltage of 2 V causes the ramp to reach the switching level of the comparator in half the time required for a 1-V level. Thus, 2 V produces pulses at twice the frequency of 1 V. The V/F converter is not practical when it is necessary to measure very small voltage levels, as the time between pulses would become excessive. To overcome this problem, a V/F converter often has a minimum output frequency, even at zero volts, that increases linearly with increasing values of V_{in}. This minimum output frequency must be subtracted from the final count before the result is displayed. The V/F con-

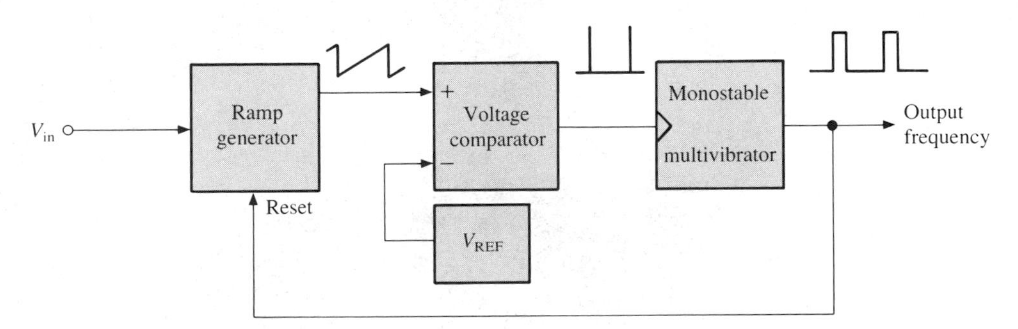

Figure 7–21 Voltage-to-frequency converter.

verter is often used where the voltage to be measured is located some distance from the display, as it is easier to transmit the pulse train reliably than it is a low-level DC signal.

Staircase Ramp A/D Converter

One of the simplest A/D converters uses the staircase ramp technique illustrated in Figure 7–22. The principle is similar to the ramp A/D converter discussed earlier in this chapter, but the ramp generator is replaced in this case with a D/A converter. The output of the D/A converter is a staircase serving the same purpose as the analog ramp in the ramp converter.

At the start of a conversion cycle, the counter is reset to zero; the output of the D/A converter is also zero. At this point, the output of the comparator is high, and the clock pulses go through the gate to the counter. As the count increases, the staircase waveform proceeds in increments towards V_{in}. When the staircase voltage level exceeds V_{in}, the output of the comparator goes low,

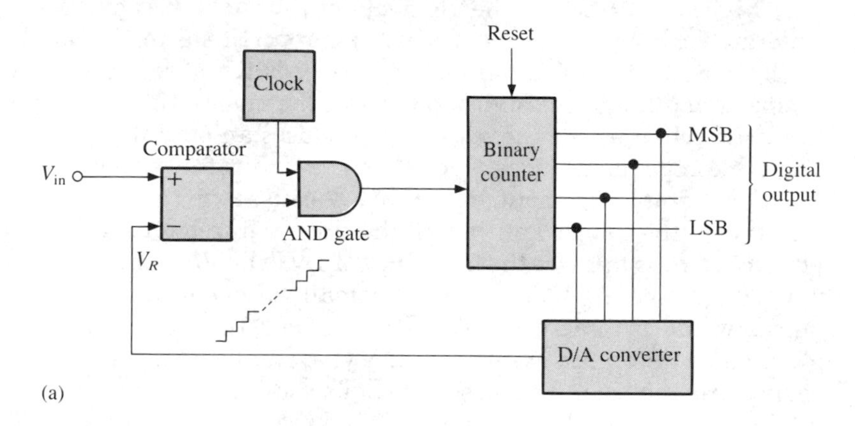

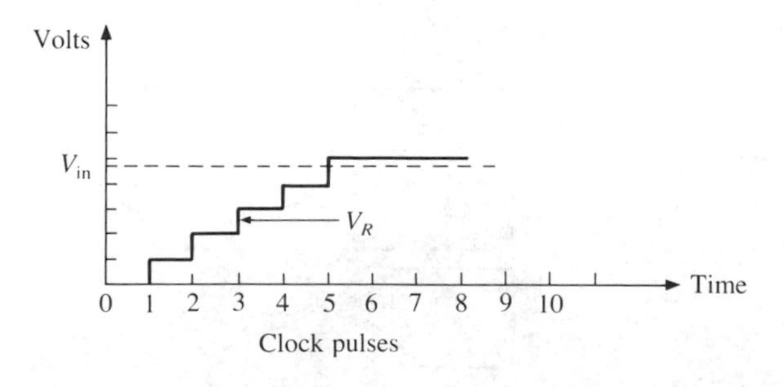

Figure 7–22 A 4-bit staircase ramp A/D converter. (a) Block diagram. (b) The binary counter stops when V_R exceeds V_{in}.

and the count stops. The count is converted to decimal form, and is used to update the display. The cycle then repeats itself.

Successive Approximation A/D Converter

The two major drawbacks of the V/F converter, dual-slope, and staircase ramp A/D converters are that (1) they are relatively slow and (2) the conversion time is a function of the magnitude of V_{in}. The successive approximation A/D converter overcomes both these problems to a considerable degree. It makes conversions in microseconds rather than the several milliseconds required by the ramp, dual-slope, and V/F converters.

Although the block diagram for the successive approximation A/D converter in Figure 7–23 looks very much like that of a staircase ramp A/D converter, it differs in the control logic section. Instead of allowing the staircase to climb from zero through all sequential steps up to V_{in}, it first tries a step equivalent to the most significant bit of the converter. If this step does not exceed the value of V_{in}, it is retained in the storage register. Otherwise, the storage register is set to zero. The next largest bit is then tried, and so on until all bits have been tried in order of decreasing size. A significant feature of this system is that it takes only N steps to produce an N-bit output, regardless of the magnitude of the input signal.

The conversion time of the successive approximation converter is a constant and is given by

$$T = \frac{N}{f} \tag{7–17}$$

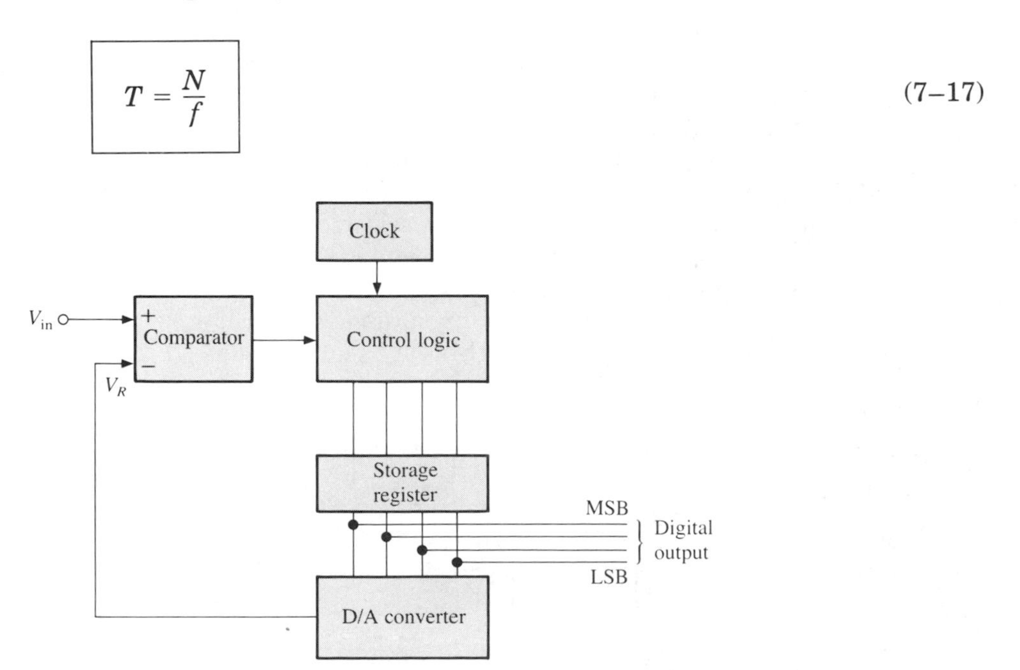

Figure 7–23 A 4-bit successive approximation A/D converter.

where N is the number of bits in the converter, and f is the clock frequency. Thus, a 10-bit converter with a 1-MHz clock makes a conversion every 10 μs.

Charge Subtraction Converters

Figure 7–24 shows the block diagram for a charge subtraction converter, also referred to as the **charge balance converter**. The input voltage continuously charges a capacitor through an integrating circuit. When the voltage on the capacitor rises above zero, a comparator triggers a unit-charge generator that removes a fixed quantity of charge from the capacitor and forces the voltage on the capacitor to return to negative. Each time this occurs, one count is sent to the counter.

A unit-charge generator causes a known current to flow for a fixed period of time. Current multiplied by time is equal to charge. The higher the input voltage, the faster the capacitor discharges back to zero, and the higher the frequency at which the unit-charge generator is triggered. Thus, this system produces a string of pulses the frequency of which is a linear function of the input voltage. The rest of the circuit counts these pulses for a fixed period of time and displays the result. This circuit is essentially a form of voltage-controlled oscillator combined with a frequency counter.

Range Selector

An analog-to-digital converter is capable of handling only a limited range of DC input voltages. To handle voltages outside the basic range of the A/D converter, it is necessary to attenuate the voltage being measured to match the basic range of the A/D converter.

A *signal attenuator* for DC voltages normally consists of some sort of resistive voltage divider network. Figure 7–25 shows a typical decade-range selector that provides six manually selected ranges. Some of the more advanced multirange meters incorporate automatic range selection through the use of solid-state switching devices such as FETs; this instrument turns on combinations of such switches for a given range of overflow, bringing the input voltage of the A/D converter to within its basic range.

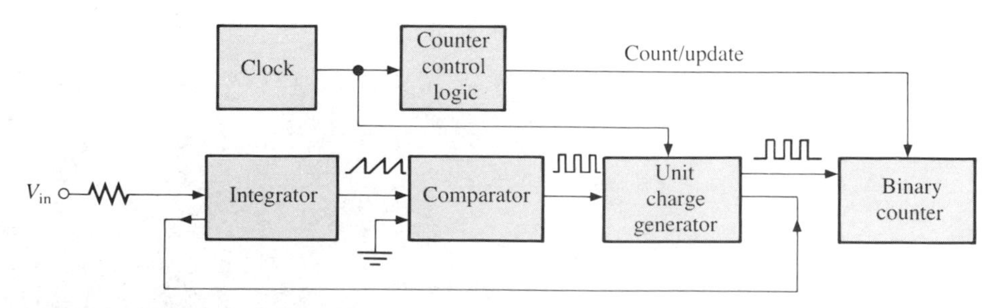

Figure 7–24 Block diagram of a charge subtraction A/D converter.

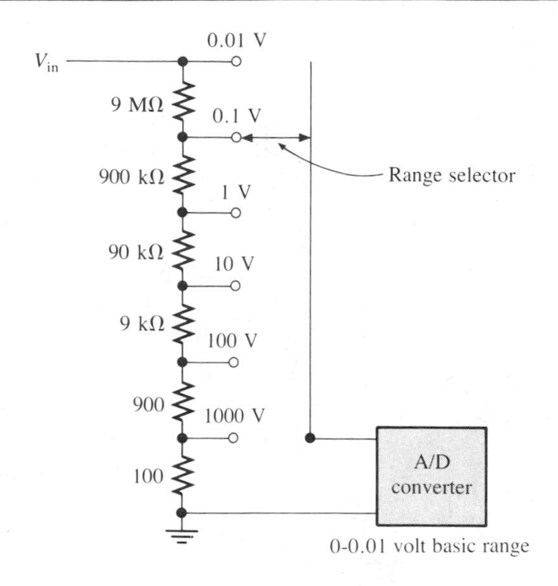

Figure 7–25 Manually selectable 6-range resistive DC attenuator for use with an A/D converter.

AC Converter

Since most A/D converters are only capable of operating with DC inputs, AC signals must be converted to DC to be usable. There are two basic types of AC/DC converters in current use; these are the *average responding* and the *true RMS* converters.

The average responding DC converter often uses a precision rectifier, as shown in Figure 7–26. The circuit shown behaves as a half-wave rectifier, but eliminates the effect of diode drop by placing the diodes inside the feedback loop of an operational amplifier.

The output of this circuit is not taken from the output of the operational amplifier, but rather from the junction between R_1 and D_1. When the input signal is positive, all feedback current flows through D_2, and the output of the circuit is zero. When the input is negative, the feedback current flows through both R_1 and D_1, and a signal appears at the output. As soon as the input goes

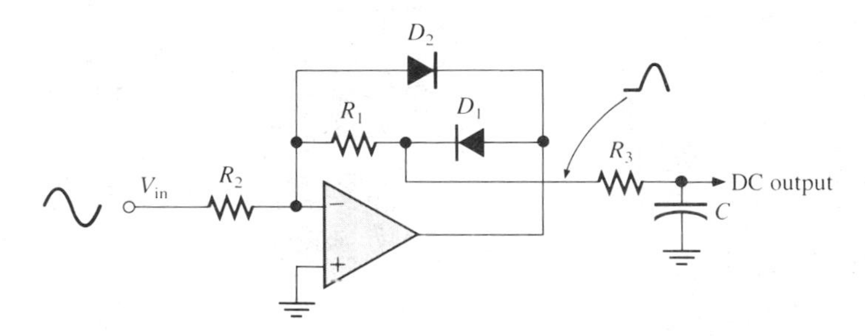

Figure 7–26 Precision half-wave rectifier and low-pass filter.

negative, the output of the operational amplifier rises to the positive level required to cause D_1 to conduct, and, in effect, eliminates the diode voltage drop from the output signal. Once diode D_1 begins to conduct, the circuit behaves as an inverting amplifier with unity gain. The opposite polarity of output could be obtained by merely reversing the direction of both diodes. The lowpass filter is formed by R_3 and C smooths out the pulsating DC signal before it is applied to the A/D converter.

A major disadvantage of the precision rectifier is that it is accurate only with a pure sine-wave input signal. Since the meter's scale factor is different for other waveforms (as discussed in Section 6–3), the meter's calibration is also different.

On the other hand, the true RMS converter gives an accurate RMS reading regardless of the waveform. One method used to implement true RMS conversion is to use a thermocouple network (as described in Section 6–5). An alternate method is to simulate the actual calculation of an RMS value by using operational amplifiers in an active network. Although true RMS converters are superior for most applications, they normally cost significantly more than the average responding type.

Ohms Converter

One of the most commonly used techniques to measure DC resistance with a digital meter is to pass a constant current through the unknown resistance and measure the resulting voltage drop, as illustrated in Figure 7–27.

For example, if the basic DVM were placed on the 1-V range and a constant current source of 1 mA were used, the meter would have a resistance range of 0 to 1000 Ω. All that is necessary to obtain a readout directly in ohms is to move the decimal point. To measure higher resistance values, a smaller current is switched in. A range from 0 to 10 kΩ requires a current source of 100 μA.

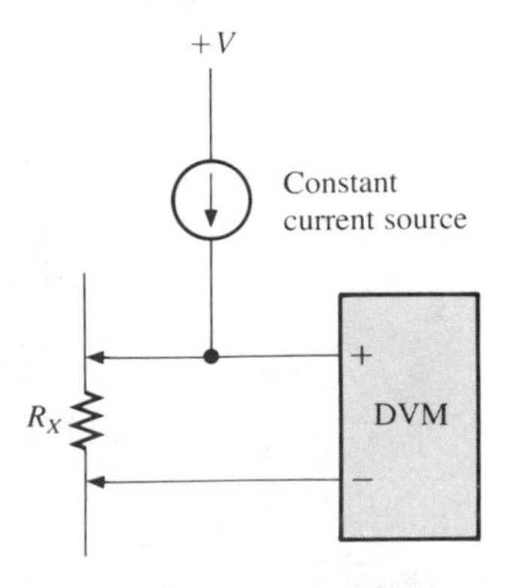

Figure 7–27 Simple resistance measurement conversion scheme used with a digital voltmeter (DVM).

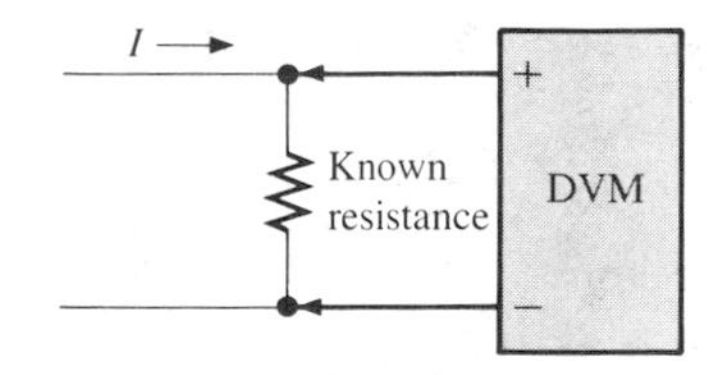

Figure 7–28 Conversion of a digital voltmeter (DVM) to measure current.

Current Converter

To measure current with a digital voltmeter, a known value resistance is placed in series with the current, and the voltage drop is measured with a standard digital voltmeter circuit (as shown in Figure 7–28). For example, a 10-Ω series resistor and a converter range of 1 V provide a full-scale current range of 100 mA. Again, all that is necessary to have a readout directly in the proper units is to move the decimal point.

DVM Specifications

Like most instruments, DVMs have parameters that indicate to the user the limits of its capabilities. Among the most important specifications are the following: accuracy, input impedance, resolution, sensitivity, and overrange capability.

Accuracy – The accuracy of DVMs used for measuring DC voltages range from 0.2 to 0.001% of the reading ±1 digit. The same meter used for measuring AC voltages, resistances, or currents has a lower degree of accuracy.

Input impedance – The input impedance of most digital voltmeters is typically 10 MΩ, shunted by about 50 pF of capacitance.

Resolution – The resolution of a DVM is taken as a ratio of the minimum value that can be displayed on a given range to the maximum value that can be displayed on that same range.

Sensitivity – The sensitivity of a DVM is the smallest change in voltage that the meter can respond to.

Overrange – Overrange capability of 100% is typical of most DVMs. For example, a 3-digit meter can read from 0 V to 1999 V. Without overrange, the maximum reading for a 3-digit meter is "999." However, a 3-digit meter usually has an extra "half" digit which reads either 0 or 1, so that the maximum reading is then "1999," which is twice that or 100% greater than without the "half" digit. Overrange on most DVMs is indicated by the addition of a "1" in the most significant position. This additional "1" is called a *half digit*. Thus, a four-digit meter with 100% overranging is called a *4½ digit meter*.

Disadvantages

Despite the many advantages offered by instruments having digital meters, the analog meter display is still preferred for certain measurements. Since most service-grade digital meters have a sampling rate ranging from 1 to 5 readings per second, sudden peaks or changes in the measured parameter may not be displayed if the reading returns to the average value before the next sample is taken. In addition, the direction and relative magnitude of a signal's change during an adjustment may be more readily apparent using an analog meter, which provides the user with better visual feedback than the digital type.

7–7 THE DIGITAL MULTIMETER

The **digital multimeter** (DMM), such as the Fluke Model 73 shown earlier in Figure 7–13, is similar to the analog multimeter discussed in Section 7–5 in that it shares a single DVM circuit and display between the three basic functions of voltmeter, ammeter, and ohmmeter (Figure 7–29). Although the function selection is manual, the range selection may be either manual or automatic. The actual circuits for measuring AC and DC voltage, AC and DC current, and resistance operate identically to those described previously in Section 7–6 and consist essentially of a DVM section combined with suitable current-to-voltage and resistance-to-voltage transducer circuits.

The input impedance on the voltage ranges is usually about 10 MΩ in parallel with 50 pF. The accuracy is about 0.5% of the reading on the DC ranges, and about 2–3% of the reading on the AC ranges. The frequency response on the AC ranges varies from 45 Hz to 1 kHz, and in the case of less

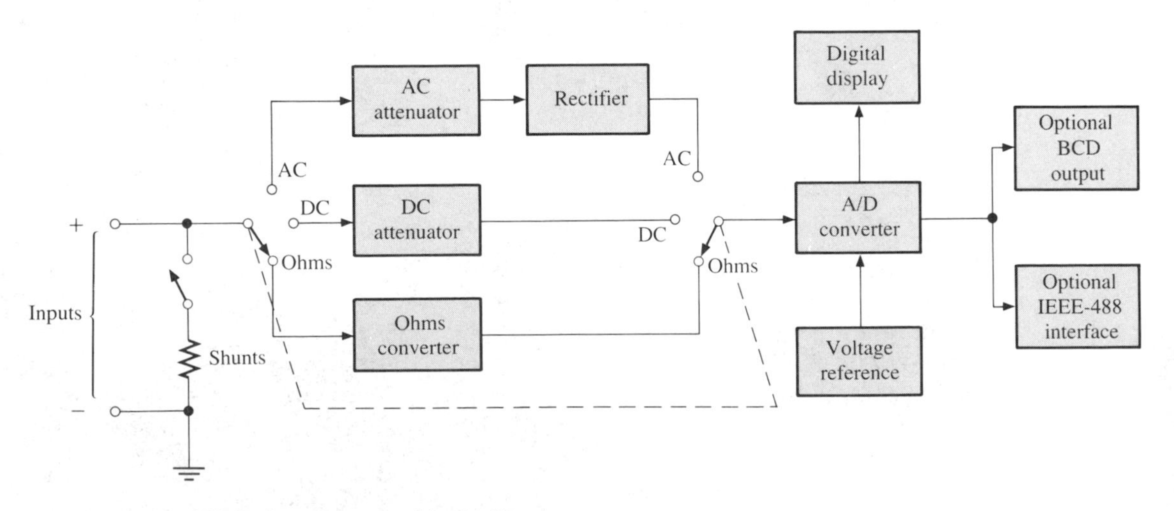

Figure 7–29 Digital multimeter block diagram.

expensive instruments, the AC ranges are calibrated for sine waves only. This type of AC meter uses an average-sensing AC-to-DC converter, and is intended only for use in sine-wave measurements where little or no distortion is present. Some of the more expensive instruments are equipped with true RMS converters and can accurately read any waveform. (The frequency response of the basic DMM is low compared to the analog meter discussed earlier.) A quality bench DMM might be expected to have a basic DC accuracy of better than 0.1% of reading and an AC accuracy of 1.5% of reading from 30 Hz to 100 kHz. On the resistance ranges, the open-circuit lead voltage is usually less than a few volts and is quite safe for most semiconductor measurements, although some bench DMMs have open-circuit voltages of over 10 V.

Diode Test

A useful feature found on many DMMs is the diode test position, usually indicated by a diode symbol. In this position, a current source of from 0.1 to 1 mA is connected between the leads, and the voltage between the leads (up to 2 or 3 V) is displayed. Thus, a good forward-biased junction will read approximately 0.7 volts, while a good reversed-biased junction will read "OL," for "overload." This is also a quick way to test the base-emitter and base-collector junctions in a bipolar transistor.

Touch Hold

Another feature available on some DMMs is **touch hold**. This feature permits the meter to capture and hold a reading, so that the user may remove the leads from the circuit under test before reading the display. This is particularly useful when taking readings in an awkward position or a dangerous location where it is prudent to watch the test probes throughout the procedure. Accessory probes are available to add touch hold to meters that do not come equipped with it.

Peak Hold

Peak hold is a feature that permits the meter to store and hold the peak measured value until cleared by the user. It also enables the meter to display a peak value that may otherwise have been missed due to the normal sample/update rate.

Bar Graph Display

Some DMMs also incorporate a bar graph feature on the display, giving the operator some of the benefits of an analog display, as well as the digital. This analog display is particularly useful for peaking and nulling tuned circuits and potentiometers, as well as for observing rapidly changing signals that are not always readily apparent with a digital display.

Digital Interface

There are a number of digital multimeters available that provide a digital interface. This is useful both for connection to data loggers or to interface with a computer for data reduction. For example, to measure temperature, a user can measure the output of a temperature-to-voltage transducer with a DVM, use a computer to linearize the output of the transducer, and display the results in degrees Celsius or Fahrenheit on the display of the voltmeter. A DMM-computer combination can also be used to display decibels referenced to any power or impedance level.

A DMM employing a digital interface designed for system use can become part of a large measuring and data-logging network containing a number of DMMs controlled and interrogated by a central computer. Such an arrangement may be used to automate measurements in the laboratory or production environment, or to act as part of a process control system. The three most commonly provided interfaces are the following: IEEE–488, RS–232C, and parallel/BCD. Chapter 16 discusses the IEEE–488 and RS–232C interfaces.

Accessory Probes

There are normally available a number of accessory probes that increase the versatility of the DMM.

1. Touch-hold probe provides a sample and hold feature for meters not so equipped.
2. High-frequency probe converts a DC voltmeter into a high-frequency voltmeter with a typical upper frequency limit of up to 500 MHz. The probe's DC output is normally calibrated to be equivalent to the RMS value of a sine wave input.
3. Temperature probe enables a digital meter to be used as a thermometer. A typical probe would produce 1 mV per degree to a DC meter.
4. High-voltage probe extends the voltage-measuring capability of an AC/DC voltmeter up to as much as 40 kV. This is a precision voltage divider in a specially molded plastic housing to protect the user from high voltage.
5. Clamp-on AC/DC current probe uses a clamp-on Hall-effect device to provide DC, AC, or composite (AC/DC) current measurements. This feature becomes particularly useful by making it necessary to break the circuit to make a current measurement.
6. Clamp-on AC current probe uses a clamp-on current transformer to extend the AC current-measuring range of a DMM. A coil on the clamp-on core serves as the secondary of a current transformer. The current-carrying conductor being measured serves as the primary.

7. Current shunt permits the measurement of AC or DC currents using a sensitive voltmeter. This is a precise resistance with a typical voltage drop of about 10 mV per A. This device is rugged and relatively inexpensive.

LED/LCD Displays

Digital meters currently use either LED or LCD displays. Light-emitting diode (LED) displays are popular for use in power line-powered, bench instruments because LEDs require simple drive circuitry and provide good visibility in low ambient light levels. However, poor visibility in direct sunlight coupled with relatively high power consumption make instruments with LED displays a poor choice for portable instruments.

On the other hand, liquid-crystal displays (LCD) are better suited for use in portable service instruments. They have low power consumption and are easily viewed in direct sunlight. However, they do require more complex drive circuitry, and have a slower response time than LED displays.

Microprocessor-Based Features

Some of the more advanced DMMs, which have their own internal microprocessor, are said to be *microprocessor-based.* Microprocessor-based DMMs often provide semiconductor memories for accumulating measurements over various time intervals. In addition, some units can further process the stored data to compute an average, make a limit comparison, or scale a reading. Such meters can also provide an offset function that converts the most recent input to a new zero level. This permits nulling out test lead resistances and facilitates differential measurements.

Another advanced feature of microprocessor-based DMMs is that they can be calibrated either manually, from their front panel, or automatically, under computer control. This capability can significantly speed up and simplify the calibration of several similar instruments.

Accuracy

The accuracy of a DDM is not necessarily guaranteed by the number of digits of its display; rather, it is the internal circuitry that determines the accuracy of a DMM. The accuracy ranges from about 3–5% of full scale for low-cost, hand-held DMMs to 0.002% for laboratory-grade meters. High-performance service-grade instruments fall in the range of 0.01 to 0.5% for DC voltage and resistance, and up to as high as 1% for high current levels. AC accuracy on frequency. Most DMMs attain their maximum accuracy in the range of 40 Hz to 20 kHz.

Resolution

The **resolution** of the DMM is the lowest variation that can be indicated by the display; assuming similar circuitry, the resolution of a DDM is limited by the number of digits in its display. For example, most service-grade DMMs have 3½ digit displays, while better units have 4½ digits. Table 7–1 lists the resolutions possible for various voltage ranges.

Table 7–1 DMM resolution for 3½ digit displays

Voltage range	Resolution
0.000 to 1.999 V	1 mV
2.00 to 19.99 V	10 mV
20.0 to 199.9 V	0.1 V
200 to 1999 V	1 V

DMM resolution for 4½ digit displays

Voltage range	Resolution
0.0000 to 1.9999 V	0.1 mV
2.000 to 19.999 V	1 mV
20.00 to 199.99 V	10 mV
200.0 to 1999.9 V	0.1 V

7–8 SUMMARY

The basic PMMC meter can be improved by adding an active device to improve the meter's input impedance. It is then referred to as an electronic meter. Furthermore, the PMMC movement can be wired to measure DC resistance as well as current and voltage. The series ohmmeter has a scale indicating increasing resistance from right to left; this type is most commonly found. On the other hand, the shunt ohmmeter, best suited for low resistance measurements, has a scale reading in the conventional manner, from left to right. In either type, however, the meter scale is nonlinear, being crowded at the higher-resistance values. Unlike voltmeters and ammeters, an ohmmeter should never be used when the test circuit has power in it.

As test instruments, meters are generally packaged as *multimeters* capable of measuring AC and DC voltage, AC and DC current, and DC resistance. Although these measurement functions are available as separate meters, it is usually more inexpensive and convenient to combine them in a single case, called a multimeter.

Current technology has made the digital meter more popular than the analog type, as it reduces measurement errors and is easy to use. Despite these advantages, however, the analog meter still enjoys a preference for certain types of measurements. This chapter discussed the fundamental building blocks of any digital meter: ramp generator, comparator, A/D converter, V/F converter, and D/A converter.

7–9 GLOSSARY

analog-to-digital (A/D) converter An electronic device in which the magnitude of an analog input signal is converted to its equivalent binary value.

charge balance converter A digital-to-analog converter using a unit charge generator to remove a fixed quantity of charge from a capacitor so that its voltage will be zero.

chopper stabilized amplifier An AC amplifier capable of amplifying DC voltages. The DC input signal is converted to a square wave by a chopper, is amplified, and is finally rectified to obtain a DC output.

digital-to-analog (D/A) converter An electronic device in which a digital signal in binary form is converted to its equivalent analog value.

digital multimeter Abbreviated DMM, a test instrument having a digital display combining the functions of a voltmeter, ammeter, and ohmmeter in a single case; frequently called a multimeter.

digital voltmeter Abbreviated DVM, a voltmeter having a digital rather than analog display.

electronic voltmeter A meter circuit built around an active device, such as a vacuum tube or a transistor, and having a high input impedance.

multimeter A test instrument measuring a variety of parameters, such as AC and DC voltage, AC and DC current, and resistance. When an analog meter is used, it is usually referred to as a volt-ohm-milliammeter (VOM). If the meter is digital, it is called a digital multimeter (DMM).

ohmmeter A test instrument using a PMMC meter movement to measure DC resistance.

resolution The lowest variation that can be indicated by digital display.

series ohmmeter A meter circuit for measuring DC resistance in which (1) the current through the meter is inversely proportional to the unknown resistance, and (2) the unknown resistance is connected in series with the meter movement. Current flows only when a resistance is being measured.

shunt ohmmeter A meter circuit for measuring low DC resistance in which (1) the current through the meter is inversely proportional to the unknown resistance, and (2) the unknown resistance is connected in parallel with the meter movement. Current always flows in a shunt ohmmeter.

single-slope converter An analog-to-digital converter that uses a steadily increasing ramp voltage as a comparison with the input voltage being measured.

touch-hold probe A probe used with a digital multimeter that allows the user to capture and hold a reading so that the user may remove meter leads from the circuit before reading the meter.

voltage-to-frequency (V/F) converter A linear voltage-controlled oscillator device that changes a DC input voltage to an output pulse train, the frequency of which is proportional to the DC input voltage.

volt-ohm-milliammeter Abbreviated VOM, a test instrument having an analog meter that combines the functions of a voltmeter, ammeter, and ohmmeter in a single case; frequently called a multimeter.

7–10 PROBLEMS

1. A series ohmmeter using the circuit of Figure 7–7 with a 100-μA, 1000-Ω PMMC movement, has a battery voltage of 1.5 V. Determine the value of R_{ADJ} to produce a mid-scale reading of 15 kΩ.
2. For the ohmmeter circuit of Problem 1, determine the percentage of full-scale reading if the unknown resistance being measured is 10 kΩ.
3. A series ohmmeter similar to that of Figure 7–9 uses a 50-μA, 1.5-kΩ PMMC movement and has a 22-kΩ resistor for R_{ADJ}. Determine the shunt resistance, R_S, required to zero the meter scale when the ohmmeter's leads are shorted together if the battery voltage is
   (a) 1.5 V
   (b) 1.42 V
4. For the ohmmeter of Problem 3, what is the minimum battery voltage for R_S still able to zero the meter scale when the ohmmeter's leads are shorted together?
5. Determine the percent resolution of an A/D converter having
   (a) 8-bits
   (b) 12-bits
   (c) 13 bits
6. If the three A/D converters of Problem 5 are used in digital voltmeters, what is the voltage resolution of the least significant bit if the meter's full-scale voltage is 20 V in each case?
7. Determine the conversion time of a 12-bit successive approximation A/D converter if the clock rate is 4 MHz.
8. If the conversion time of an A/D converter having an internal clock frequency of 1 MHz is 30 μs, determine the percent resolution of the least significant bit.
9. Determine how many bits are required for a D/A converter to provide output increments of less than 25 mV if the meter's full-scale voltage is 10 V.
10. A ramp A/D converter has a slope of 2 V/ms and a clock frequency of 50 kHz. How many clock cycles will be counted at the end of each conversion cycle if the input voltage is 6.88 V?

8

POTENTIOMETERS AND ANALOG RECORDERS

8–1 INSTRUCTIONAL OBJECTIVES

After completing this chapter, you will be able to

- Explain the operation and limitations of a simple potentiometer.
- Define the term "null."
- Describe the differences between the manual slide-wire potentiometer (including the Nalder, Varley, and Feussner circuits) and the self-balancing potentiometer.
- Explain how to calibrate ammeters and voltmeters and to measure resistances using a potentiometer.
- Describe the operations of and differences between the galvanometer, self-balancing, and *X–Y* recorders.
- Recognize how damping affects the recorded signal.
- Explain how an *X–Y* recorder can be used to determine the characteristic curves of diodes and transistors.

8–2 INTRODUCTION

This chapter first addresses the operation and use of potentiometers. Their purpose is to make very accurate voltage measurements, using a null technique similar to that of a Wheatstone resistance bridge. Potentiometers, whether manual or self-balancing, are also frequently used to calibrate DC voltmeters and ammeters, and to perform accurate resistance measurements.

Also covered here is the analog recorder. The analog recorder provides a graphical record of any measured quantity, such as voltage, current, temperature, flow rate, pressure, and others; essentially, it uses a heavy-duty PMMC arrangement to which a stylus has been attached. This stylus may be a light beam, pressure, ink, or thermal type. The reason for discussing both instruments in a single chapter is that the self-balancing potentiometer actually belongs to one class of recorders.

8–3 POTENTIOMETERS

A **potentiometer** is a device used to make very accurate DC voltage measurements. In its simplest form, as shown in Figure 8–1, a potentiometer consists of a length of resistance wire with a momentary sliding contact connected across an accurate voltage source; this forms an adjustable voltage divider. A sensitive galvanometer is connected between the sliding contact and the voltage to be measured, and the sliding contact is adjusted so that the meter reads zero, or is **"nulled,"** when the contact is closed. When this null condition is established,

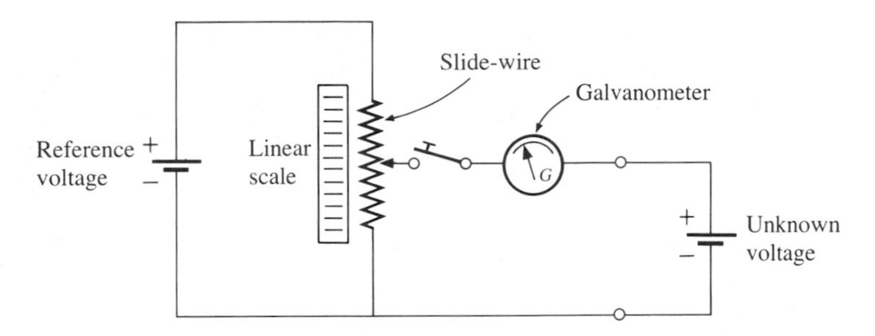

Figure 8–1 Simple slide-wire potentiometer with scale.

the voltage at the sliding contact is equal to the voltage being measured, and no current flows through the meter.

In this simple instrument, the slide wire is mounted above a scale, enabling the operator to determine what percentage of the reference voltage appears between the sliding contact and the end of the wire used as the zero reference. For example, if the scale were calibrated from zero to 100 and the slider to null the meter at a reading of 60 units, the measured voltage would equal 60% of the reference voltage.

The slide wire itself is a length of resistance wire with a cross section made uniform, so that any given increments of equal length have equal resistances. This provides a linear voltage drop throughout the length of the wire.

Compared with analog DC voltmeters, potentiometers have the following advantages:

1. Because no current flows from the source being measured, the potentiometer does not load this source when nulled. Hence, "meter loading" is not a problem.
2. The potentiometer permits comparison between an unknown voltage and an accurate reference (such as a standard cell).
3. The potentiometer all but eliminates most errors due to the mechanical components of the PMMC movement (such as bearing friction).

Manual Potentiometers

The simple slide-wire potentiometer described is limited to the measurement of voltages less than the internal reference voltage. Also, it is cumbersome due to its long, necessarily straight resistance wire and its scale. A more practical form of potentiometer uses a stable working voltage source to make the actual comparison measurements, rather than using the reference voltage directly. As a result, this permits the working voltage to be *much higher* than the reference, permitting the measurement of higher voltages. The straight resistance wire is replaced with a switch-selectable decade resistor network for coarse adjustments, and by a small section of rotary slide-wire controls for finer adjustments. Adjustments made to the resistor network must not alter the total resistance of the potentiometer; otherwise, the current through the divider changes and produces a corresponding change in calibration. This requirement is inherent in the slide-wire potentiometer, and the dial resistance network must use a circuit that provides the same effect.

The rotary controls are calibrated in units of voltage, and they provide greater resolution and convenience than straight slide-wire controls. Of these rotary types, the Nalder, Feussner, and Varley slide wires are most frequently used. The **Nalder potentiometer,** or *resistance dial* potentiometer (Figure 8–2), substitutes many adjusted resistance units for the slide wire. Although shown in Figure 8–2 as straight line resistance strings for clarity, the resistances are actually arranged in a circular pattern with a rotary switch. In the Feussner circuit of Figure 8–3, double dials with double contact arms insert resistance between the contact points Y and Z from one-half of each dial. At the same time an equal resistance is cut out of the other parts of the potentiometer in the lower half of each dial. The total resistance, therefore, remains constant.

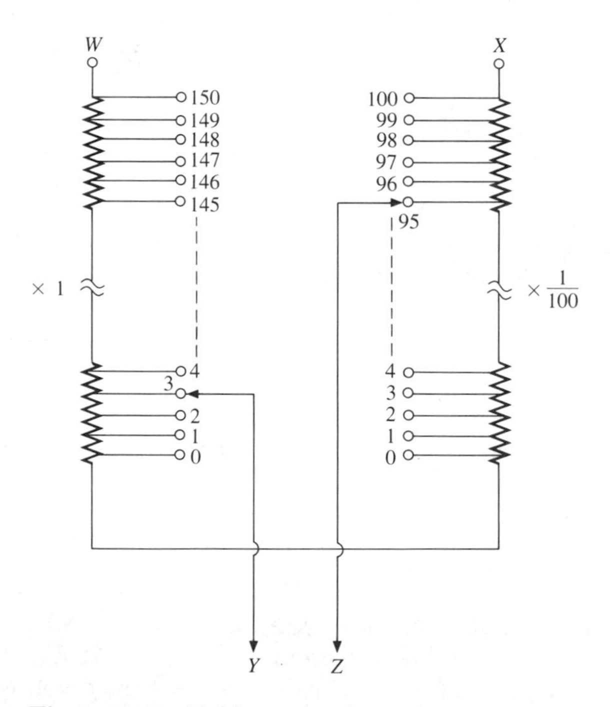

Figure 8–2 Nalder manual resistance network.

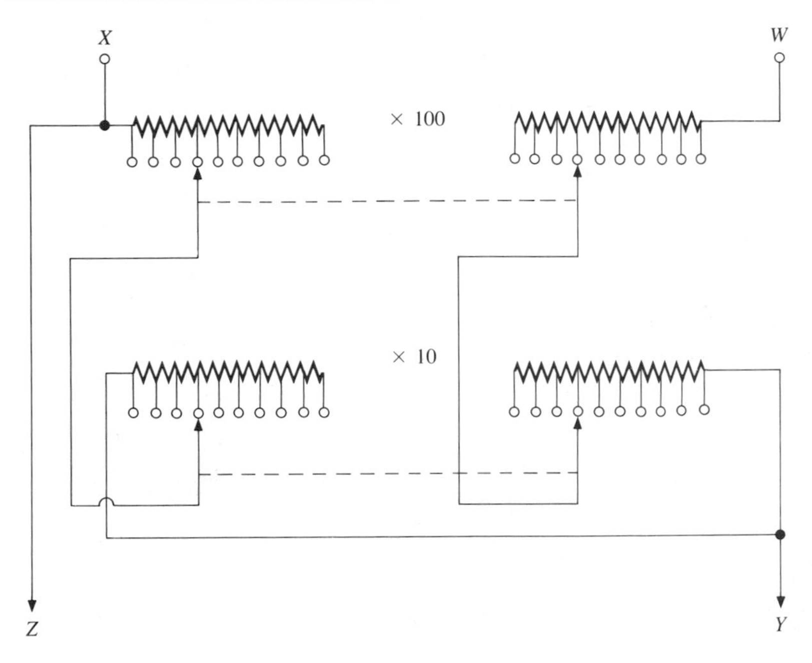

Figure 8–3 Feussner manual resistance network.

On the other hand, the **Varley slide arrangement** of Figure 8–4 uses a pair of sliding contacts moving together, continuously bridging any two of the resistance coils of a given circularly arranged group. A second circular dial of 11 positions, with a total resistance equivalent to two units of the main dial, is in parallel with the sliding contacts. The parallel resistance is therefore equal to one unit of the main dial, and this unit is subdivided into 10 parts by the second dial. In all three systems, terminals W and X are connected to the working voltage standard circuit, while terminals Y and Z are connected to the galvanometer and the unknown voltage to be measured.

Figure 8–5 shows the essential elements of a standard manual potentiometer; however, for clarity, the divider network is represented as a single resistor. To calibrate the instrument, switch A is turned to the "CALIBRATE" position and the switch across the current limiting resistor is opened. The dials of the precision voltage divider are set to the value of the reference cell (1.019 V for a standard cell), and the rheostat is set to null the galvanometer when the key is closed. The switch across the current-limiting resistor is closed, increasing the sensitivity of the meter, and the null is further refined. This establishes the correct voltage across the divider. Finally, switch A is turned to the "MEASURE" position; this is the last step in preparing the instrument.

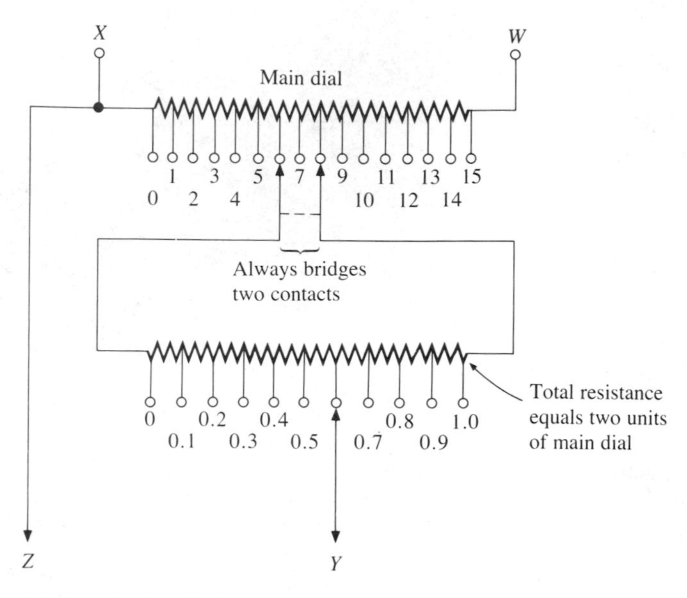

Figure 8–4 Varley slide manual resistance network.

Once the calibration rheostat has been set, it must not be moved during the measurement. Otherwise, the calibration is invalidated. The switch across the current-limiting resistor is again opened to decrease meter sensitivity. When the key is closed, the dials are again used to null the meter. The current-limiting resistor is switched out and the null refined. The value of the unknown voltage can now be read from the dials of the potentiometer. Commercial portable manual potentiometers (Figure 8–6) typically have an accuracy of 0.05%.

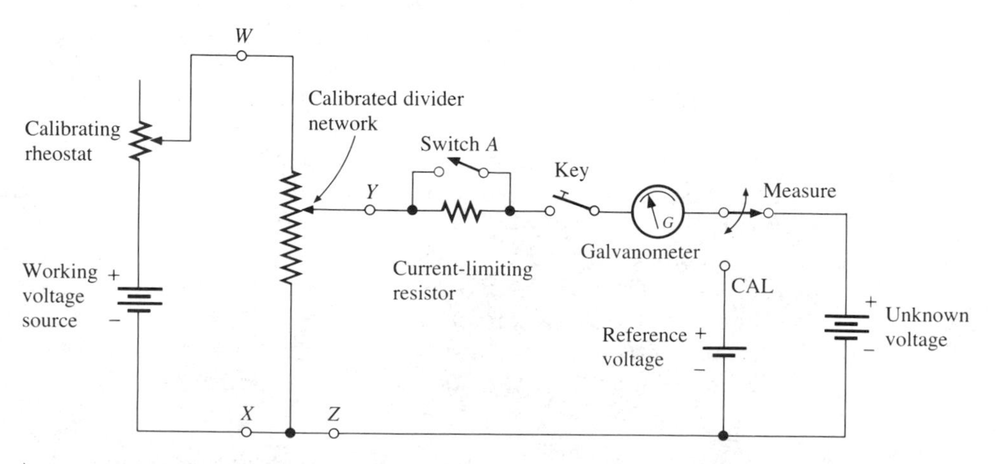

Figure 8–5 Standard manual potentiometer circuit.

Figure 8–6 Commercial manual potentiometer. Courtesy Leeds and Northrup.

Self-Balancing Potentiometers

The potentiometer can be made to balance itself by replacing the galvanometer with a servo amplifier and motor, as shown in Figure 8–7. Any voltage difference between the voltage at the slider of the motor-driven slide-wire and the unknown voltage, called the **error voltage,** is applied to the input of the servo

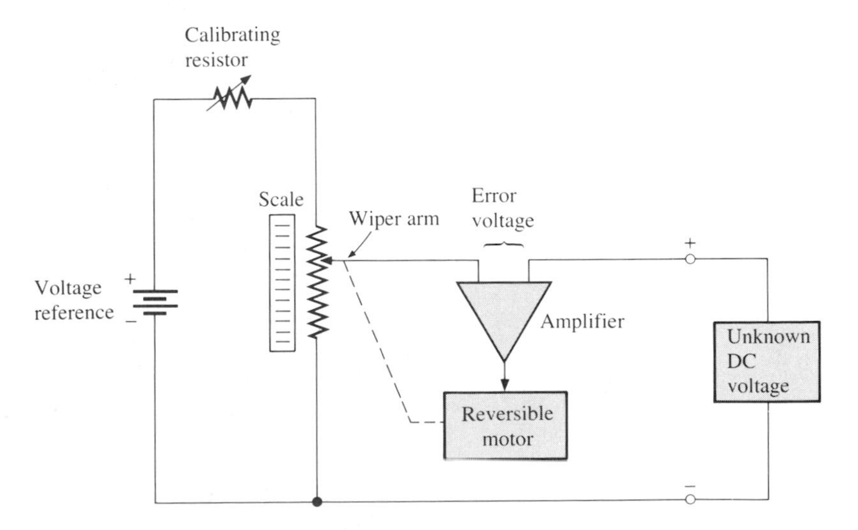

Figure 8–7 Block diagram of a self-balancing potentiometer.

amplifier. This difference is amplified and used to drive the slider motor in the proper direction, balancing the voltages and nulling the input to the amplifier.

As the voltage difference continues to shrink, the motor slows, eventually stopping when the error approaches zero. This is a popular system for chart recorders, in which the motor is also used to drive a marking pen across the face of a moving chart. The pen position is a function of the slider position; thus, it can be used to mark a chart calibrated directly in voltage or any other suitable unit.

EXAMPLE 8–1

The basic slide-wire potentiometer circuit of Figure 8–8*a* is used to measure the voltage of a thermocouple. The voltage varies with temperature. When adjusted, the 150-Ω galvanometer reads zero current when the upper portion of the slide wire is 350 Ω. The entire slide wire has a resistance of 600 Ω. Determine the EMF voltage at the terminals of the thermocouple if the reference voltage of the potentiometer is 1.5 V.

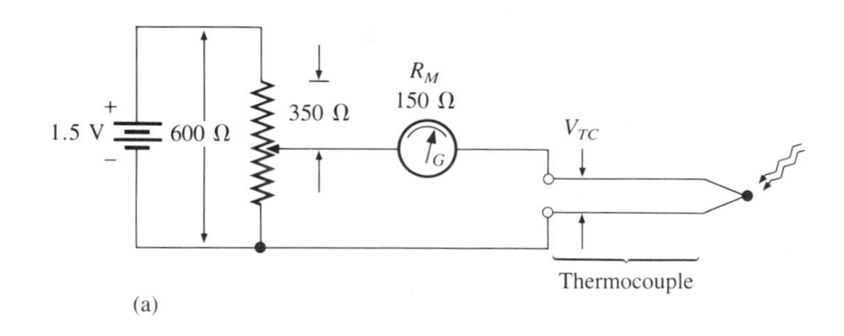

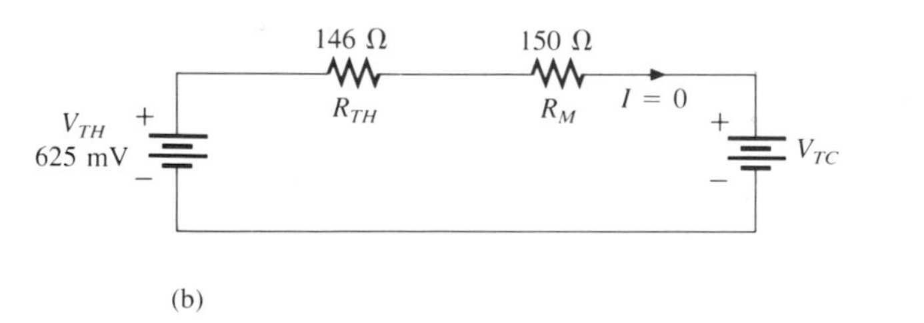

Figure 8–8 Circuit for Example 8–1.

Solution:

By Thevenizing the potentiometer-thermocouple circuit, as shown in Figure 8–8*b*, the Thevenin voltage across thermocouple and galvanometer is

$$V_{TH} = (1.5\text{ V})\,\frac{250\ \Omega}{600\ \Omega}$$

$$= 625 \text{ mV}$$

while the Thevenin resistance is

$$R_{TH} = 350\ \Omega \parallel 250\ \Omega$$
$$= 146\ \Omega$$

Since no current flows through the meter, the thermocouple EMF equals the Thevenin voltage, or

$$V_{TC} = 625 \text{ mV}$$

By knowing the calibration curve of the thermocouple, the temperature that results in an EMF voltage of 625 mV can be determined.

Meter Calibration and Resistance Measurement

A potentiometer can be used to calibrate DC voltmeters and DC ammeters, and to measure resistance. Figure 8–9 shows a method for calibrating a DC voltmeter. This method also works when the voltmeter's full-scale voltage is beyond the range of the potentiometer.

When two precision resistors are connected as a voltage divider, making the potentiometer measure the voltage across R_2,

$$V_2 = \frac{R_2}{R_1 + R_2} V_{DC} \tag{8–1}$$

Since the voltmeter under test measures the power supply voltage,

$$V_{meter} = V_{DC} \tag{8–2}$$

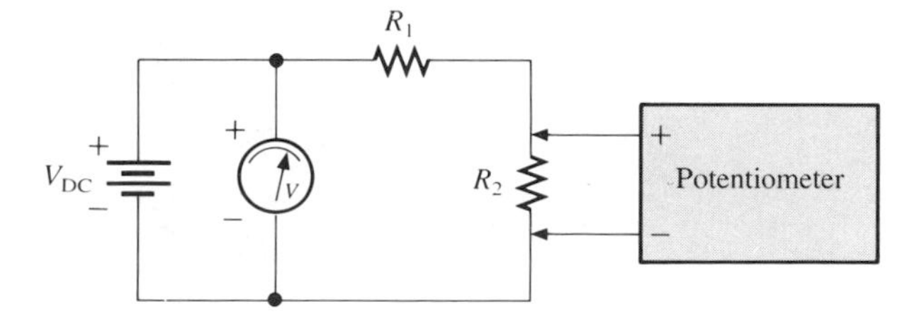

Figure 8–9 Potentiometric calibration of DC voltmeters.

The DC power supply is adjusted at specified voltages over the meter's range (usually at each major division of the meter's scale). At each reading, the potentiometer is used to measure the voltage across R_2; then, Equation 8–1 is used to determine the expected reading of the voltmeter under test. Either a calibration curve or a regression line can be determined to express the relationship between the meter's measured value against the expected value.

EXAMPLE 8–2

A DC voltmeter with a full-scale voltage of 10 V is to be calibrated using the calibration circuit of Figure 8–10, with $R_1 = 10\ \Omega$ and $R_2 = 75\ \Omega$. Determine the expected reading of the voltmeter under test when the potentiometer reads

(a) 247.1 mV
(b) 1.000 V

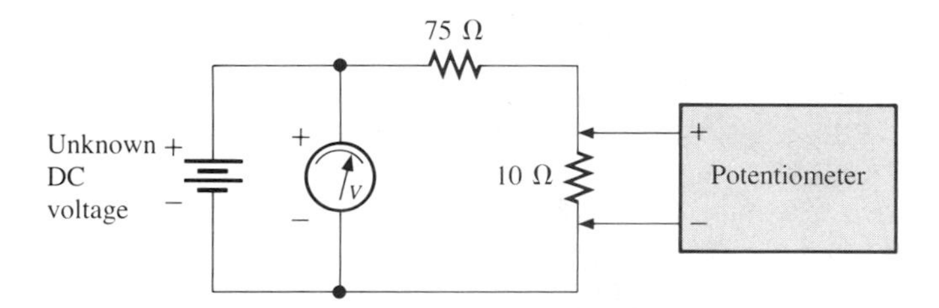

Figure 8–10 Circuit for Example 8–2.

Solution:

(a) When the potentiometer reads 247.1 mV, the DC voltmeter under test should read

$$V_{\text{meter}} = (247.1\ \text{mV})\left(\frac{85\ \Omega}{10\ \Omega}\right)$$
$$= 2.1\ \text{V}$$

(b) At 1.000 V, the meter should read

$$V_{\text{meter}} = (1.000\ \text{V})\left(\frac{85\ \Omega}{10\ \Omega}\right)$$
$$= 8.5\ \text{V}$$

DC ammeters are calibrated by using the arrangement shown in Figure 8–11; this is nothing more than a simple application of Ohm's law. The voltage across a precision resistor is measured with the potentiometer. The resulting current is determined by Ohm's law. As with the voltmeter calibration, the DC power supply is adjusted so that measurements are made at convenient points over the meter's entire range.

EXAMPLE 8–3

A DC ammeter is calibrated using the circuit of Figure 8–11, with $R_1 = 10\ \Omega$. If the potentiometer reads 61.4 mV and the ammeter under test reads 6.0 mA, determine the percent error, if any, of the ammeter at this value.

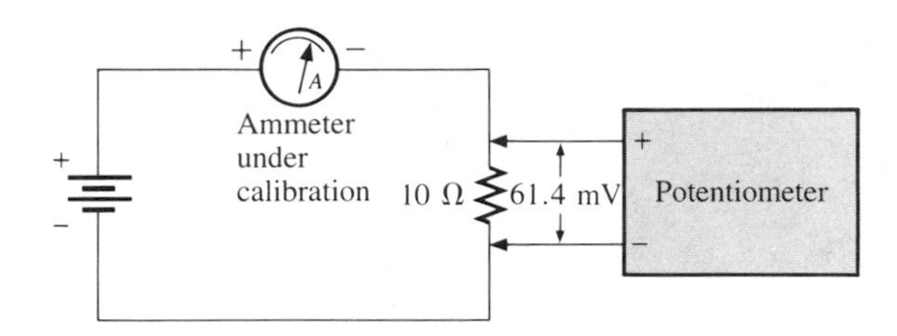

Figure 8–11 Potentiometric calibration of DC ammeters.

Solution:

Using Ohm's law, the current through the 10-Ω resistor is the true current that should be read by the meter under test, or

$$I = \frac{61.4\ \text{mV}}{10\ \Omega}$$

$$= 6.14\ \text{mA}$$

Since the meter under test actually indicates 6.0 mA when it should be reading 6.14 mA, the percent error is

$$\%\ \text{error} = \left(\frac{6.00 - 6.14\ \text{V}}{6.14\ \text{V}}\right) \times 100\%$$

$$= -2.28\%$$

This indicates that the meter reading is 2.28% less than the true value.

Using the arrangement shown in Figure 8–12, the potentiometer can be used to perform resistance measurements with a high degree of accuracy. The voltage across the unknown resistor R_U is measured, and the power supply adjusted so the voltage across R_U is at a convenient level. Since the voltage across the known precision resistor R_K can also be measured. the current through both resistors is the same; thus, $V_U/R_U = V_K/R_K$. Rearrangement yields

$$R_U = \left(\frac{V_U}{V_K}\right)R_K \tag{8–3}$$

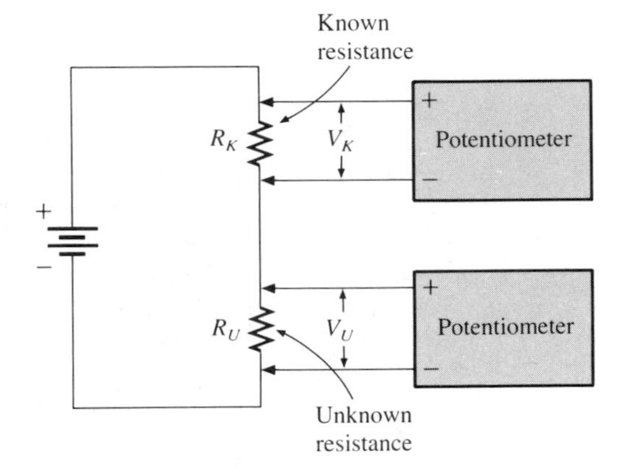

Figure 8–12 Potentiometric calibration of DC resistance.

8–4 ANALOG CHART RECORDERS FOR ELECTRICAL SIGNALS

Analog recorders are used to provide a continuous, permanent graphic record of a signal's amplitude as a function of either time or a second signal. There are three types of graphic recording instruments that are in common use.

1. the galvanometer recorder
2. the self-balancing recorder
3. the *X–Y* recorder

In the *X–Y* recorder, the marking pen or stylus is simultaneously moved along the *X* and *Y* (Cartesian) coordinates of a stationary rectangular chart by two separate and independent signals. In the self-balancing and galvanometeric recorders, the marking device is moved across a chart moving at a constant speed in a direction perpendicular to the path of the pen. Consequently, this type records signals as a function of time. Self-balancing and galvanometer recorders use either a long rolled strip or circular disc chart (Figure 8–13), while *X–Y* recorders use a rectangular sheet of a fixed size.

In all three cases, a suitable scale is usually printed on the chart paper.

Figure 8–13 Strip-chart recorder. Courtesy Leeds and Northrup.

The actual marking of the chart can be done with an inked stylus, a pressure stylus, a light beam, or a heated stylus. In the case of heat, light, or pressure styli, special sensitized paper must be used. Although the inkless marking systems tend to require less attention and eliminate the hazard of ink spillage and clogging, they are more complex and require expensive paper. The ink marking system is the least expensive to operate, but requires frequent cleaning of the pen and attention to the ink reservoir. Strip chart recorders frequently include a second stylus, called an *events marker,* which is switch activated. It is used to produce a small rectangular pulse near the bottom of the chart paper, signifying the start or end of some particular event, or marking units of time.

In strip and circular type recorders, a chart-drive motor either rotates the circular chart or advances the strip chart at a constant speed. This speed may or may not be selectable. In the case of AC powered instruments, the chart-drive motor is usually synchronous to ensure a constant speed.

Galvanometer Recorders

In **galvanometer recorders** (Figure 8–14), the marking stylus or pen is mounted at the end of the pointer of a heavy-duty d'Arsonval meter movement. Galvanometer recorders are more compact and have a higher frequency response than self-balancing recorders, but can not be made as sensitive. The

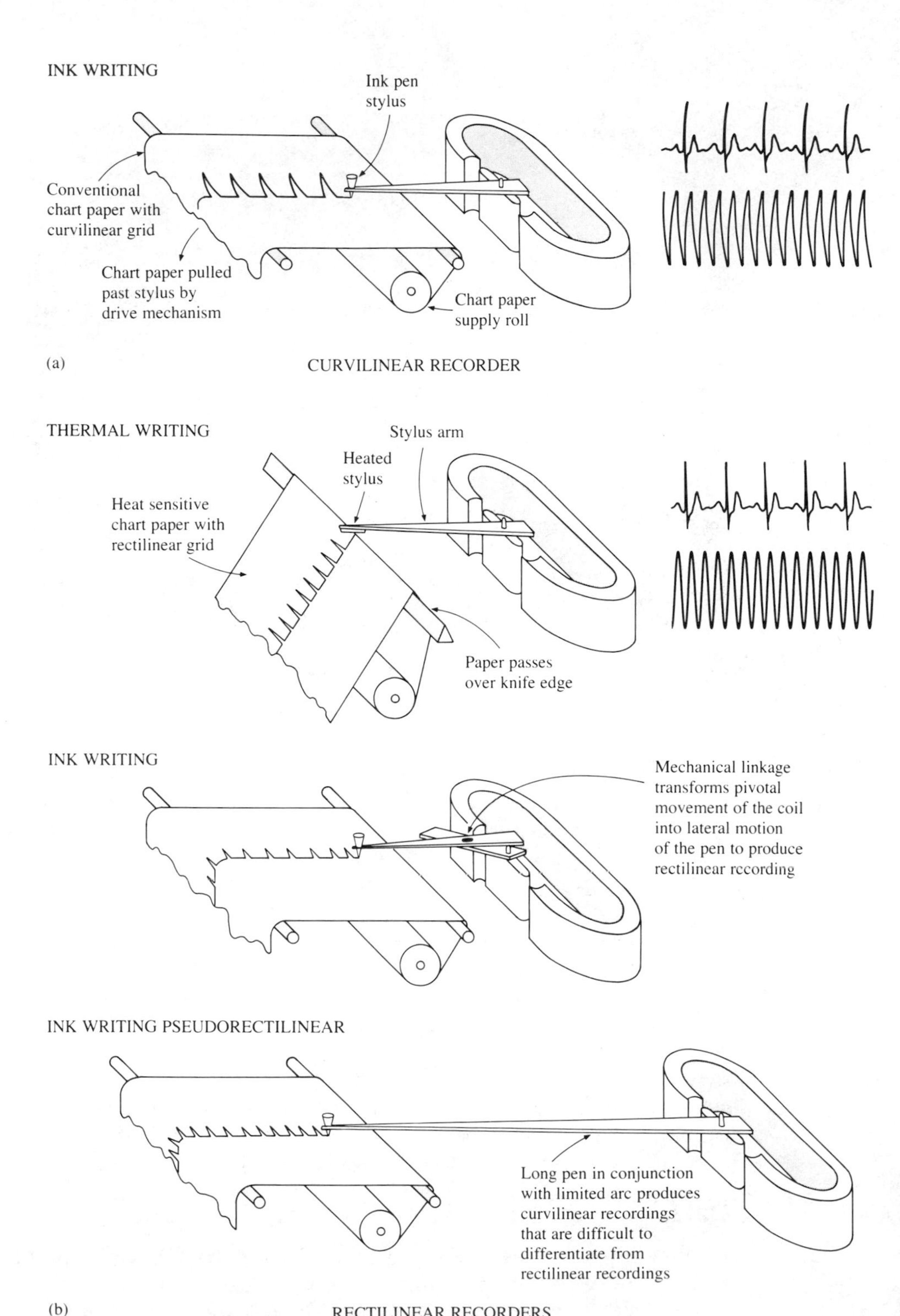

Figure 8–14 Galvanometer type recorder where the marking stylus is at the end of the pointer of a heavy-duty d'Arsonval meter movement. Reproduced with permission of Tektronix, Inc. Copyright 1970.

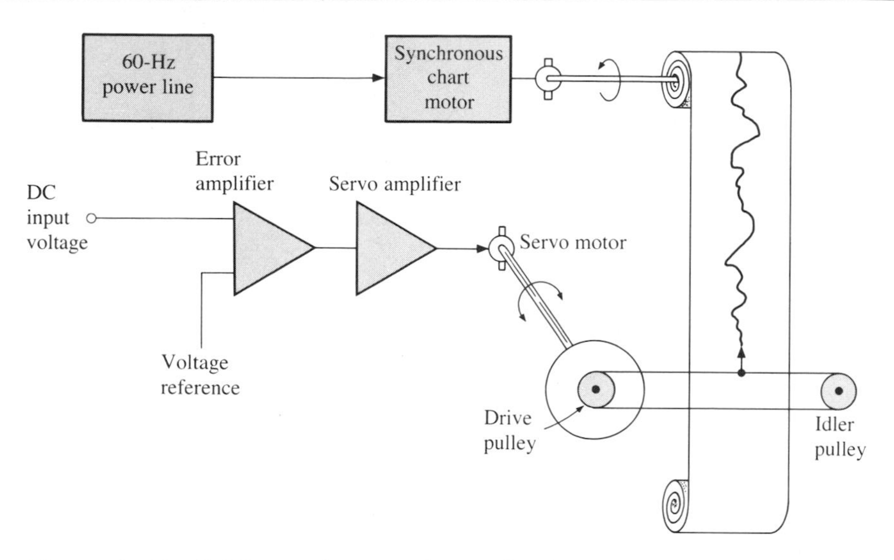

Figure 8–15 Self-balancing recorder mechanism.

maximum frequency response is about 100 Hz, and the maximum accuracy approximately ±1 to ±2%.

This type of movement can be used with both strip and circular charts. Because of their size, galvanometer recorders are often used in multi-channel recorders to provide a simultaneous record of two or more time-varying signals.

Self-Balancing Recorders

Instead of a galvanometer and a manual balance control, the **self-balancing recorder** uses an electronic circuit that drives a motorized balance control, continuously maintaining the potential balance (Figure 8–15). The same motor also drives a marking stylus through a linkage or cable arrangement. Thus, the self-balancing recorder has all the advantages of the potentiometer while eliminating most of the inaccuracy inherent in the drag of a marking stylus. On the other hand, the self-balancing mechanism tends to be larger, slower, and more expensive than its galvanometer counterpart.

Commercially available thermocouples are easily adapted for use with self-balancing recorders (both rectangular and circular). As mentioned briefly in Chapter 6, the thermocouple, using two dissimilar metals, is based on the Seebeck effect: the EMF developed at its terminals is proportional to temperature. Thermocouples are further discussed in Chapter 10.

Figure 8–16 shows the basic scheme of temperature measurement using a thermocouple. The "hot" junction is at the temperature being measured, while the "cold" junction is maintained at a reference temperature (32 °F or 0 °C, the

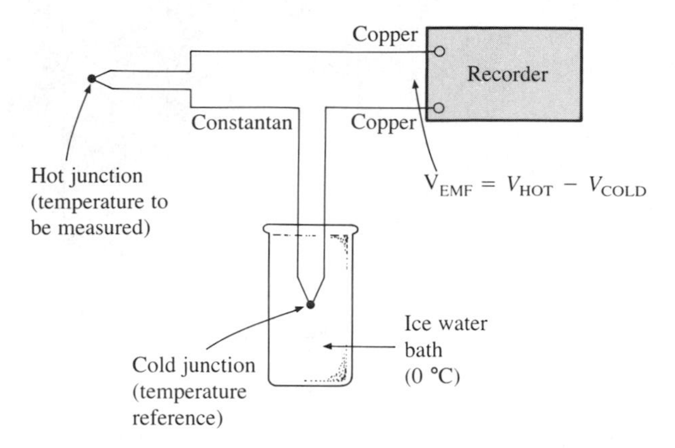

Figure 8–16 Thermocouple-type recorder setup for measuring temperature.

freezing point of water) that must be used for absolute temperature measurements. The EMF of the cold junction is known. The recorder measures the difference between the two junctions.

$$V_{EMF} = V_{HOT} - V_{COLD} \tag{8-4}$$

By knowing the relationship between junction EMF and temperature, the temperature of the "hot" junction can be determined. In a practical situation, however, the maintenance of a cold junction ice bath is awkward. Commercial recorders used specifically for thermocouples have an electronic reference voltage included for a given thermocouple (such as copper-constantan).

Chart Speed

Strip chart recorders have a drive motor that advances the chart paper at a constant speed. The rate at which the recording paper of the strip chart recorder moves is called the **chart speed.** It is generally expressed in units of length per time (for example, millimeters per second). If the chart speed is known, time or frequency measurements of the recorded data can be determined. For circular chart recorders, the chart speed is expressed in revolutions per unit time.

X–Y Recorders

The ***X–Y* recorder** is used to plot the variation of two signals against each other. In the *X–Y* recorder (Figure 8–17), two closed-loop servomechanisms drive the marking stylus with a wire and pulley arrangement through a Car-

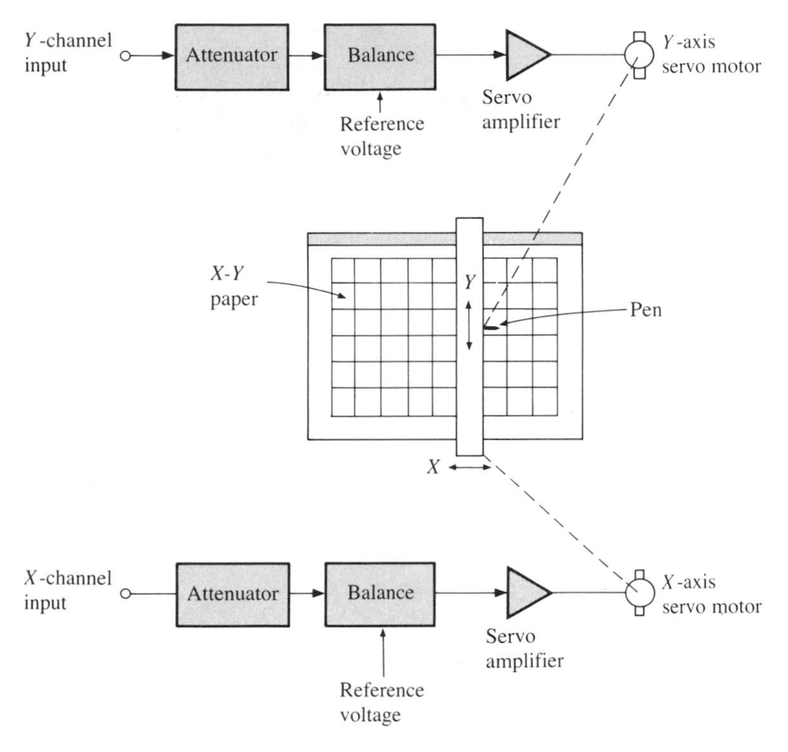

Figure 8–17 X-Y recorder block diagram.

tesian coordinate system. The stylus, under the control of two separate input signals, can move simultaneously in both the X and Y directions. Unlike strip chart recorders that feed moving paper from a roll onto sprockets, the paper used on an X–Y recorder is stationary, held in place either by vacuum or electrostatic attraction.

To provide a hard copy record of the output, X–Y recorders are often used as output display devices in conjunction with other electronic instruments such as spectrum analyzers, impedance analyzers, polar displays, and semiconductor parameter analyzers (i.e., transistor curve tracers). Although the X–Y recorder is primarily under the control of two separate input signals, many X–Y recorders include a time base so that a single signal can be plotted against time.

For an X–Y recorder, typical parameters include the following:

input impedance: 1 MΩ, each channel
accuracy: ±0.5% of full scale
resettability: 0.2%
overshoot: less than 2% of full scale
slewing speed: greater than 20 inches per second

The recorder's **resettability** refers to how well the servomechanism repositions the marking stylus to its "zero" position when the plotting is finished. Since there is frequently a cable pulley or mechanical linkage arrangement, system backlash is a primary cause of resettability error. The **slewing speed** is similar in concept to the slew rate of an operational amplifier. A slewing speed of greater than 20 inches per second means that the servomechanism and stylus can respond by moving greater than 20 inches in one second. This provides the user with a measure of how fast the recording system responds to sudden changes.

Other Considerations

As the pen recorder is a mechanical system, the servomechanism and pen assembly must overcome a certain amount of inertia *before* moving. In strip chart recorders, this movement is perpendicular to the motion of the chart paper. In X–Y recorders, this movement can be in any direction. The range of signals insufficient to overcome inertia is referred to as the **deadband** zone. In terms of an electrical signal, the pen does not move until a certain minimum input voltage is present. As shown in Figure 8–18, once this minimum level is exceeded, the pen's position is linearly proportional to the applied input voltage. For low-level signals, the deadband phenomenon causes a distortion of the signal traced by the pen onto the paper. Manufacturers generally specify deadband as a percentage of the recorder's full-scale value; a percentage less than 0.1% indicates a good instrument.

Once the servomechanism and pen assembly are moving, their momentum must be controlled so that, when subjected to rapidly changing signals, the pen position is an accurate representation of the input signal. That is, it must neither undershoot nor overshoot its intended position. The amount of overshoot or undershoot in a mechanical system is characterized by the term **damping.** As shown in Figure 8–19, the amount of system damping may be characterized by the three terms, **underdamped, critically damped,** and **overdamped.**

Since the leading edge of a square wave has the fastest changing signal known (an infinite slope), it is used to determine the amount of damping

Figure 8–18 Deadband zone characteristic.

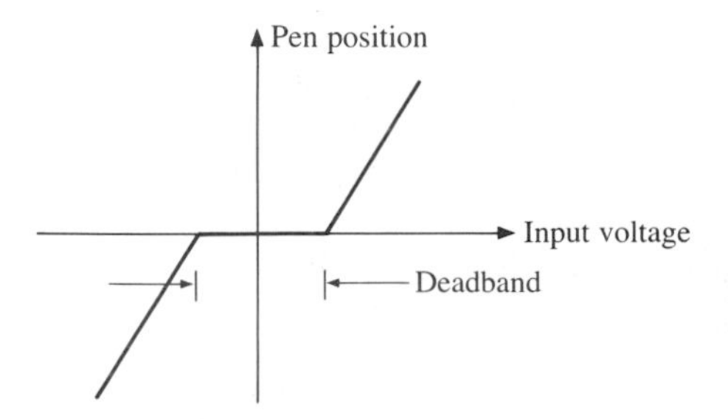

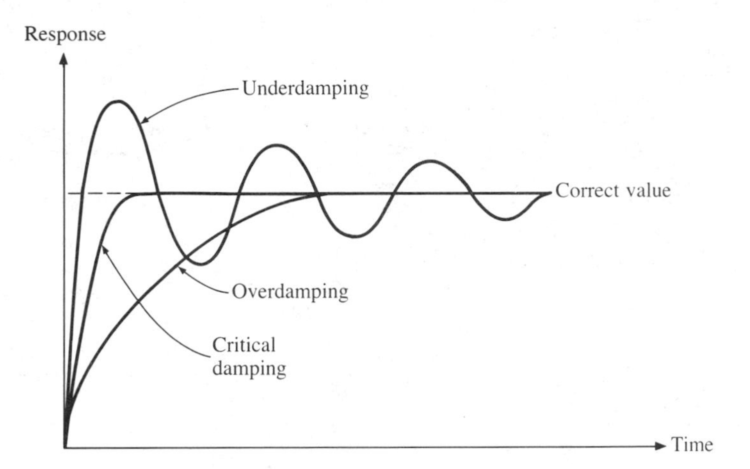

Figure 8–19 System damping.

present. This is sometimes referred to as *transient analysis*. When the square wave changes from its minimum value to its maximum at a virtually instantaneous rate, the recorder's servomechanism and pen assembly also try to follow this change. An underdamped system overshoots the correct value (position) and oscillates about the correct value until decaying to that value. The oscillation that occurs while the pen position decays to the correct value is termed **ringing.** The amount of overshoot is usually specified as a percentage above the correct value. The **settling time** is the amount of time it takes before the pen, having first exceeded the correct value, can remain within a given percentage of the correct value. It is desirable that both the overshoot and settling time be as small as possible.

An overdamped system is one that reaches its correct value very slowly, and without any overshoot. The movement of the servomechanism and pen assembly appear sluggish, so rapidly changing signals tend to have "rounded edges." A critically damped system, which is difficult to achieve, has no overshoot and virtually no undershoot. It is characterized by an exact representation of the input signal.

8–5 RECORDING SEMICONDUCTOR PARAMETERS

The characteristic curves of semiconductor devices, such as diodes and transistors, can be graphed using an X–Y recorder to provide a permanent record for later reference.

Figure 8–20 shows the arrangement used for plotting the characteristic curve of an ordinary diode. The vertical (Y-axis) input measures the voltage across the 1-kΩ resistor. Using Ohm's law, this voltage is actually displayed in terms of current. If the Y-axis sensitivity were set at 1 V/in (1 V/cm, or 1 V/division), the diode current would be 1 V/in divided by 1 kΩ, or 1 mA/in of

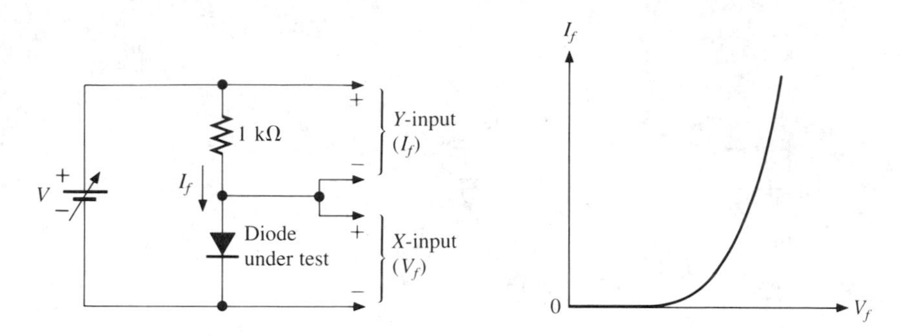

Figure 8–20 Plotting a semiconductor diode's forward characteristic curve using an *X-Y* recorder.

vertical pen deflection. The horizontal (*X*-axis) measures the voltage drop across the diode; its sensitivity is usually 0.1 V/in.

The DC supply voltage is slowly varied from zero to a level sufficient to trace out the characteristic curve of diode forward current as a function of diode voltage. The forward diode current must be kept below its rated maximum value. Because the voltage across the 1-kΩ resistor is a *differential* voltage, both terminals of the resistor are "floating" with respect to ground. If the *X*- and *Y*-axis inputs are not true differential inputs, then the circuit of Figure 8–20 must not be used. Rather, a differential amplifier, like the op-amp circuit of Figure 8–21, should be used with the vertical input. As shown in Figure 8–22, the characteristic curve of the breakdown region of a zener diode is accomplished in a manner similar to that of the rectifier diode, though the horizontal sensitivity may have a different value, appropriate to the particular situation.

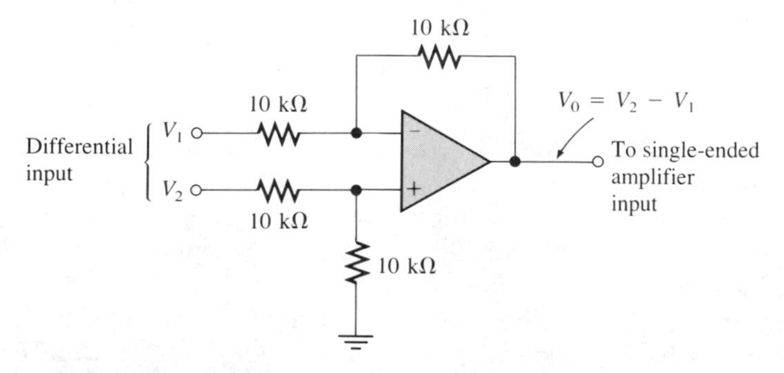

Figure 8–21 Using a unity-gain difference amplifier to convert a single-ended recorder channel input to a differential input.

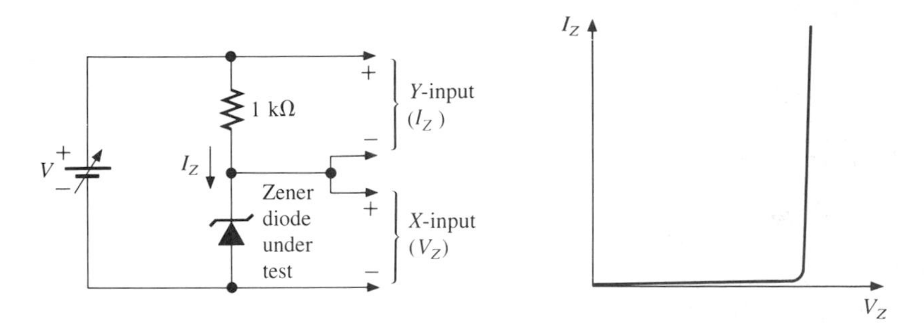

Figure 8–22 Plotting a zener diode's breakdown characteristic curve using an *X-Y* recorder.

Two power supplies are required to trace the family of collector curves of a bipolar (junction) transistor. Figure 8–23*a* is used for NPN transistors, and Figure 8–23*b* for PNP transistors. The transistor's collector current (I_C), measured in terms of the voltage across the 1-kΩ resistor, is applied to the recorder's vertical amplifier. The collector-to-emitter voltage (V_{CE}) is connected to the horizontal input.

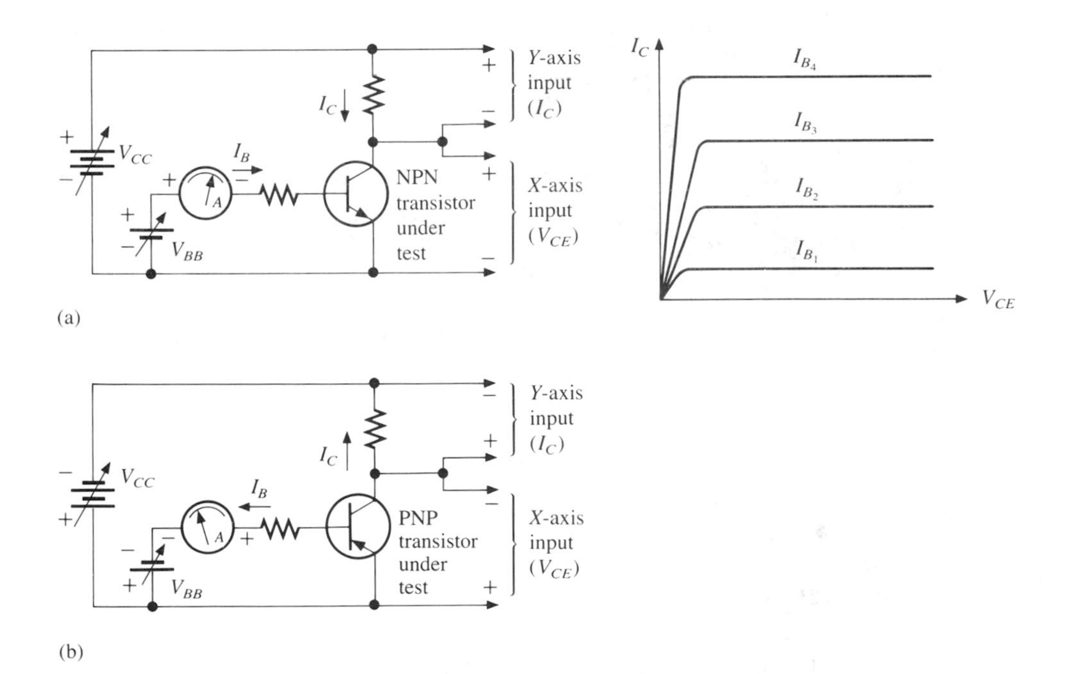

Figure 8–23 Plotting a bipolar transistor's collector characteristic curve using an *X-Y* recorder. (a) NPN transistor. (b) PNP transistor.

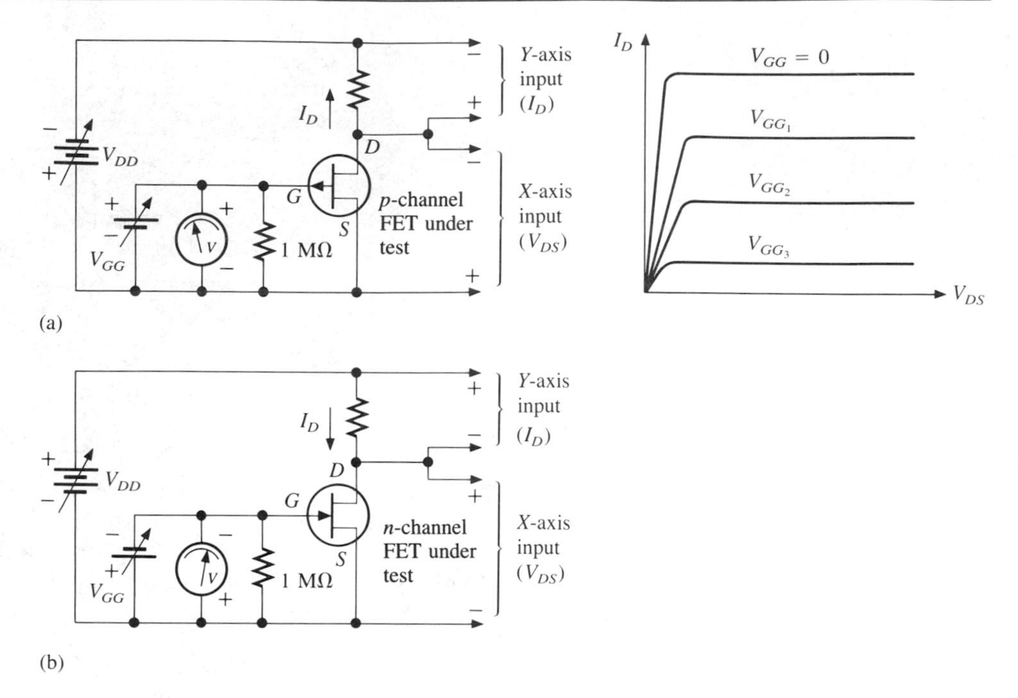

Figure 8–24 Plotting a FET's drain characteristic curve using an *X-Y* recorder. (a) *p*-channel FET. (b) *n*-channel FET.

The DC base voltage supply (V_{BB}) is adjusted so that the ammeter reads a convenient base current (I_B), such as 10 μA. The DC collector supply voltage (V_{CC}) is varied from zero to a level sufficient to trace out the characteristic curve. V_{CE} must be kept below the maximum rating of the particular transistor under test.

After this first curve is traced, the collector supply is returned to zero, and the base supply voltage again adjusted for a different value of base current, usually a fixed increment of the first value. The collector supply voltage is varied as before. This process is repeated for as many levels of base current as required. The resultant graph is a plot of I_C vs. V_{CE}. Figure 8–24 shows a similar arrangement for *n*-channel and *p*-channel field effect transistors.

8–6 SUMMARY

Although voltmeters are frequently used to measure voltage, the potentiometer (manual or self-balancing) is capable of accurately measuring voltage levels where circuit loading by the voltmeter is a problem. Using the null technique, the potentiometer at balance draws no current from the circuit under test, thus eliminating loading errors. Besides measuring low-level voltages, potentiometers are capable of calibrating DC voltmeters and ammeters, and of measuring DC resistances.

Analog recorders are galvanometric, self-balancing, or *X–Y* types; they provide a permanent graphic record of the variations of a measured quantity with respect either to time or another parameter. Their applications include the measurement of temperature using thermocouples and the ability to graph the characteristic curves of diodes and transistors.

8–7 GLOSSARY

analog recorder An instrument that provides a continuous permanent graphic record of a signal's amplitude as a function of time.

chart speed The speed at which the chart paper moves past the stylus.

critically damped The degree of damping required to give the most rapid transient response without either overshoot or oscillation.

damping Any action that extracts energy from a vibrating system, minimizing its vibration or oscillation.

deadband The range of signal amplitudes too small to cause movement of the pen or stylus in a recorder.

error voltage In a servo system, the voltage that results when the input and output voltages are not equal.

galvanometric recorder A pen-recording instrument that has its marking pen or stylus mounted at the end of the pointer of a heavy-duty d'Arsonval meter movement.

Nalder potentiometer Also called a resistance-dial potentiometer, a potentiometer which substitutes a large number of adjusted resistance units in place of a slide wire.

null A condition of circuit balance so that the current or output is zero.

overdamped A high degree of damping characterized by the absence of overshoot or oscillation.

potentiometer A instrument producing a known voltage that can be balanced against an unknown voltage.

resettability A measure of how well the recorder's servomechanism repositions the marking stylus to its "zero" position when plotting is finished.

ringing An oscillatory transient that occurs in the output of a system as a result of a sudden change in input.

settling time The time required for the transient response to decrease to a given percentage of its peak value

self-balancing recorder A pen-recording instrument, using a servo amplifier and motor in place of a galvanometer (d'Arsonval) movement.

slewing speed A measure of the response of a servo mechanism as a function of how far its recording stylus can move in a given amount of time.

slide-wire potentiometer A potentiometric instrument that uses a resistance wire whose resistance is proportional to its length.

underdamped A low level of damping that is unable to prevent a circuit from oscillating after the application of a sharply rising or falling input signal.

Varley potentiometer A potentiometer having a pair of sliding contacts moving together to continuously bridge any two of the resistance coils of a circularly-arranged group.

***X–Y* recorder** A recorder used to graph the variation of two signals against each other on a Cartesian coordinate system.

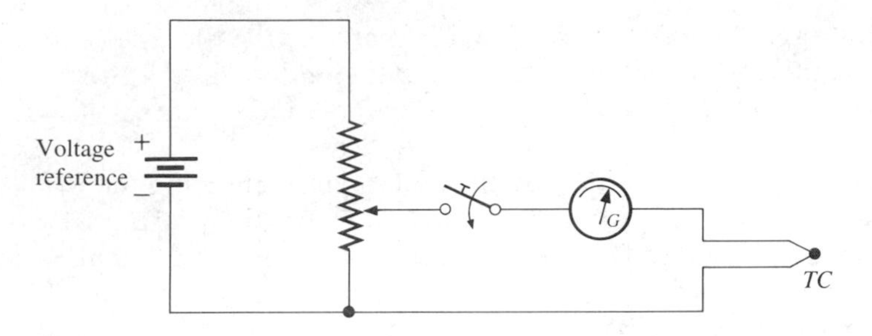

Figure 8–25 Circuit for Problem 1. Slide-wire potentiometer circuit with galvanometer to measure voltage of thermocouple.

8–8 PROBLEMS

1. The slide-wire potentiometer circuit with a 100-Ω galvanometer in Figure 8–25 is used to measure the voltage of a thermocouple. When nulled, the potentiometer's meter reads zero, the resistance of the upper portion of the slide wire is 759 Ω, and the entire slide wire has a resistance of 1 kΩ. Determine the generated EMF at the thermocouple's terminals if the reference voltage of the potentiometer is 1.019 V.
2. A DC voltmeter with a full-scale voltage of 10 V is to be calibrated using the calibration circuit of Figure 8–9, with $R_1 = 20\ \Omega$ and $R_2 = 150\ \Omega$. Determine the expected readings of the voltmeter under test for each of the following readings: 212, 495, 701, 947, and 1118 mV.
3. For the data of Problem 2, assuming there is a linear relationship between the meter's scale and the true value, determine the regression line relating the true voltage as a function of the voltmeter's reading if the intended readings were to be 2.0, 4.0, 6.0, 8.0, and 10 V, respectively.
4. A DC ammeter is calibrated, using the circuit of Figure 8–11, with $R_1 = 10\ \Omega$. If the potentiometer reads 98.3 mV, and the ammeter under test indicates a full-scale current of 10.0 mA, determine the percent error of the ammeter at this value.
5. Two potentiometers are used to measure an unknown resistance using the circuit of Figure 8–12. Resistor R_K is a precision 100-Ω resistor. If the voltages measured by the two potentiometers across the known and unknown resistors are, respectively, 851.7 mV and 0.567 V, determine the value of the unknown resistor.
6. The chart speed of an analog chart recorder is 10 in/s. At this speed, one cycle of the signal being recorded covers 7.3 in. Determine the period and frequency of the recorded signal.

9

AC AND DC BRIDGES

9–1 INSTRUCTIONAL OBJECTIVES

At the completion of this chapter, you will be able to

- Determine unknown resistances using either a Wheatstone or Kelvin bridge.
- Explain the use of a guard on a Wheatstone bridge for measuring very high resistances.
- Determine the location of faults in cables using either the Murray or Varley loop test methods.
- Identify and compare the characteristics of measuring capacitance with the series-resistance, parallel-resistance, and Schering bridges.
- Identify and compare the characteristics of measuring inductance with the Maxwell, Hay, Owen, and inductance comparison bridges.
- Explain the operation of the Wagner ground connection.

9–2 INTRODUCTION

Bridge circuits are commonly used to measure many quantities in electronics, such as resistance, capacitance, inductance, impedance, and admittance. A bridge circuit usually takes the form of a two-port network with a source of excitation connected to one port and a detector to the other. The component to be measured is inserted in one branch of the network, and the network is adjusted until the detector indicates no output. At this point, the bridge is said to be *balanced;* the unknown value may be determined from known values of the circuit. Because the unknown value is determined on the basis of known standard values, bridges are often considered *comparison instruments*. Though all based on the simplest of all bridges, the Wheatstone bridge, which measures DC resistance, other forms of this simple circuit can measure capacitance, inductance, and complex impedances.

9–3 RESISTANCE BRIDGES

Depending on the magnitude of the quantity to be measured, several resistance bridges are available as alternatives to the voltmeter-ammeter method and ohmmeter for the measurement of DC resistance. The Wheatstone bridge is used for most resistances greater than several ohms. The Kelvin bridge is used for low values ranging from 1 μΩ to 1 Ω. Modification of the Wheatstone bridge allows for the location of faults in cables. Although bridges of these types can be excited by either a DC or AC source, they are primarily DC excited. This increases their versatility and portability.

Wheatstone Bridge

The four-arm, diamond-shaped circuit shown in Figure 9–1 is called a **Wheatstone bridge.** It is used to measure the value of an unknown resistance in terms of the known values of the other three resistances. Although its principle was first proposed in 1833 by S. Christie, its full potential as an accurate method for the measurement of resistance was not shown until Sir Charles Wheatstone did so fourteen years later. It has a typical accuracy of 0.1% for resistance measurements greater than several ohms, as opposed to the 3% error exhibited by ordinary ohmmeters.

To understand its operation, it is helpful to look at the Wheatstone bridge arrangement as two voltage dividers in parallel with the same voltage source. Figure 9–1*a* shows the Wheatstone bridge as it is usually drawn, while Figure 9–1*b* shows it redrawn as two voltage dividers. If the voltage drops across R_2 and R_X are equal, the galvanometer detector indicates zero current; this situation is referred to as a *null,* since there is no voltage across the output of the bridge. Using the rule for a voltage divider, the voltages across R_2 and R_X, respectively, are

$$V_2 = \frac{R_2}{R_1 + R_2} V_S \tag{9–1a}$$

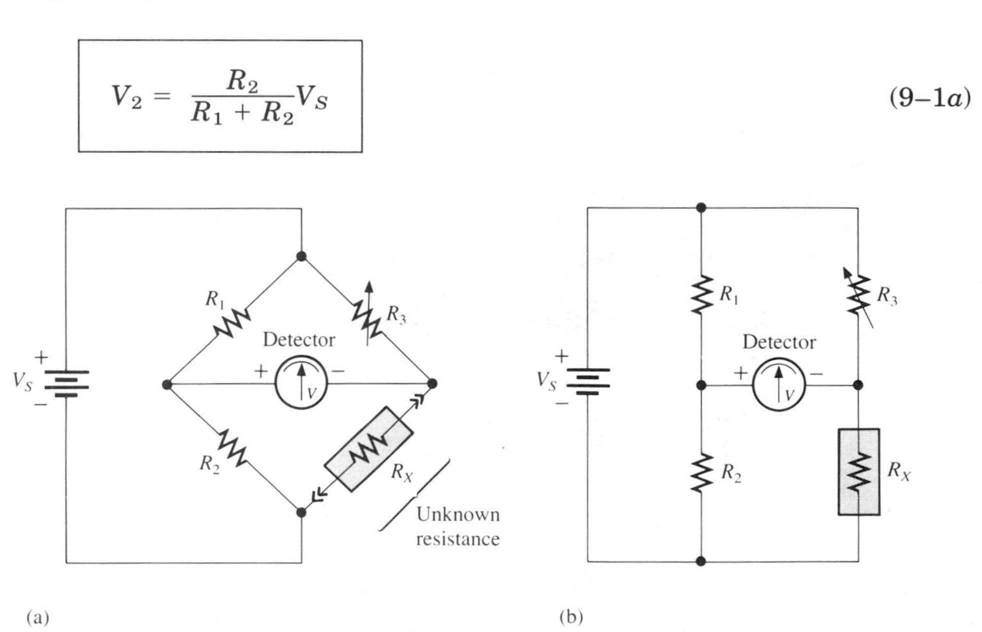

Figure 9–1 DC Wheatstone bridge. (a) Normally drawn circuit. (b) Bridge circuit drawing showing two voltage dividers.

and

$$V_X = \frac{R_X}{R_3 + R_X} V_S \tag{9–1b}$$

Since $V_2 = V_X$, the resistance values needed to balance the bridge can be determined by setting Equations 9–1*a* and 9–1*b* equal to each other:

$$R_1 R_X = R_2 R_3 \tag{9–2}$$

In practice, the ratio of R_2/R_1 is controlled by a selector switch that changes this ratio by decades (factors of 10). Thus, the ratio R_2/R_1 can be selected to be 0.001, 0.01, 0.1, 1, 10, 100, 1000, and so on. Resistor R_3 is a continually variable standard resistor with a calibrated scale (Figure 9–2). When a null is achieved with this arrangement, the value of the unknown resistance can be read directly from the dials. As an example, if the variable standard resistor dial reads 123, and the ratio (multiplier) dial reads 10, the value of the unknown resistor would be 123 × 10, or 1230 Ω.

Commercial versions of the Wheatstone bridge usually include a means of varying the sensitivity of the detector, usually a galvanometer in a DC bridge,

Figure 9–2 A commercial Wheatstone DC resistance bridge. Courtesy Leeds and Northrup.

and a switch for connecting and disconnecting the source. In general, the procedure for making a measurement is as follows:

1. Connect the unknown resistor, R_X, to the terminals of the bridge.
2. Set the scale of the detector (galvanometer) to the least sensitive range to prevent damage to the detector (galvanometer) if the bridge is severely unbalanced.
3. Connect the excitation source and adjust the resistor dials until a null is obtained (zero deflection of the galvanometer).
4. If not at the most sensitive scale, increase the sensitivity of the galvanometer and repeat Step 3; otherwise, proceed to Step 5.
5. Calculate the unknown resistance with Equation 9–2. By substituting an AC source and an AC detector, it is possible to measure the value of a resistor at the frequency at which it will be used.

EXAMPLE 9–1

For the circuit of Figure 9–1, used to determine an unknown resistance R_X, the following resistances are noted when the bridge is balanced:

$$R_1 = 10\ \text{k}\Omega$$

$$R_2 = 5\ \text{k}\Omega$$

$$R_3 = 2786\ \Omega$$

Determine the unknown resistance, R_X.

Solution:

From Equation 9–2,

$$R_X = \frac{(5000\ \Omega)(2786\ \Omega)}{(10{,}000\ \Omega)}$$

$$= 1393\ \Omega$$

Although a sensitive galvanometer is generally used as the detector, a voltmeter can be used, usually with an operational amplifier connected as a differential amplifier, as shown in Figure 9–3. The amplifier expands the voltage difference between nodes A and B. An oscilloscope can be used in place of a voltmeter, but should not be used without the additional differential amplifier circuit. Unless an oscilloscope with a differential input is used, one portion of the input lead of most oscilloscopes is always connected to ground. Accidental grounding of either node A or B of the bridge leads to incorrect operation and possible damage of the equipment. Naturally, op-amps used in this manner should have a high common mode rejection (CMR > 100 dB) for best performance. The sensitivity can be as high as practical considerations dictate.

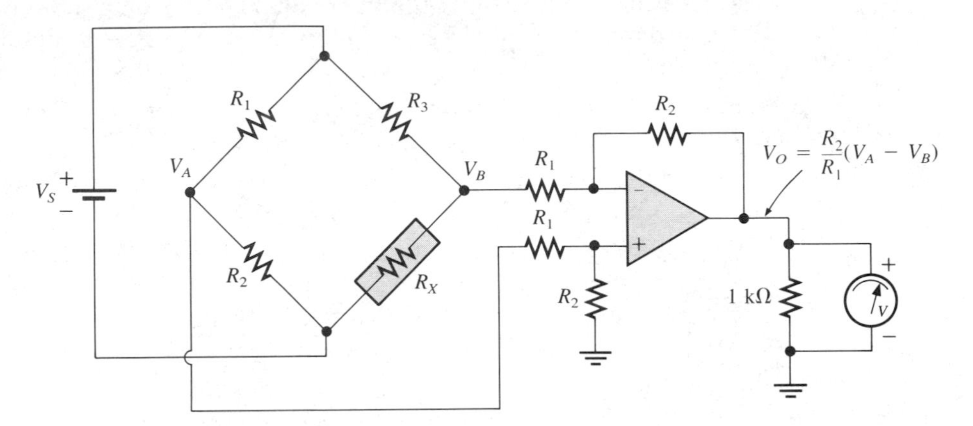

Figure 9–3 Using a differential amplifier to increase the output voltage of a Wheatstone bridge.

Guarded Wheatstone Bridge

While the basic Wheatstone bridge circuit of Figure 9–1 is useful in most cases, its performance deteriorates substantially when measuring high resistances, such as the insulation of cables and other materials. Resistances of this type are generally greater than 1 MΩ. The major problem of high resistance measurements is the **leakage current** that travels around and over the binding post of the resistor being measured, as illustrated in Figure 9–4*a*. Because these currents can bypass the resistor being measured and flow through the detector branch, they affect the measurement.

To reduce this measurement error, the leakage current is diverted away from the unknown resistance by using a *guard wire* and *guard ring*. The guard ring surrounds the positive terminal post of the bridge circuit (Figure 9–4*b*) and minimizes the leakage current by minimizing the potential difference. Only the current flowing through R_X flows through the detector; as a result, a true reading is obtained.

Kelvin Bridge

When measuring the value of small resistances (below 1 Ω), the small but finite resistance of leads and electrical contacts can cause a significant error if a Wheatstone bridge is used. To eliminate this error when measuring very low resistances, a modification of the basic Wheatstone bridge is used. Called a

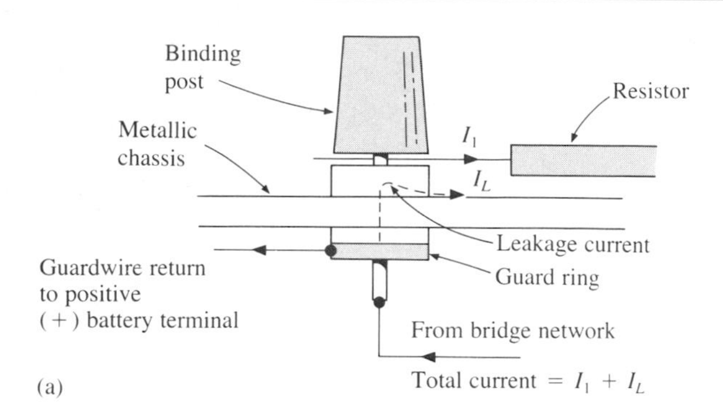

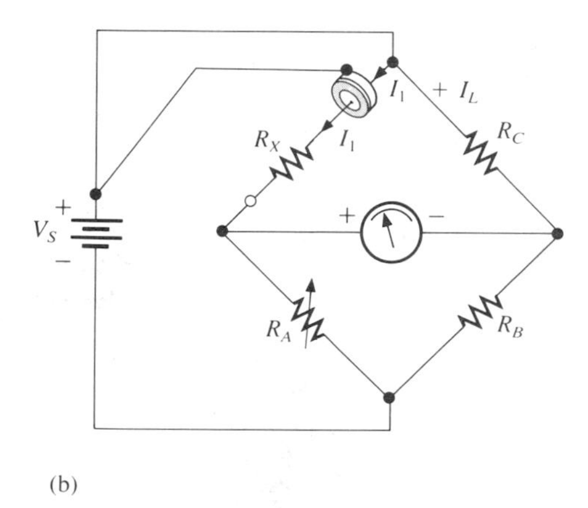

Figure 9–4 Guarded Wheatstone bridge. (a) Guard wire used to eliminate leakage current near binding posts. (b) Bridge circuit showing guard ring and terminal.

Kelvin bridge, it allows accurate resistance measurements from 1 Ω down to 1 μΩ.

Figure 9–5 shows the circuit of a Kelvin bridge, sometimes referred to as a **Thompson bridge.** Resistance R_W represents the resistance of the conductor (wire) connecting the unknown resistor, R_X, to R_3. Resistance R_S limits the current. By bridging R_W with two resistors, R_5 and R_6, that have the same ratio as R_1 and R_2, the effect of the voltage drop across R_W is eliminated. Because the circuit contains a second pair of ratio arms (R_5 and R_6), the circuit is often called a **Kelvin double bridge.** Figure 9–6 shows the physical arrangement of the Kelvin bridge. It is not necessary to compensate for the connection at the other end of R_X; if the value of R_W is negligible, the Kelvin bridge reduces to a Wheatstone bridge with the parallel combination of R_5 and R_6 in series with the galvanometer.

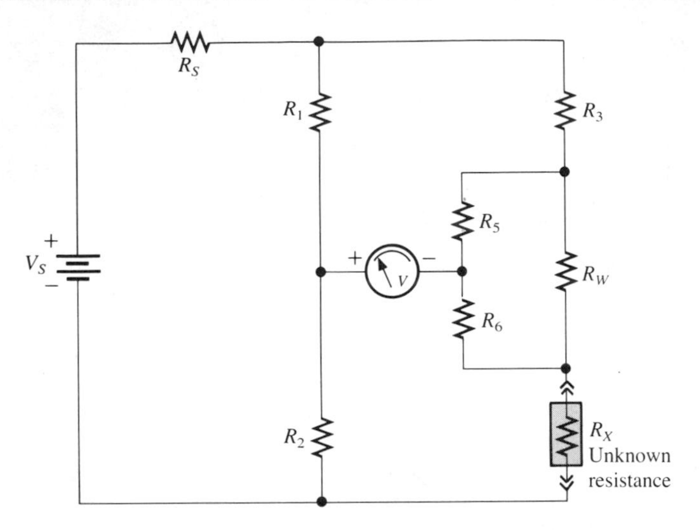

Figure 9–5 Kelvin bridge for measurement of small resistances down to 1 Ω.

At balance, the voltage across R_2 equals the voltage across the series combination of R_6 and R_X. The unknown resistance can then be determined from

$$R_X = \frac{R_2 R_3}{R_1} + \frac{R_5 R_W}{R_5 + R_6 + R_W}\left(\frac{R_2}{R_1} - \frac{R_6}{R_5}\right) \tag{9–3}$$

However, Equation 9–3 is greatly simplified by requiring that $R_6/R_5 = R_2/R_1$, in which case

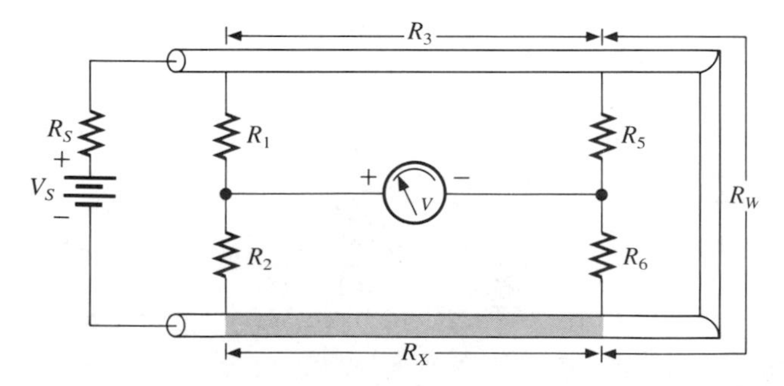

Figure 9–6 Arrangement of Kelvin double bridge components.

$$R_X = \frac{R_2 R_3}{R_1} \tag{9–4}$$

The requirement that $R_6/R_5 = R_2/R_1$ must be maintained for the balance equation. This is generally done by varying R_2 and R_1 with the switch closed until a null is obtained. The switch is then opened, and R_5 and R_6 are adjusted for null. The process is repeated until a null is maintained with the switch either open or closed.

Varley and Murray Loop Test Methods

By a slight modification of the Wheatstone bridge, the resultant circuit can be used to locate faults such as a short to ground in wire lines and cables. This is particularly helpful when the cable is buried underground. Instead of unearthing the entire cable, repair workers can locate the precise position of a fault with this method, thus making repairs only where necessary.

Figure 9–7*a* shows the circuit for what is known as the **Murray loop test** method. The faulted line is connected with a jumper wire to a good line at one end, forming a loop. The resistance of the loop is $2R_a$, and is measured with a Wheatstone bridge. The loop is connected, forming a resistance bridge in which one arm contains the resistance R_x between the test set and the fault, and the adjacent arm contains the remainder of the loop resistance between the test set and the fault, $2R_a - R_x$. When R_1 is adjusted to achieve balance,

$$(2R_a - R_x)R_1 = R_2 R_x \tag{9–5a}$$

so that solving for R_x is as follows:

$$R_x = \frac{2R_a R_1}{R_1 + R_2} \tag{9–5b}$$

Since the length, L_a, of the line is proportional to its resistance, R_a, and since R_x is proportional to the distance of the fault from the test set (L_x), $L_x/L_a = R_x/R_a$. Figure 9–7*b* shows this same method applied to a fault between two conductors of the same cable.

A third resistor is added in the **Varley loop** method (Figure 9–8). As before, $2R_a$ represents the resistance of the wire or cable from the test set to the location of the fault. When the switch is at position 1, the circuit functions as a Wheatstone bridge. Thus, by adjusting R_2 for balance,

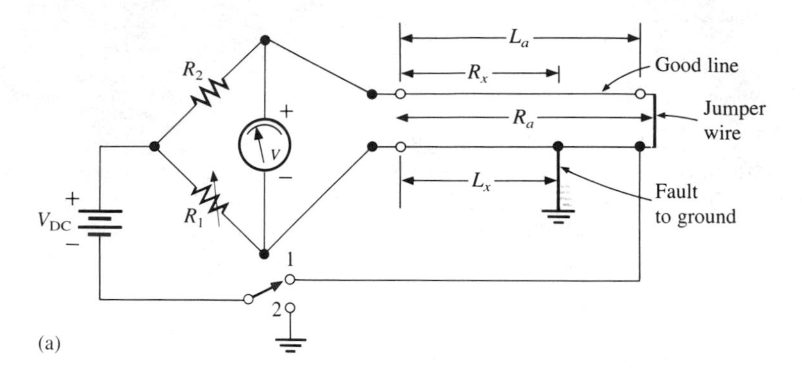

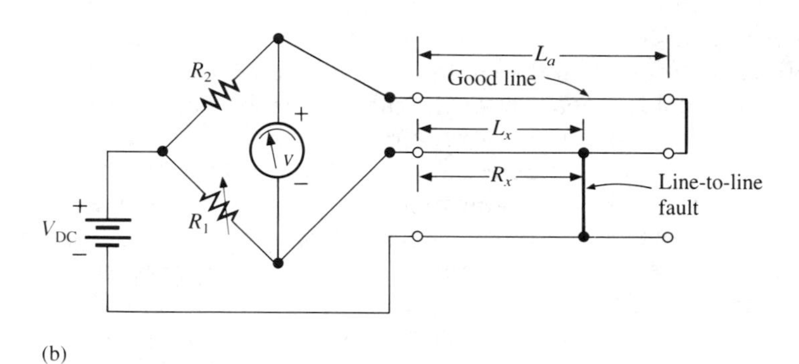

Figure 9–7 Murray loop test. (a) Test circuit for a single-wire line. (b) Finding the fault between two conductors of the same cable.

$$R_1R_2 = 2R_aR_3 \tag{9–6a}$$

$$R_a = \frac{R_1R_2}{2R_3} \tag{9–6b}$$

The switch is turned to position 2 and the bridge again balanced, so that

$$R_1(R_2 + R_x) = (2R_a - R_x)R_3$$

By solving for R_x

$$R_x = \frac{2R_aR_3 - R_1R_2}{R_1 + R_3} \tag{9–7}$$

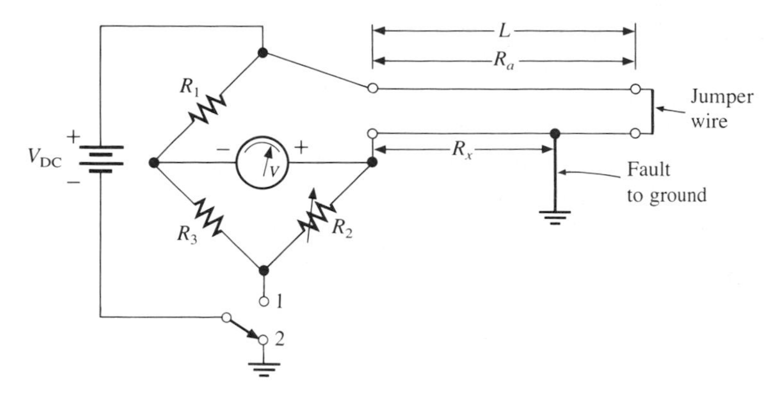

Figure 9–8 Varley loop test.

The value for R_2 is not the same in both situations. Computations associated with the Varley loop method are illustrated by the following example.

EXAMPLE 9–2

A fault is located somewhere on a 256-foot line. Use the Varley loop test circuit of Figure 9–8, with $R_1 = 100\ \Omega$ and $R_3 = 100\ \Omega$. In position 1, R_2, at balance, was 36.5 Ω. In position 2, R_2 was found to be 20.4 Ω. Determine the distance, L_x, of the fault from the test set.

Solution:

In position 1, the resistance of one-half the loop is found from Equation 9–6*b*:

$$R_a = \frac{(100\ \Omega)(36.5\ \Omega)}{(2)(100\ \Omega)}$$

$$= 18.25\ \Omega$$

In position 2, the resistance of the line between the fault and test set (Equation 9–7) is

$$R_x = \frac{[(2)(18.25\ \Omega)(100\ \Omega)] - [(100\ \Omega)(20.4\ \Omega)]}{100\ \Omega + 100\ \Omega}$$

$$= 8.05\ \Omega$$

Since R_x is 8.05/18.25, or 44.1% of the resistance of the 256-foot line, the distance of the fault from the test set is 0.441 times 256 feet, or 112.9 feet.

9–4 SIMPLE AC BRIDGES

When the four resistive arms of the basic Wheatstone bridge are replaced by impedances and the bridge excited by an AC source (Figure 9–9), the result is referred to as an **AC bridge.** It is capable of measuring complex impedances. Unlike the Wheatstone bridge, however, *two balance conditions must be satisfied.* Such a bridge must balance both the resistive and reactive components before the detector indicates a null. In general, the balance equations, in polar form, of the bridge are

$$(Z_1\angle\theta_1)(Z_X\angle\theta_X) = (Z_2\angle\theta_2)(Z_3\angle\theta_3) \tag{9–8a}$$

or

$$Z_1Z_X\angle(\theta_1 + \theta_X) = Z_2Z_3\angle(\theta_2 + \theta_3) \tag{9–8b}$$

Balance for the resistive component in an AC bridge is usually obtained by the adjustment of a precision variable resistance. Balance for the reactive component is obtained from a similar reactance in an adjacent arm of the bridge, or from an opposite reactance in an opposite arm. In most circuits, the reactance is in the form of a fixed precision capacitor, either in series or parallel with a variable resistance. Two conditions must be met for the AC bridge to be balanced. The first condition requires that the magnitudes of the four impedances satisfy this relationship:

$$Z_1Z_X = Z_2Z_3 \tag{9–9a}$$

The second condition relates the impedance angles:

$$\angle\theta_1 + \angle\theta_X = \angle\theta_2 + \angle\theta_3 \tag{9–9b}$$

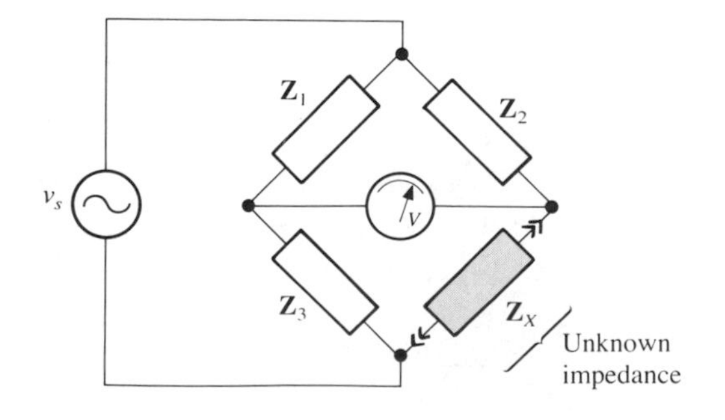

Figure 9–9 Generalized AC bridge.

As a consequence, when the bridge is balanced, the voltages across Z_X and Z_2 must be equal both in amplitude and phase.

EXAMPLE 9–3

At balance, the three known impedances of the AC bridge of Figure 9–9 are

$$Z_1 = 200\ \Omega\ \angle 60°$$

$$Z_2 = 150\ \Omega$$

$$Z_3 = 100 + j300\ \Omega$$

Determine the unknown impedance, Z_X, in polar form (j is a mathematical operator used by electrical engineers to represent $\sqrt{-1}$).

Solution:

For this problem, it is more convenient to express impedances in polar form. Consequently

$$Z_3 = 100 + j300\ \Omega$$

$$= 316\ \Omega \angle 71.6°$$

Using Equation 9–8*a*, the magnitude of Z_X is

$$Z_X = \frac{(150\ \Omega)(316\ \Omega)}{200\ \Omega}$$

$$= 237\ \Omega$$

Using Equation 9–8*b*, the impedance angle is

$$\angle \theta_X = 0° + 71.6° - 60°$$

$$= 11.6°$$

The unknown impedance in polar form is 237 Ω ∠11.6°.

AC bridges must be able to determine the resistive component of the impedance under test. As shown in Figure 9–10, the resistive component may be either in series or parallel with the reactive component when testing capacitors and inductors.

As shown in Figure 9–10*a*, the typical equivalent of a capacitor consists of a parallel resistor, R_P, called the **leakage resistance,** and the actual capacitance, C_P. Figure 9–10*b* shows the equivalent series circuit. Although, in a particular situation, a given equivalent circuit may be more convenient than the other, either arrangement can be used in a bridge circuit.

Figure 9–10 Capacitor and Inductor equivalent circuits. (a) Capacitor with parallel resistance. (b) Capacitor with equivalent series resistance. (c) Inductor with series resistance. (d) Inductor with equivalent parallel resistance.

C_P R_P

(a)

$R_S = R_P \dfrac{D^2}{1 + D^2}$

$C_S = C_P(1 + D^2)$

(b)

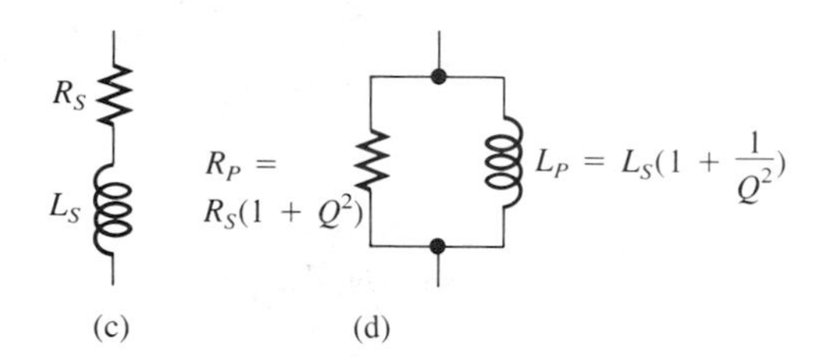

Most AC bridges do not measure the resistive component of a capacitor or inductor in direct terms, but, rather, as the ratio of reactance to resistance of the component under test. For the equivalent circuit of a capacitor having a parallel leakage resistor, this ratio is called the **dissipation factor, *D*,** where

$$D = \frac{X_P}{R_P} \qquad (9\text{–}10a)$$

$$= \frac{1}{\omega R_P C_P} \qquad (9\text{–}10b)$$

It is directly proportional to power loss per cycle. A capacitor with a very low leakage resistor is a high quality capacitor. The resistance shown for the parallel equivalent circuit (Figure 9–10*a*) is involved in the calculation of the dissipation factor. Ideally, the parallel resistance should be much larger than the capacitive reactance. When the series equivalent circuit (Figure 9–10*b*) is used, the dissipation factor is

$$D = \frac{R_S}{X_S} \qquad (9\text{–}11a)$$

$$= \omega C_S R_S \qquad (9\text{–}11b)$$

In an inductor, the ratio of reactance to its DC coil resistance is called the **storage factor** or **quality factor, *Q*** where

$$Q = \frac{X_S}{R_S} \tag{9–12a}$$

$$= \frac{\omega L_S}{R_S} \tag{9–12b}$$

The quality factor is directly proportional to the energy stored per cycle. As with capacitors, the equivalent circuit for an inductor is represented as either a series L–R (Figure 9–10*c*) or a parallel circuit (Figure 9–10*d*). Equation 9–12 applies to the Q of a series L–R circuit. The Q of a parallel circuit can be shown to be

$$Q = \frac{R_P}{X_P} \tag{9–13a}$$

$$= \frac{R_P}{\omega L_P} \tag{9–13b}$$

Inspection of Equations 9–11 and 9–12 (or 9–10 and 9–13) yields the result that $Q = 1/D$.

As with the Wheatstone bridge, the galvanometer frequently used as the detector for an AC bridge may be replaced with an active circuit, like that of Figure 9–11. Here, an op-amp differential amplifier is connected to the output of the bridge. When a PMMC meter is used, the diode half-wave rectifier con-

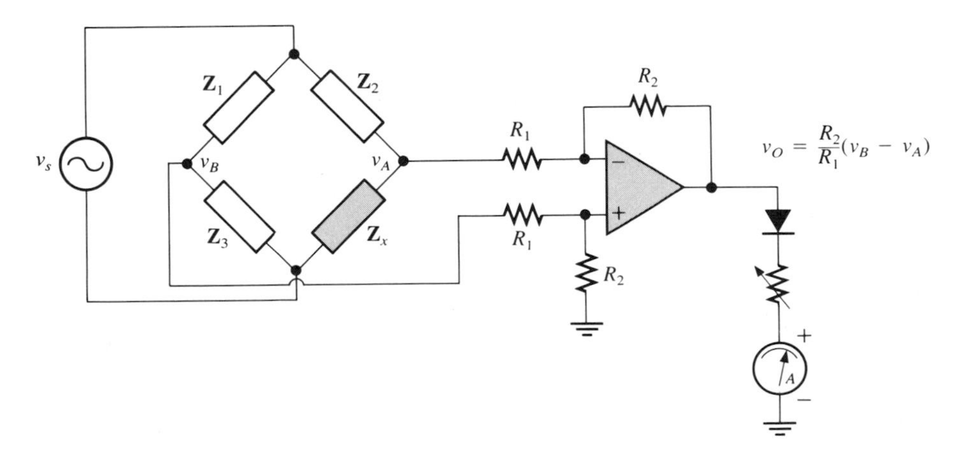

Figure 9–11 Using a differential amplifier with a diode rectifier-PMMC detector.

nected to the output of the amplifier produces a unidirectional current flow that is read by the meter. If an oscilloscope is used as the detector, the oscilloscope is connected directly to the amplifier's output without the diode rectifier.

9–5 CAPACITANCE BRIDGES

Although the measurement of capacitors is now frequently done with self-contained digital capacitance meters similar to digital ohmmeters, capacitance measurements done with a capacitance bridge are still preferred. It uses the basic AC bridge circuit (Figure 9–9) with two capacitive arms and two resistive arms. Figure 9–12 shows the circuit for the basic capacitance bridge. At balance,

$$C_X = C_1 \frac{R_1}{R_2} \qquad (9\text{–}14)$$

The unknown capacitor is actually more a function of the value of the known standard capacitor than the values of R_1 and R_2, because only the *ratio* of the resistances is involved. Because of its simplicity, the basic circuit has no provision for balancing the resistive component of a capacitor. It is intended for pure capacitive elements with no leakage (i.e., infinite parallel resistance).

EXAMPLE 9–4

For the capacitance bridge of Figure 9–12, the standard capacitor is 0.01 μF. $R_1 = 100\ \Omega$. If R_2 can be adjusted from 50 to 500 Ω, determine the range of the unknown capacitor that can be measured with the bridge.

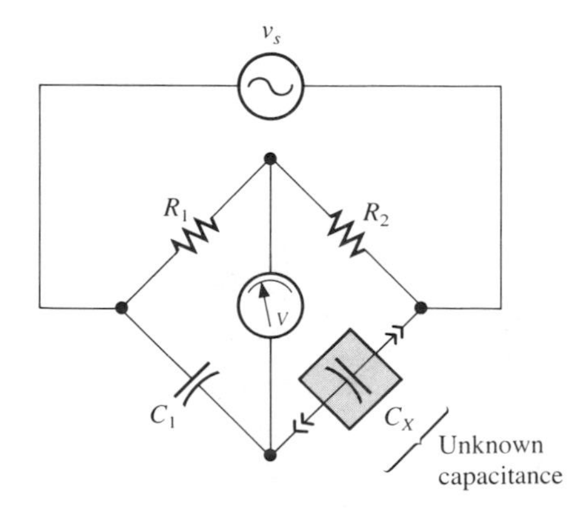

Figure 9–12 Basic capacitance bridge.

Solution:

Using Equation 9–14 when $R_2 = 50\ \Omega$,

$$C_X = \frac{(100\ \Omega)(0.01\ \mu\text{F})}{50\ \Omega}$$

$$= 0.02\ \mu\text{F}$$

When $R_2 = 500\ \Omega$,

$$C_X = \frac{(100\ \Omega)(0.01\ \mu\text{F})}{500\ \Omega}$$

$$= 0.002\ \mu\text{F}$$

As R_2 varies from 50 to 500 Ω, the bridge can measure any unknown capacitance from 0.002 to 0.02 μF.

Capacitance bridges discussed in the following sections have the capability for balancing the associated resistive component, determining its value on the basis of the dissipation factor. Although there are many variations of the basic circuit of Figure 9–12, the most common types are the following: series-resistance capacitance comparison bridge, parallel-resistance capacitance bridge, and Schering bridge.

Series Resistance-Capacitance Comparison Bridge

The **series resistance-capacitance comparison bridge,** shown in Figure 9–13, is primarily used for capacitors having dissipation factors between 0.001 and 0.1. It is also frequently called the **similar-angle bridge.** Based on the series resistance equivalent circuit of a capacitor, it uses a series R–C network in an adjacent arm of the bridge. At balance, the unknown capacitance and series resistance are determined from

$$C_S = C_1 \frac{R_2}{R_3} \tag{9–15a}$$

$$R_S = \frac{R_1 R_3}{R_2} \tag{9–15b}$$

In both equations, there is no dependence on frequency; the actual excitation frequency is not of major importance. However, the excitation frequency is usually either 1 kHz or a frequency at which the capacitor normally operates.

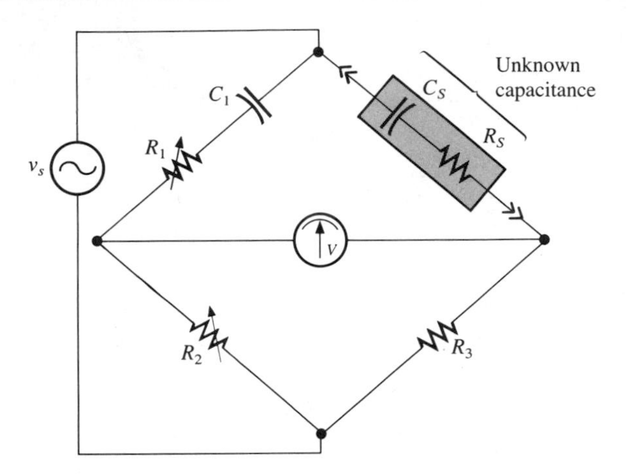

Figure 9–13 Series resistance-capacitance (similar angle) bridge for capacitors having low dissipation factors.

Resistors R_1 and R_2 are variable, while R_3 and C_1 are fixed-value standards. When used in a balanced commercial bridge with calibrated dials, both the capacitance and the dissipation factor are read directly from the dial settings. Using the dissipation factor of Equation 9–11*b*, the series leakage resistance can be found from

$$R_S = \frac{D}{\omega C_S} \tag{9–16}$$

EXAMPLE 9–5

A balanced similar-angle bridge like that of Figure 9–13 has the following values using a 100-kHz sine-wave source:

$C_1 = 500\ \text{pF}$

$R_1 = 176\ \Omega$

$R_2 = 8524\ \Omega$

$R_3 = 10\ \text{k}\Omega$

Determine (a) the unknown capacitance, (b) the unknown series resistance, and (c) the dissipation factor.

Solution:

(a) From Equation 9–15*a*,

$$C_S = \frac{(8524\ \Omega)(500\ \text{pF})}{(10\ \text{k}\Omega)}$$

$$= 426\ \text{pF}$$

(b) From Equation 9–15*b*,

$$R_S = \frac{(175\ \Omega)(10\ \text{k}\Omega)}{(8524\ \Omega)}$$

$$= 206.5\ \Omega$$

(c) The dissipation factor at 100 kHz, using Equation 9–11*b*:

$$D = (6.28)(100\ \text{kHz})(426\ \text{pF})(206.5\ \Omega)$$

$$= 0.055$$

Parallel Resistance-Capacitance Comparison Bridge

The parallel resistance-capacitance comparison bridge of Figure 9–14 is more advantageous for larger values of D (0.05 to 50). Resistors R_1 and R_2 are variable, while R_3 and C_1 are fixed-value standards. At balance,

$$C_P = C_1\frac{R_2}{R_3} \qquad (9\text{–}17a)$$

$$R_P = \frac{R_1R_3}{R_2} \qquad (9\text{–}17b)$$

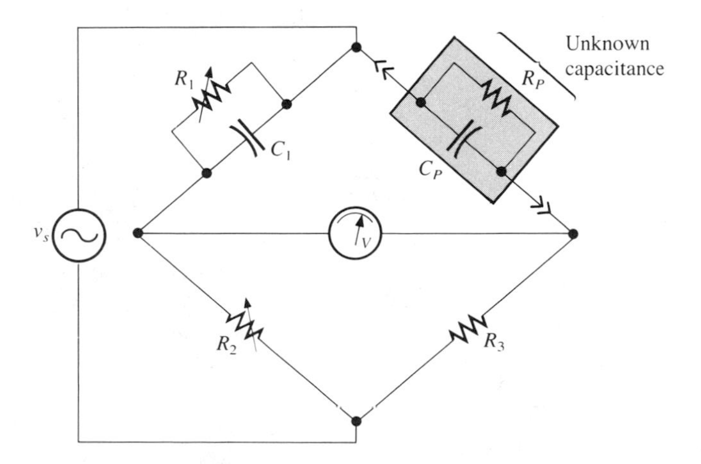

Figure 9–14 Parallel resistance-capacitance comparison bridge.

As with the similar-angle bridge, neither balance equation has any dependence on frequency; the actual excitation frequency is not of major importance. When used in a balanced commercial bridge with calibrated dials, both the values of capacitance and dissipation factor are read directly from the dial settings. Using the dissipation factor for a parallel R–C equivalent circuit (Equation 9–10*b*), the parallel leakage resistance can be found from

$$R_P = \frac{1}{\omega C_P D} \tag{9–18}$$

Schering Bridge

For measuring capacitors with a very low dissipation factor (with an impedance angle very close to 90°), the **Schering bridge** of Figure 9–15 is superior to the other two circuits. Besides capacitances and dissipation factors, it also measures the insulating properties of electrical cables and equipment. At balance the equations are

$$C_S = C_3 \frac{R_1}{R_2} \tag{9–19a}$$

$$R_S = R_1 \frac{C_1}{C_3} \tag{9–19b}$$

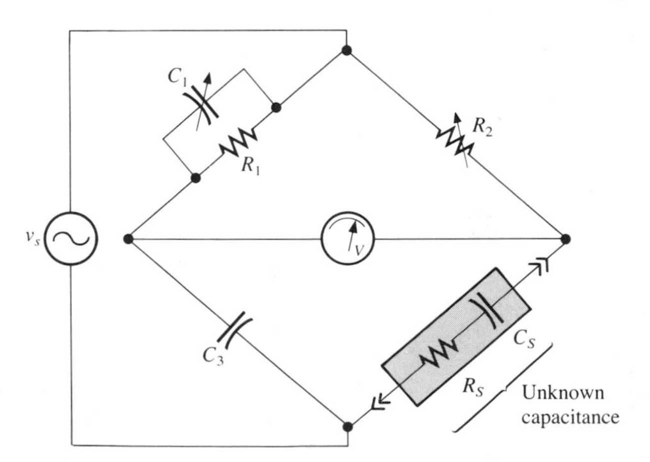

Figure 9–15 The Schering capacitance bridge.

Again, the balance is independent of frequency.

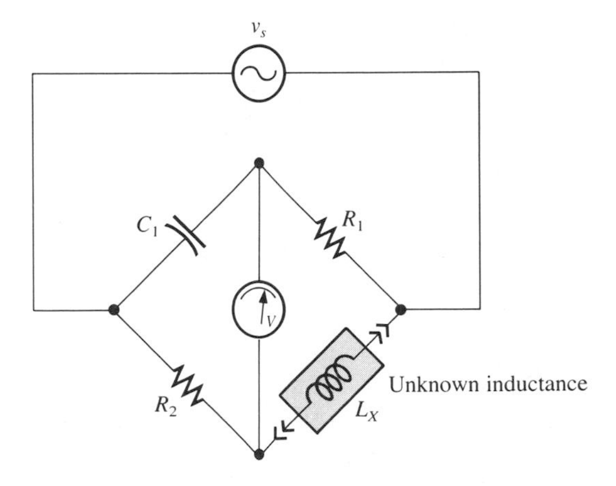

Figure 9–16 Basic inductance bridge.

9–6 INDUCTANCE BRIDGES

Unknown inductances are usually measured using a bridge circuit. Figure 9–16 shows the circuit for a simple inductance bridge. At balance,

$$L_X = C_1 R_1 R_2 \tag{9–20}$$

Like the simple capacitance bridge of Figure 9–12, the inductance bridge has no provision for balancing the resistive component of an inductor. It is intended for pure inductive elements having no series resistance.

Inductance bridges have the capability for balancing the associated resistive component, generally determining its value on the basis of the Q factor. Although there are many variations of the basic circuit of Figure 9–16, the most common types are the Maxwell, Hay, Owen, and inductance comparison bridges, which are discussed in the following sections.

Maxwell Bridge

The **Maxwell bridge,** or **Maxwell-Wien bridge** (Figure 9–17), is used to measure both a given inductance (with a Q between 1 and 10) and its series resistance by comparison to a standard capacitance. Using a capacitance as a standard offers several important advantages. Capacitors are easy to shield and produce almost no external field of their own. In addition, they are compact and fairly inexpensive.

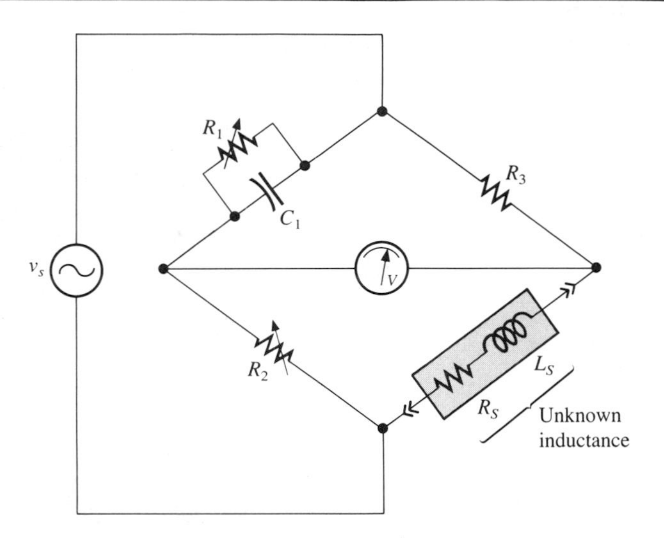

Figure 9–17 The Maxwell bridge for inductances having quality factors between 1 and 10.

For balance, resistors R_1 and R_2 are variable and R_3 and C_1 are fixed-value standards. At balance,

$$L_S = C_1 R_2 R_3 \qquad (9\text{–}21a)$$

$$R_S = \frac{R_2 R_3}{R_1} \qquad (9\text{–}21b)$$

The balance is independent of frequency. When used in a balanced commercial bridge with calibrated dials, both the values of the inductance and Q factor are read directly from the dial settings. Using the Q factor for a series L–R equivalent circuit (Equation 9–12b), the series resistance can be found from

$$R_S = \frac{\omega L_S}{Q} \qquad (9\text{–}22)$$

EXAMPLE 9–6

A balanced Maxwell bridge similar to that of Figure 9–17 has the following values when using a 1-kHz sine-wave source:

$C_1 = 0.01\ \mu\text{F}$

$R_1 = 1471\ \Omega$

$R_2 = 602\ \Omega$

$R_3 = 500\ \Omega$

Determine (a) the unknown inductance, (b) the unknown series resistance, and (c) the Q factor.

Solution:

(a) From Equation 9–21*a*,

$$L_S = (0.01\ \mu\text{F})(602\ \Omega)(500\ \Omega)$$
$$= 3.01\ \text{mH}$$

(b) From Equation 9–21*b*,

$$R_S = \frac{(602\ \Omega)(500\ \Omega)}{1471\ \Omega}$$
$$= 204.6\ \Omega$$

(c) The Q factor at 1-kHz, using Equation 9–22 is as follows:

$$Q = \frac{(6.28)(1\ \text{kHz})(3.01\ \text{mH})}{204.6\ \Omega}$$
$$= 0.092$$

Hay Bridge

The **Hay bridge,** or **opposite-angle bridge** (Figure 9–18), is used for the measurement of high-Q inductors ($Q > 10$). The bridge arm opposite the unknown inductance contains a capacitive reactance. The Hay bridge differs from the Maxwell in that it has a standard capacitor in series with a variable resistor, as opposed to a parallel combination in the arm opposite the unknown inductor. Since these two opposite arms contain opposite reactance types, they have opposite impedance angles (hence the "opposite" in the name). Like the Maxwell bridge, it has the advantages of using a known standard capacitance and containing a capacitive reactance (opposite impedance angle) as well as using a standard capacitor to measure inductance.

For balance, resistors R_1 and R_2 are variable, while R_3 and C_2 are fixed-value standards. At balance,

$$L_S = \frac{R_1 R_3 C_2}{1 + (\omega R_2 C_2)^2} \tag{9–23a}$$

$$R_S = \frac{(\omega C_2)^2 R_1 R_2 R_3}{1 + (\omega R_2 C_2)^2} \tag{9–23b}$$

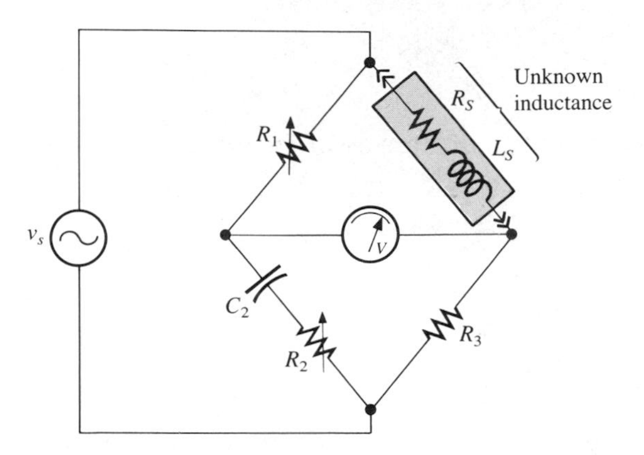

Figure 9–18 The Hay (opposite angle) bridge for inductance quality factors greater than 10.

Unlike the capacitance and Maxwell inductance bridges, balance is dependent on frequency.

EXAMPLE 9–7

A balanced Hay inductance bridge similar to that of Figure 9–18 has the following values using a 1-kHz sine-wave source:

$$C_2 = 0.1\ \mu\text{F}$$

$$R_1 = 983\ \Omega$$

$$R_2 = 94\ \Omega$$

$$R_3 = 500\ \Omega$$

Determine (a) the unknown inductance, (b) the series resistance, and (c) the Q factor.

Solution:

(a) From Equation 9–23*a*,

$$L_S = \frac{(983\ \Omega)(500\ \Omega)(0.1\ \mu\text{F})}{1 + [(6.28)(1\ \text{kHz})(0.1\ \mu\text{F})(94\ \Omega)]^2}$$

$$= 49\ \text{mH}$$

(b) From Equation 9–23*b*,

$$R_S = \frac{[(6.28)(1\ \text{kHz})(0.1\ \mu\text{F})]^2(983\ \Omega)(94\ \Omega)(500\ \Omega)}{1 + [(6.28)(1\ \text{kHz})(0.1\ \mu\text{F})(94\ \Omega)]^2}$$

$$= 18.2\ \Omega$$

(c) The Q factor at 1 kHz from Equation 9–12*b:*

$$Q = \frac{(6.28)(1 \text{ kHz})(49 \text{ mH})}{18.2 \ \Omega}$$

$$= 16.9$$

Owen Bridge

The **Owen bridge,** shown in Figure 9–19, is used to measure a wide range of inductances. At balance,

$$L_S = R_1 R_3 C_2 \quad (9\text{–}24a)$$

$$R_S = R_3 \frac{C_2}{C_1} \quad (9\text{–}24b)$$

Balance is independent of frequency in this bridge, as in the Maxwell bridge. Resistors R_1 and R_3 are variable, while C_1 and C_2 are fixed-value standards.

Inductance Comparison Bridge

The circuit for the **inductance comparison bridge** (Figure 9–20) is similar to the series-resistance capacitance (similar-angle) bridge, except that inductors replace capacitors. Unlike the Maxwell, Hay, and Owen bridges, the inductance comparison bridge determines the value of the unknown inductance on the basis of a known standard inductor, rather than a capacitor. At balance,

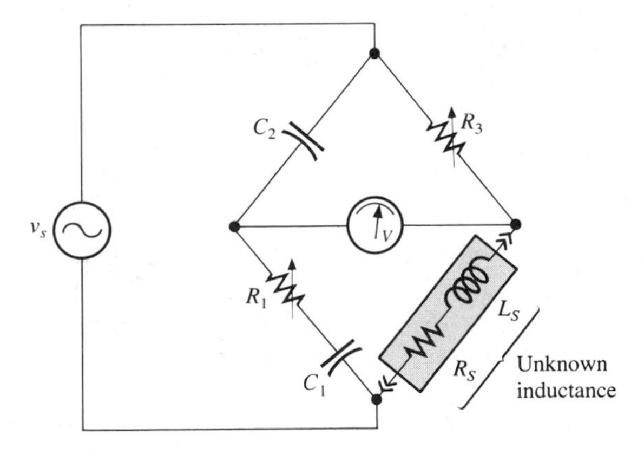

Figure 9–19 The Owen inductance bridge.

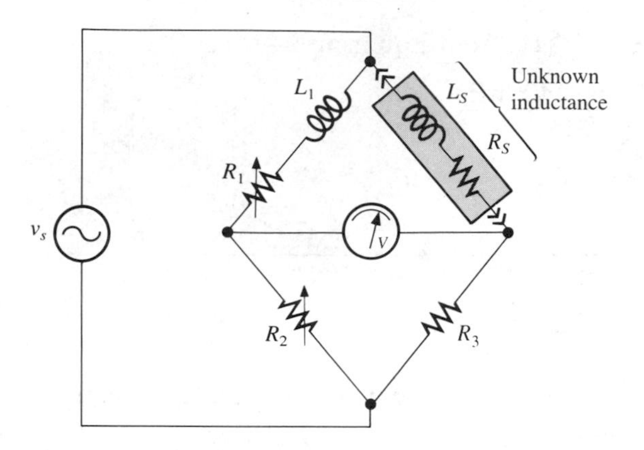

Figure 9–20 Inductance comparison bridge.

$$L_S = L_1 \frac{R_3}{R_2} \tag{9–25a}$$

$$R_S = \frac{R_1 R_3}{R_2} \tag{9–25b}$$

L_S and R_S are independent of frequency. Resistors R_1 and R_2 are variable, while R_3 and L_1 are fixed-value standards.

9–7 THE WAGNER GROUND CONNECTION

All the AC bridge circuits discussed in sections 9–5 and 9–6 are based on the ideal situation of no stray capacitances between the various bridge elements, connections, and ground. At low frequencies, this is not much of a problem; the reactance of the stray capacitances are usually large enough to be ignored. However, at higher frequencies, or with either large inductances or small capacitances, stray capacitances do cause errors in measurements.

The Wagner ground connection, using the circuit of Figure 9–21, eliminates the stray capacitances between the detector terminal and the ground. The AC signal source is removed from its usual ground connection and is bridged by a series resistor-capacitor (R_W–C_W) combination. The grounded junction of R_W and C_W is called the **Wagner ground.**

Using a series-resistance capacitance bridge as an example in Figure 9–21, the adjustment of R_W follows the procedure below:

1. Switch S_1 to position A. In this case, the bridge circuit functions normally so that the bridge can be balanced.
2. Switch S_1 to position B, which now grounds one side of the detector.

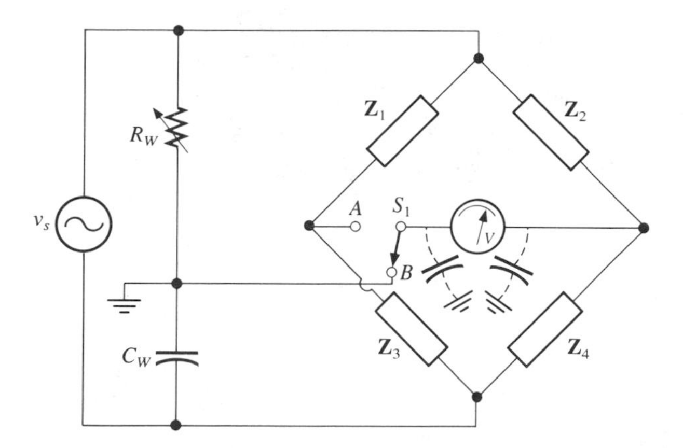

Figure 9–21 The Wagner ground connection used to minimize bridge stray capacitance errors at high frequencies.

3. Adjust resistor R_W for null.
4. Switch S_1 back to position A and again balance the bridge.
5. Repeat Steps 2, 3, and 4 until the same null occurs in both positions A and B.

After the final null is obtained, nodes A and B are both at ground potential. Stray capacitances at the detector's terminal are essentially shorted to ground, and have no effect on bridge balance.

9–8 COMMERCIAL RLC BRIDGES

Universal impedance bridges can measure resistance, capacitance, and inductance over a wide range of values. Such instruments usually contain many of the previously discussed bridge circuits in a single package. Although they normally include an internal fixed-frequency excitation source, most have provision for the connection of an external AC source to permit operation at various frequencies.

9–9 AUTOMATIC BRIDGES

All the bridge circuits discussed are manually operated. With the advent of digital circuitry and microprocessors, it is possible to automate the steps necessary to select the proper range and establish balance. In addition, like most other digital instruments, these automatic bridges (Figure 9–22) provide a digital readout, and may be interfaced with computers via RS–232C or IEEE–488 interface buses (Chapter 16).

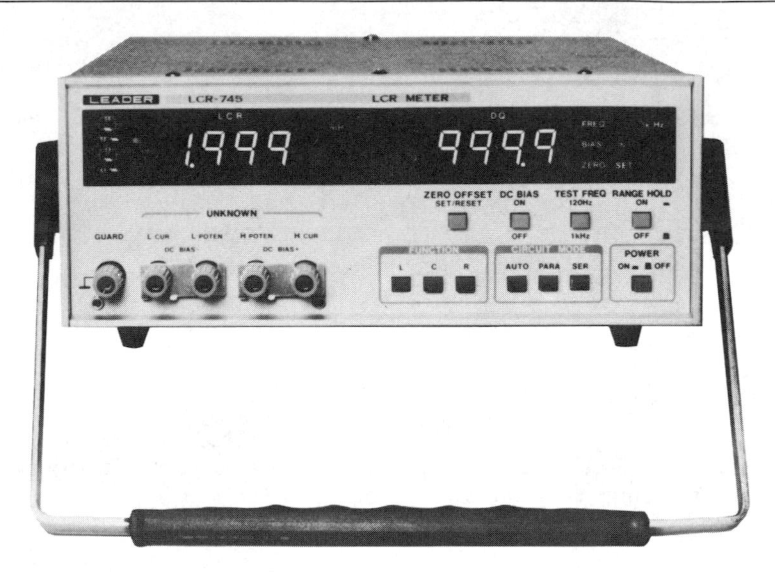

Figure 9–22 Model LCR-745G Digital LCR Meter. Courtesy of Leader Instruments Corporation.

9–10 SUMMARY

This chapter discusses various bridge circuits that measure resistance, capacitance, inductance, the quality factor (Q), and the dissipation factor (D). The basic DC Wheatstone bridge can be expanded to permit the measurement of very low resistances (Kelvin bridge) and very high resistances (guarded Wheatstone bridge), and to determine the location of faults in both wire and cable (either the Murray or Varley loop test).

Because they use an AC source, capacitance and dissipation factors are primarily measured using either series-resistance, parallel-resistance, or Schering bridges. Inductance and Q factor are measured with Maxwell, Hay, Owen, or inductance comparison bridges. The Wagner ground is a useful method for reducing stray capacitances, which cause errors when dealing with small capacitances at high frequencies or values of large inductors in AC bridge circuits. Although, in their basic form, all bridge circuits are manually balanced, sophisticated RLC bridges now use microprocessor technology to automate the process, including digital readout and the ability to interface with other computers.

9–11 GLOSSARY

AC bridge A bridge circuit excited by an AC source that measures either capacitance or inductance.

bridge A measuring circuit that has four series arms in the shape of a diamond. Two opposite nodes of the bridge serve as input terminals, while the two remaining nodes are the output terminals.

dissipation factor A measure of the quality of a capacitor. For capacitors with parallel leakage resistance, it is the ratio of the capacitive reactance to the parallel resistance.

guard ring A metal ring that surrounds the battery's positive terminal post of a Wheatstone bridge; used to minimize the leakage current that flows through the terminal.

Hay bridge Also called an **opposite angle bridge,** it is an AC bridge for measuring high-Q inductances in terms of resistance, capacitance, and frequency.

inductance comparison bridge An AC bridge for measuring an unknown inductor and its series resistance in terms of known resistances and a standard inductor.

Kelvin bridge Also called a **Thompson,** or **Kelvin double bridge,** it is a seven-arm adaptation of the DC Wheatstone bridge. It measures very low resistances by offsetting the parasitic resistance of its leads.

Kelvin double bridge See **Kelvin bridge.**

leakage current An unwanted but small flow of current.

leakage resistance The resistance path through which leakage current flows. In a capacitor, the finite DC resistance that exists between its terminals.

Maxwell bridge Also called a **Maxwell-Wien bridge,** it is an AC bridge used for measuring low-Q inductances in terms of resistance and capacitance.

Maxwell-Wien bridge See **Maxwell bridge.**

Murray loop test A method of finding a fault in either a cable or a wire by connecting the damaged line with a good one as adjacent arms of a Wheatstone bridge.

opposite angle bridge See **Hay bridge.**

Owen bridge An AC bridge used for measuring inductance in terms of resistance and capacitance.

parallel resistance-capacitance comparison bridge An AC bridge for measuring capacitors having high dissipation factors in terms of known resistances and capacitance.

quality factor The Q factor (also sometimes called storage factor) of an inductor is the ratio of the inductive reactance to the series coil resistance.

Schering bridge An AC bridge for measuring capacitance in terms of resistance and a standard capacitance.

series resistance-capacitance comparison bridge An AC bridge for measuring capacitors having low dissipation factors in terms of known resistances and capacitances.

similar angle bridge An AC bridge for measuring capacitance and its series leakage resistance in terms of resistance and a standard capacitance.

storage factor See **quality factor.**

Thompson bridge See **Kelvin bridge.**

Varley loop test While similar to the Murray loop test, it differs in having a third resistor. The cable under test makes up the fourth arm of a Wheatstone bridge.

Wagner ground A ground connection used with AC bridges to minimize errors due to stray capacitances. The oscillator is removed from its usual ground connection and bridged by a series RC network.

Wheatstone bridge A bridge with four resistor arms that is excited by either an AC or DC voltage and used for the measurement of resistance.

9–12 PROBLEMS

1. The Wheatstone bridge of Figure 9–1 is used to measure the unknown resistance, R_X. At balance, $R_1 = 581\ \Omega$, $R_2 = 100\ \Omega$, and $R_3 = 743\ \Omega$. Determine the unknown resistance.
2. Using the Wheatstone bridge of Problem 1, if R_3 varies from 100 Ω to 1 kΩ, determine the range of the unknown resistance that can be measured with the bridge.
3. The Kelvin double bridge of Figure 9–5 is used to measure resistance. The resistance ratio R_6/R_5 is maintained as equal to $R_2/$

R_1. At balance, $R_1 = 100\ \Omega$, $R_2 = 9.8\ \Omega$, and $R_3 = 0.1\ \Omega$. Determine the unknown resistance.

4. The Murray loop test circuit of Figure 9–7*a* is used to determine the location of a fault in a 590-foot wire buried underground. A Wheatstone bridge measures the resistance of the loop as 15.4 Ω. With $R_2 = 100\ \Omega$, R_1 is 28.6 Ω at balance. Determine the distance of the fault from the test set.

5. The location of a fault in a 105-meter cable is to be determined using the Varley loop test circuit of Figure 9–8 with $R_1 = 500\ \Omega$, and $R_3 = 400\ \Omega$. When the switch is in position 1, R_2 is 22.5 Ω at balance. When the switch is moved to position 2 and the circuit rebalanced, $R_2 = 10.1\ \Omega$. Determine the distance of the fault from the test set.

6. Using the basic AC impedance bridge of Figure 9–9, the three known impedances at balance are

$$Z_1 = 32 - j60$$
$$Z_2 = 97\ \Omega\ \angle -90°$$
$$Z_3 = 100\ \Omega$$

(a) Determine the unknown impedance, Z_X, in both polar and rectangular forms.

(b) Is the reactance of the unknown impedance capacitive or inductive?

7. For the simple capacitance bridge of Figure 9–12, the values of the three known arms are the following: $R_1 = 1\ \text{k}\Omega$, $R_2 = 372\ \Omega$, and $C_1 = 0.5\ \mu\text{F}$. Determine the value of the unknown capacitor, C_X.

8. For the bridge of Problem 7, if R_2 varies from 50 Ω to 950 Ω, determine the range of the unknown capacitance that can be measured with the bridge.

9. Use the similar-angle bridge circuit of Figure 9–13 with a 1-kHz source and a 0.1-μF capacitor for C_1. At balance, $R_1 = 115.6\ \Omega$, $R_2 = 1825\ \Omega$, and $R_3 = 1\ \text{k}\Omega$. Determine (a) the unknown capacitance, (b) the series leakage resistance, and (c) the dissipation factor.

10. Using Figure 9–10, calculate the parallel equivalent circuit components for the unknown capacitor of Problem 9.

11. Use the parallel-resistor capacitor bridge circuit of Figure 9–14 with a 1-kHz source and a 0.1-μF capacitor for C_1. At balance, $R_1 = 462\ \Omega$, $R_2 = 579\ \Omega$, and $R_3 = 1\ \text{k}\Omega$. Determine (a) the unknown capacitance, (b) the parallel leakage resistance, and (c) the dissipation factor.

12. Use the Schering bridge circuit of Figure 9–15 with a 10-kHz source to determine the value of an unknown capacitor. At balance, $R_1 = 1\ \text{k}\Omega$, $R_2 = 14.2\ \Omega$, $C_1 = 100$ pF, and $C_3 = 7.3$ pF. Determine (a) the unknown capacitance, (b) the series leakage resistance, and (c) the dissipation factor.

13. Use the Maxwell bridge circuit of Figure 9–17 with a 1-kHz source to determine the value of an unknown inductor. At balance, $R_1 = 2649\ \Omega$, $R_2 = 1565\ \Omega$, $R_3 = 1\ \text{k}\Omega$, and $C_1 = 0.1\ \mu\text{F}$. Determine (a) the unknown inductance, (b) the series coil resistance, and (c) the Q factor.

14. Use the opposite angle (Hay) bridge circuit of Figure 9–18 with a 1-kHz source to determine the value of an unknown inductor. At balance, $R_1 = 873\ \Omega$, $R_2 = 184\ \Omega$, $R_3 = 100\ \Omega$, and $C_2 = 0.5\ \mu\text{F}$. Determine (a) the unknown inductance, (b) series coil resistance, and (c) the Q factor.

15. Use the Owen bridge circuit of Figure 9–19 with a 1-kHz source to determine the value of an unknown inductor. At balance, $R_1 = 1\ \text{k}\Omega$, $R_3 = 14.2\ \Omega$, $C_1 = 100$ pF, and $C_2 = 7.3$ pF. Determine (a) the unknown inductance, (b) series coil resistance, and (c) the Q factor.

16. Use the inductance comparison bridge circuit of Figure 9–20 with a 1-kHz source to determine the value of an unknown inductor. At balance, $R_1 = 395\ \Omega$, $R_2 = 604\ \Omega$, $R_3 = 1\ \text{k}\Omega$, and $L_1 = 100$ mH. Determine (a) the unknown inductance, (b) the series coil resistance, and (c) the Q factor.

10

TRANSDUCERS

10–1 INSTRUCTIONAL OBJECTIVES

At the completion of this chapter, you will be able to

- Define the term transducer.
- List 12 electrical phenomena that can be the basis for a transducer.
- List 13 characteristics of transducers which must be selected for any given application.
- Explain how resistance-changing transducers, such as the RTD, thermistor, strain gage, and photoconductive devices work, and also how their electrical changes can be measured.
- Explain how self-generating transducers, such as thermocouple, piezoelectric, magnetic-induction, and photovoltaic devices work, and also how their electrical changes can be measured.
- Describe the construction and explain the principle of the electromagnetic flow meter.
- Explain how inductive and capacitive transducers work, and also how their electrical changes can be measured.
- Describe the operation of phototransistors and photodiodes, and describe suitable circuits for their use.

10–2 INTRODUCTION

In general terms, a **transducer** is any device that converts energy from one form to another. For electrical measuring systems, a transducer develops a usable electrical output signal in response to a specific physical phenomenon. Commercial transducers are available for the measurement of mechanical forces, acceleration, pressure, physical position, temperature, light intensity, and any other commonly encountered physical parameter.

Transducers may be broadly classified as either *self-generating* or *externally powered.* Self-generating transducers develop their own voltage or current, while externally powered devices require power from an external source. Inside these two major classifications, there exist twelve different types of electrical phenomena. These are the basis of a transducer:

1. Capacitive
2. Electromagnetic
3. Inductive
4. Ionization
5. Photoresistive/photoconductive
6. Photoelectric/photoemissive
7. Photovoltaic
8. Piezoelectric
9. Potentiometric
10. Resistive
11. Thermovoltaic
12. Variable permittivity or resistivity

Though there are many types of transducers used in instrumentation systems, which can be the subject of a book in itself, most transducers fit into one of the following four categories:

1. Resistance-changing
2. Self-generating
3. Inductance-changing
4. Capacitance-changing

This chapter discusses these four. One or more of these basic types covers most measurement applications.

10–3 SELECTION CONSIDERATIONS

To select the right transducer requires careful consideration of each of the following specifications:

1. *Sensitivity* Defined as the ratio of the output per unit input, the sensitivity must be great enough for the resolution of the system.
2. *Range* The transducer must be able to respond over the minimum to maximum values of the parameter to be measured.
3. *Physical properties* The methods for mounting the transducer in the system to be measured, for protecting it, and for making and shielding its electrical connections must be considered.
4. *Loading effects and distortion* Because all transducers absorb some energy from the physical phenomenon being measured, the user must be certain that the transducer will not significantly distort, or "load down," the measured quantity.
5. *Frequency response* The transducer system must be able to accurately respond to the maximum rate of change of the phenomenon being studied.
6. *Electrical output format* The form of the output signal must be compatible with the rest of the measuring system. If not, a means must be available for conversion to a compatible form.
7. *Output impedance* The transducer's output impedance must be compatible with the rest of the measuring system.
8. *Power requirements* It is important that the proper supply voltage or voltages are provided for externally powered devices.
9. *Noise* Noise is any unwanted signal present in the system. The output signal from a transducer should be, as far as possible, unaffected by noise.
10. *Accuracy or error* The manufacturer usually evaluates the accuracy at which a transducer is expected to respond. Because this is quoted under

specified conditions, the user must make sure that the conditions under which the transducer is to be used are applicable.

11. *Calibration* The properties of many transducers can drift with time and aging. This must be compensated for by periodic recalibration. The frequency and desirability of this must be evaluated.
12. *Environment* A transducer's performance is critically affected by such environmental factors as temperature, humidity, and dust.
13. *Cost* This is perhaps the most difficult to judge. In general, you get what you pay for. However, relaxation of one or more of the above considerations may allow for a less expensive type of transducer without a serious loss in measurement performance.

Some of these requirements are more important than others in any given application.

10–4 RESISTANCE-CHANGING TRANSDUCERS

Resistance-changing transducers, while simple and inexpensive, have a wide range of applications in measurement systems. The intrinsic resistance of the transducer is normally changed by either mechanical linkage or the direct application of some physical parameter such as strain, pressure, or temperature. To produce an electrical output, the resistance-changing element is made part of a voltage divider, bridge, or similar circuit.

As an example, Figure 10–1 shows a simple resistance-changing element, the slide-wire potentiometer. As shown, it is connected across a constant voltage source to act as a **relative position transducer.** The movable wiper arm on the potentiometer can be linked to a moving part in either a mechanism or an instrument. As the wiper arm moves towards the top of the circuit, the output voltage increases, and vice versa. Mathematically, this relation is

$$V_o = \frac{R_1}{R_1 + R_2} V_{DC} \tag{10–1}$$

$R_1 + R_2$ equals the total resistance of the slide wire (R_T). Since the ratio of the instantaneous distance of the wiper arm from the bottom (x) to the total distance of the slide wire (L) is equal to the ratio of the instantaneous resistance from the bottom (R_1) to the total resistance (R_T), Equation 10–1 can be expressed as

$$V_o = \frac{x}{L} V_{DC} \tag{10–2}$$

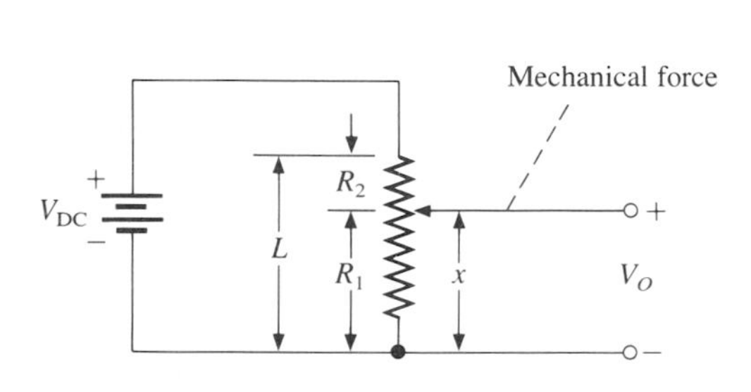

Figure 10–1 Voltage divider variable resistance transducer using a slide-wire potentiometer.

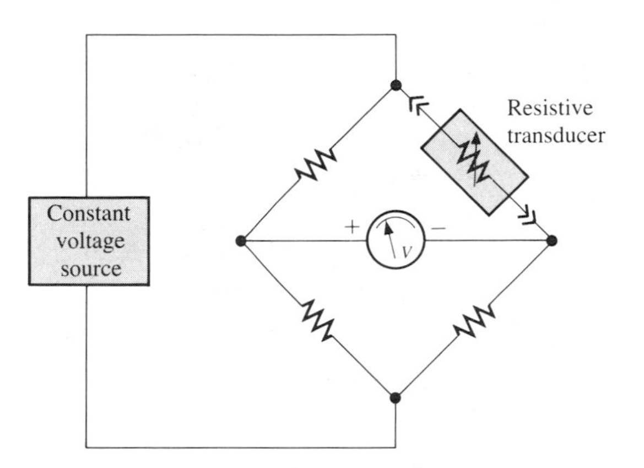

Figure 10–2 Using a resistance-changing transducer as one arm of a DC Wheatstone bridge.

In this case, the output voltage is a *linear* function of the physical position of the wiper arm. Figure 10–2 shows the slide-wire potentiometer connected as one arm of a Wheatstone bridge circuit.

The Resistance-Temperature Detector (RTD)

Figure 10–3 shows a device, called an **RTD,** or **resistance-temperature detector,** connected in a series circuit driven by a constant-current source. An RTD has a positive temperature coefficient that is virtually linear over a certain range of temperatures.

In the circuit shown, the output voltage is equal to the resistance of the RTD multiplied by a constant (the value of the current). Thus, any change in temperature produces a proportional change in output voltage. The RTD resistance, as a function of temperature, is given by

Figure 10–3 A resistance-temperature detector (RTD) measurement circuit driven by a constant-current source.

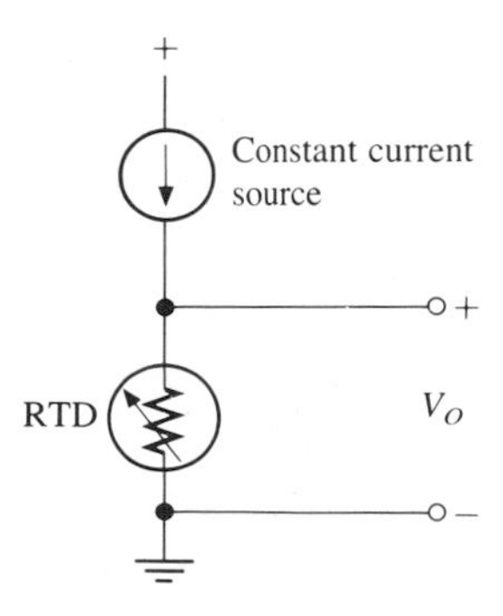

$$R_T = R_o[1 + \alpha(T - T_o)] \tag{10–3}$$

where

R_o = RTD resistance at a standard temperature reference

T_o = standard temperature reference used to determine R_o

T = RTD temperature

α = temperature coefficient per °C

Thermistors

A commonly used resistance-changing device is the **thermistor.** The thermistor is a temperature-sensitive resistor like the RTD, but more sensitive. Thermistors are usually composed of semiconductor material. Most thermistors exhibit a negative temperature coefficient, although devices having a positive temperature coefficient are available. This temperature coefficient can be as large as several percent per degree Celsius, allowing the measurement of minute changes in temperature that are unobservable with other types of devices. However, the price paid for this increased sensitivity is a loss of linearity. The thermistor is a nonlinear device and is very dependent on manufacturing process parameters. As a result, manufacturers have not standardized thermistor curves.

Despite the basic tradeoff between sensitivity and linearity in the simplest thermistors, composite thermistor assemblies that offer an almost linear response to temperature change are available. These assemblies consist of two or three thermistors mounted in the same package. When combined with specified external resistance values, they can provide a linear response over a given temperature range.

Figure 10–4 shows a typical circuit including a linear thermistor component. In general, the output voltage is given by the straight-line equation:

$$V_o = \pm mT + b \tag{10–4}$$

where

m = slope in volts/°C

b = value of V_o when $T = 0$

T = temperature in °C

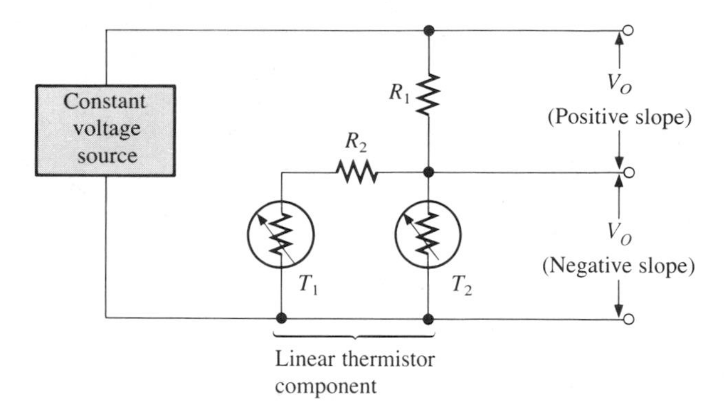

Figure 10–4 Thermistor linearization.

Because of their response speed, thermistors are frequently used in medical applications. In particular, a thermistor is placed in the nose to measure airflow due to breathing because the intake and expulsion of air cyclically warms and cools the thermistor.

Strain Gages

The **strain gage** is another example of a resistance-changing transducer. As shown in Figure 10–5*a*, a strain gage basically consists of a fine wire, about 0.001 inches in diameter, bonded to the face of a mounting plate. Normally, the wire is looped back and forth many times to allow more length. A tensile stress along the proper axis elongates the wire, thereby increasing its length. At the same time, the cross-sectional area of the wire decreases. From the formula used to calculate the resistance of a wire,

$$R = \rho\frac{L}{A} \tag{10–5}$$

where

ρ = resistivity (Ω-cm)

L = length of wire (cm)

A = cross-sectional area (cm^2)

The wire resistance is directly proportional to its length and inversely proportional to its cross-sectional area. The combined effect serves to increase the resistance of the strain gage. Figure 10–5*b* illustrates how four fine resistance wires are used to construct an unbonded strain gage having four active elements.

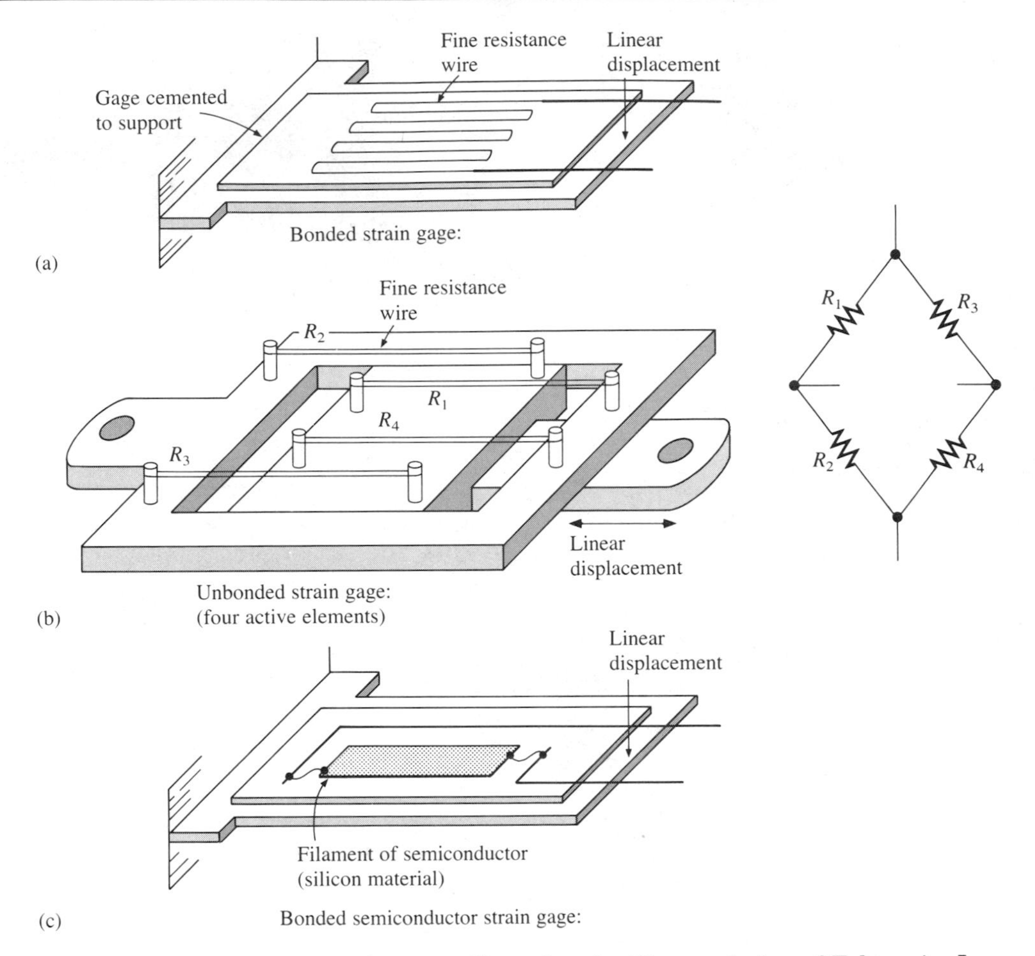

Figure 10–5 Resistive strain gages. Reproduced with permission of Tektronix, Inc. Copyright 1970.

Another type of strain gage uses lengths of doped silicon bonded to metal or some other slightly elastic surface (Figure 10–5*c*). These silicon-based resistors undergo noticeable resistance changes when stressed, because distortion of the crystal lattice structure changes both the concentration and the mobility of the charge carriers (electrons or holes). Strain gages of this type can typically be stretched or compressed up to 0.5% of their original length, producing a resultant resistance change of 25 to 100%, depending on the level of doping.

One of the shortcomings of strain gages is that their resistance also varies with temperature. To compensate for this temperature effect, a second identical strain gage can be mounted on the same structure, but *perpendicular to the direction of force*. Each gage is connected as one leg of a resistance bridge, as shown in Figure 10–6. The second (dummy) gage, because of its orientation, is

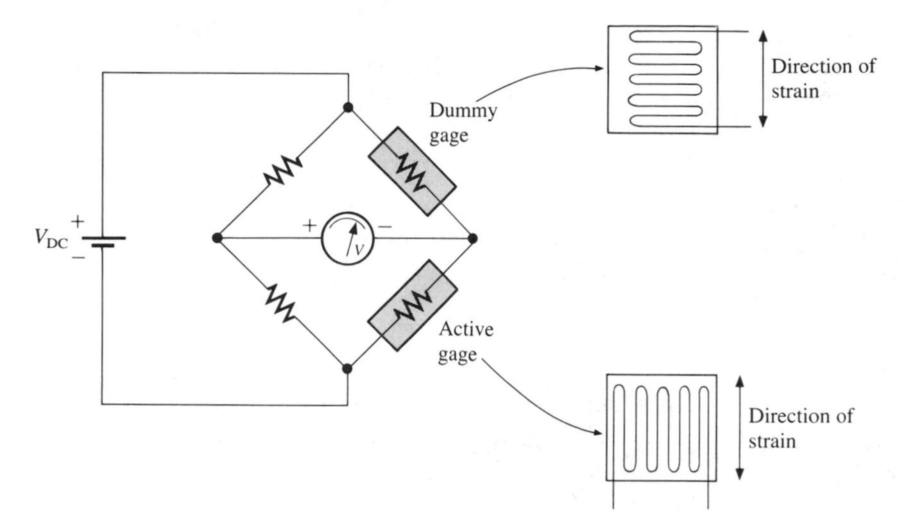

Figure 10–6 Using a second (dummy) resistive strain gage in a Wheatstone bridge to cancel temperature effects of the active gage.

affected only by temperature and not by stress. Since temperature changes are equal in each gage, temperature variations do not affect the bridge output.

Associated with strain gages is a dimensionless parameter called the **gage factor (*K*)** relating a change in resistance to a change in length:

$$K = \frac{(\Delta R/R)}{(\Delta L/L)} \tag{10–6}$$

where

R = nominal (unstrained) resistance

ΔR = change in resistance due to strain

L = nominal length

ΔL = change in length due to strain

If the wire has a cylindrical cross-sectional area with a nominal diameter d, then the gage factor is expressed as

$$K = 1 + 2\left(\frac{\Delta d/d}{\Delta L/L}\right) \tag{10–7}$$

In both Equations 10–6 and 10–7, the ratio $\Delta L/L$ is the same as the applied strain (σ). The gage factor, as well as calibration data, is provided by the manufacturer for a specific strain gage.

EXAMPLE 10–1

Determine the resistance of a strain gage mounted on a steel bar. The gage factor is 1.5, the nominal resistance of the strain gage is 150 Ω, and the applied strain is 2.5×10^{-4}.

Solution:

From Equation 10–6, the resistance change is

$$\Delta R = (1.5)(2.5 \times 10^{-4})(150\ \Omega)$$
$$= 0.05625\ \Omega$$

so the resistance of the strain gage is 150 + 0.05625, or 150.05625 Ω.

When a single strain gage is used with the circuit of Figure 10–3, in the absence of strain, the output is never zero. As a desirable characteristic of any transducer system, the output signal should be (1) zero in the absence of an applied stimulus, otherwise, (2) proportional to the stimulus. The above two criteria are satisfied using a Wheatstone bridge circuit.

The Wheatstone bridge can be used in either of two distinct modes: as a *balanced* bridge where an unknown resistance is measured by adjusting the resistance of one of the remaining arms for zero output, or as an *unbalanced* bridge whereby an unknown resistance is measured by measuring the output voltage produced by the bridge unbalance. The balanced bridge, as discussed in Section 9–3, is primarily used for *static* measurement situations, while the unbalanced bridge is used for dynamic measurements. Most transducers based on the strain gage use the unbalanced Wheatstone bridge.

As shown in Figure 10–7, a DC Wheatstone resistance bridge has strain gages in all four arms, but only one arm is active. Without any applied strain, the resistances of all four arms are equal, so the bridge is balanced and the voltage between points A and B is zero. When a strain is applied, the resistance of the strain gage becomes $R + \Delta R$. Since the bridge circuit is actually a pair of voltage dividers, the output voltage V_{AB} is

$$V_{AB} = \left(\frac{R + \Delta R}{2R + \Delta R} - \frac{1}{2}\right)V_{\text{DC}} \tag{10–8}$$

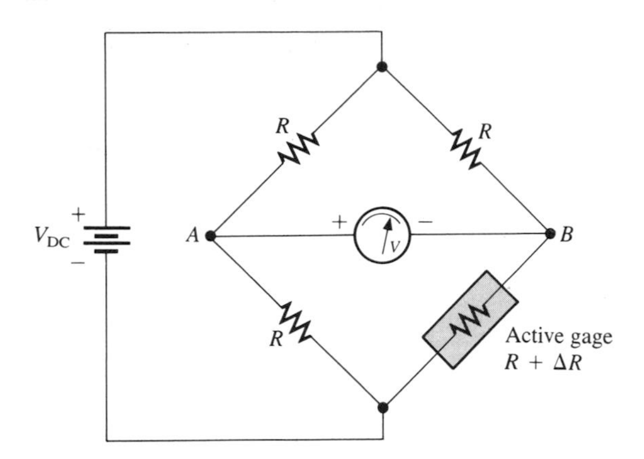

Figure 10–7 Wheatstone bridge using strain gages in all four arms, although only one arm is active.

If the resistance change ΔR is very small ($\Delta R < 0.1R$) compared to the resistance of each of the other three arms, then Equation 10–8 can be approximated by

$$V_{AB} \simeq \frac{\Delta R}{4R} V_{DC} \tag{10–9a}$$

Combining Equation 10–6 with Equation 10–9*a*, we have

$$V_{AB} \simeq \frac{KV_{DC}}{4} \text{ hr. } (\Delta L/L) \tag{10–9b}$$

The percentage error is approximately $\pm 5\%$ for $\pm 10\%$ changes in R. Consequently, the unbalanced Wheatstone bridge with one active element is accurate to $\pm 5\%$ for values of $\Delta R/R$ from -0.1 to $+0.1$. Figure 10–8 graphs both the actual output (Equation 10–8) from an unbalanced Wheatstone bridge and the ideal output (Equation 10–9*a*). The graph confirms the accuracy, though for large ratios of $\Delta R/R$, the unbalanced bridge with one active arm is substantially in error, and Equation 10–9*b* no longer valid.

EXAMPLE 10–2

A strain gage having a gage factor of 2.0 is mounted on a piece of steel 5.2 inches long and is subjected to a given amount of strain. If the strain gage is used in the bridge circuit of Figure 10–7 with each arm having a resistance of 200 Ω and a DC source voltage of 10 V, determine

(a) the change in resistance in the strain gage if the output voltage of the bridge is 5 mV,
(b) the length of the steel bar.

Solution:

(a) From Equation 10–9*a*,

$$\Delta R = \frac{(4)(5\text{ mV})(200\ \Omega)}{10\text{ V}}$$

$$= 0.4\ \Omega$$

(b) From Equation 10–6,

$$\Delta L/L = (0.4/200)/2$$

$$= 0.001$$

so that $\Delta L = (0.001) \times (5.2\text{ inches}) = 0.0052$ inches. The new length of the steel bar is 5.2052 inches.

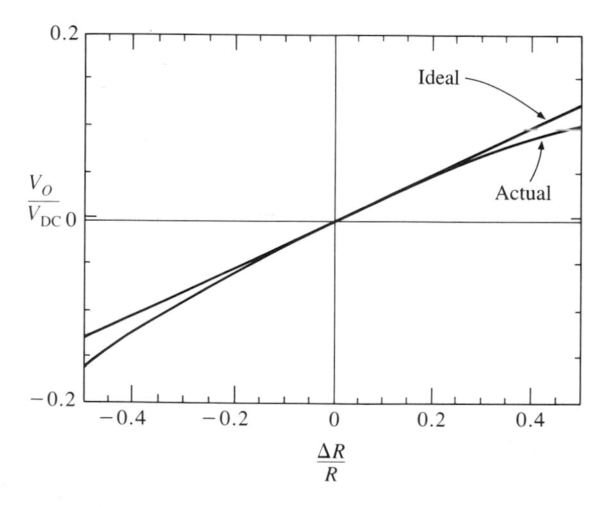

Figure 10–8 Graph of normalized output voltage (V_o/V_{DC}) of an unbalanced Wheatstone bridge as a function of the ratio of the resistance change of any one arm to its nominal value ($\Delta R/R$).

Very rarely does the Wheatstone bridge with a commercial strain gage have equal resistance in all four arms without stimulus. As a result, an offset, or error, voltage exists that should be nulled or balanced out. Figure 10–9 shows a modification of the Wheatstone bridge that is able to be balanced with zero stimulus.

Many transducer systems are arranged so that a changing physical quantity causes a resistance change in either two or all four bridge arms of the unbalanced Wheatstone bridge. Analysis of the unbalanced bridge shows that, if the physical quantity to be measured has an opposite effect on each of the two active elements in a given bridge, the output voltage is twice that of a bridge having a single active arm, or

$$V_o = \frac{\Delta R}{2R} V_{\mathrm{DC}} \tag{10–10}$$

For a bridge circuit with all four arms active, the output voltage is

$$V_o = \frac{\Delta R}{R} V_{\mathrm{DC}} \tag{10–11}$$

Strain gages mounted on a diaphragm are often used as the basis of transducers used to measure pressure, flow (in terms of pressure), mechanical vibra-

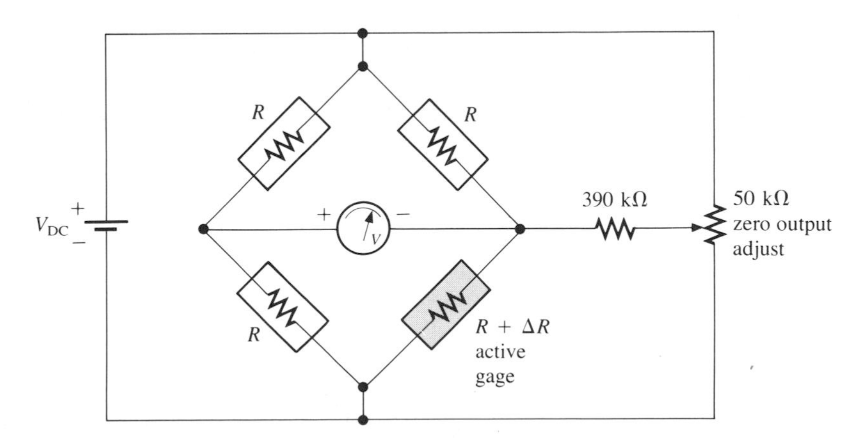

Figure 10–9 Balancing Wheatstone bridge output offset voltage with no stimulus.

tion, and deformation. Such transducers may internally consist of a Wheatstone bridge with one or more active elements; it is nevertheless unnecessary to know the characteristics of the elements of the bridge, as long as the **sensitivity factor** (F) is supplied with the transducer at the time of purchase.

The transducer sensitivity factor is normally given in terms of voltage output per voltage of excitation per unit of the physical quantity being measured. For example, a strain gage pressure transducer might produce 62 μV per 1 volt of excitation for every millimeter of mercury of pressure; the sensitivity factor is then 62 μV/V/mm Hg pressure. The output voltage of a Wheatstone bridge in terms of F is

$$\boxed{V_o = FQV_{\mathrm{DC}}} \tag{10–12}$$

where

V_{DC} = DC bridge excitation voltage

Q = transduced quantity

The sensitivity factor can be determined by subjecting the transducer to a known amount of the physical quantity the transducer is designed to measure. If the transducer is excited by a known source voltage and the output voltage measured accurately, the sensitivity factor can be determined by rearranging Equation 10–12.

EXAMPLE 10–3

A given pressure transducer has, internally, two active arms whose nominal resistance is 130 Ω each. It is excited by a 6-volt DC source. Determine the measured pressure if the transducer's sensitivity factor is 118 μV/V/cm Hg and the output voltage of the bridge is 56 mV.

Solution:

From Equation 10–12,

$$Q = \frac{56\ \mathrm{mV}}{(118\ \mu\mathrm{V/V/cm\ Hg})\ (6\ \mathrm{V})}$$

$$= 79.1\ \mathrm{cm\ Hg}$$

Photoconductive Transducers

A two-terminal device whose resistance varies inversely with the intensity of incident light is called a *photoconductive cell,* and frequently a **photo-**

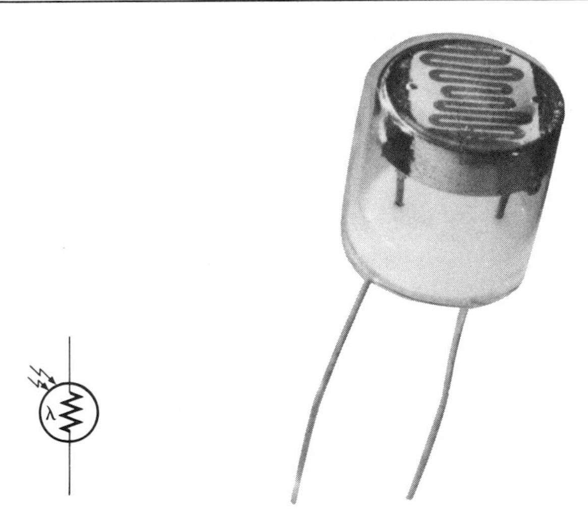

Figure 10–10 A photoconductor (photoresistor) cell.

resistor. This is not to be confused with the **solar cell,** which generates a voltage proportional to light intensity and is discussed in Section 10–5. Two of the more commonly used materials in the fabrication of photoconductive cells are cadmium sulfide (CdS) and cadmium selinide (CdSe). These are sensitive from frequencies near infrared through most of the visible light spectrum, and can respond to light intensities from as low as 0.001 to 1000 footcandles. A typical photoconductive cell and its schematic symbol are shown in Figure 10–10.

The photoconductive cell consists simply of a thin layer of material connected between two terminals. When the incident light on the device increases in intensity, the energy state of many electrons in the material also increases. This results in a larger number of free electrons and a consequent decrease in resistance. The speed of response of photoconductive cells tends to be slower than that of other types of photosensitive detectors, and, consequently, they are used for applications where high-frequency response is not necessary. A phototransistor (Section 10–9) is often a better choice for high-speed applications.

10–5 SELF-GENERATING TRANSDUCERS

Self-generating transducers, as their name implies, produce their own output voltage from an external stimulus (such as motion or heat). There is a tremendous variety of self-generating transducers in use, but many of these are merely variations of a basic circuit. The following sections discuss several of the more common types, such as thermoelectric, piezoelectric, and photovoltaic devices.

Thermocouples

As briefly discussed in Sections 6–5 and 8–4, when two wires composed of dissimilar metals are joined together in a closed loop (Figure 10–11*a*), a current flows in the loop when a junction is heated. If one junction is open-circuited, and this junction heated, as illustrated in Figure 10–11*b*, the open-circuit EMF generated between terminals *A* and *B*, known as the *Seebeck voltage,* is a function of the junction temperature and the composition of the two metals.

All dissimilar metal pairs exhibit this effect, known as the **Seebeck effect.** A transducer using this principle is called a **thermocouple,** and the small EMF voltage produced is proportional to the temperature difference between the heated junction and the cooler ends. The Seebeck voltage cannot be measured directly; if instrument leads are connected to the circuit, the leads themselves produce additional junctions, known as **cold junctions** or **reference junctions,** where they connect to the thermocouple leads. However, if these junctions are kept at a known temperature (Figure 10–12*a* and its equivalent circuit of Figure 10–12*b*), their effect can be subtracted from the total voltage, providing the hot junction voltage, so that

$$V_{\text{EMF}} = V_{\text{HOT}} - V_{\text{COLD}} \tag{10–13a}$$

$$= k_1(T_{\text{HOT}} - T_{\text{COLD}}) + k_2(T^2_{\text{HOT}} - T^2_{\text{COLD}}) \tag{10–13b}$$

where k_1 and k_2 are constants of the particular thermocouple material. The

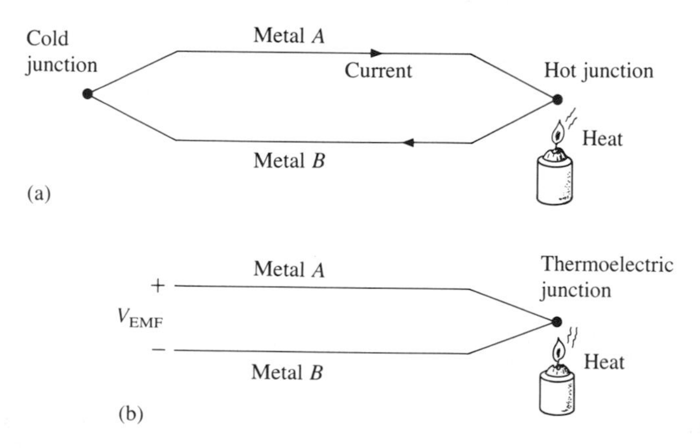

Figure 10–11 The thermoelectric (Seebeck) effect in a thermocouple formed by joining two dissimilar metals. (a) A current flows in a closed loop when one junction is heated. (b) Heating a thermocouple junction produces an EMF between the leads.

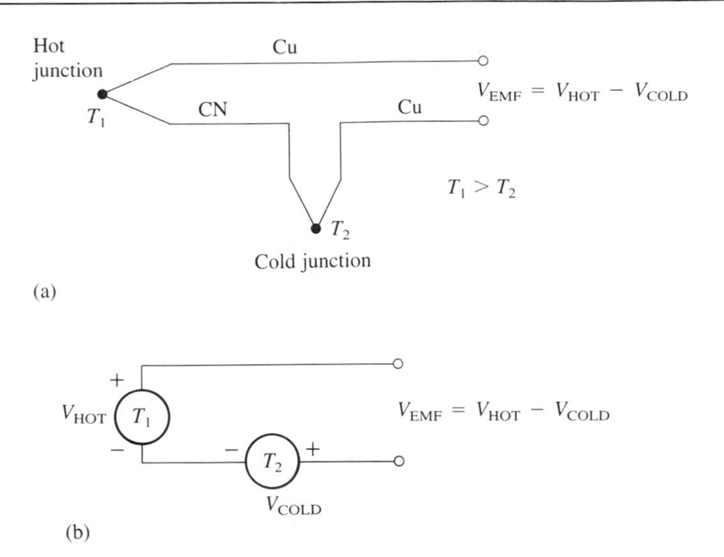

Figure 10–12 Thermocouple reference (cold) junction compensation. (a) Wiring diagram. (b) Equivalent circuit.

voltage is usually measured using either a potentiometer or a chart recorder designed for this purpose.

Another compensation technique uses an electronic circuit to measure the ambient temperature at the reference junctions, and adds a voltage to the thermocouple output equal to the voltage developed by the reference, but of opposite polarity. The resulting output from the thermocouple circuit is equal to the voltage at the measuring (hot) junction.

Table 10–1 lists a number of standard thermocouples, along with their ANSI (American National Standards Institute) letter codes. These cover the range from near absolute zero to almost 2500 °C. **Alumel** is an alloy of aluminum and nickel; **chromel,** an alloy of chromium and nickel; and **constantan** an alloy of copper and nickel. Figure 10–13 graphically compares the outputs of the various thermocouples listed in Table 10–1 as a function of temperature, when referenced to 0 °C.

The thermocouple can work in reverse, which is known as the **Peltier effect.** When a voltage is placed across terminals *A* and *B* of Figure 10–11, one junction gets hot (i.e., absorbs heat) and the other junction gets cold (i.e., loses heat).

Piezoelectric Transducers

When a mechanical pressure or vibration is applied to a crystal of barium titanate, quartz, tourmaline or Rochelle salt, the displacement of the crystals

Table 10–1 Characteristics of Common Thermocouples

Thermocouple junction materials	ANSI code	Temperature range (°C)
Platinum-6% rhodium/platinum-30% rhodium	B	38 to 1800
Tungsten-5% rhenium/tungsten-26% rhenium	C	0 to 2300
Chromel/constantan	E	0 to 980
Iron/constantan	J	−180 to 760
Chromel/alumel	K	−180 to 1250
Platinum/platinum-13% rhodium	R	0 to 1590
Platinum/platinum-10% rhodium	S	0 to 1540
Copper/constantan	T	−180 to 400

causes a potential difference. This property, known as the **piezoelectric effect,** is used in piezoelectric transducers. As illustrated in Figure 10–14, the crystal is usually placed between a solid base support and a force-transmitting member. An externally applied force exerts pressure on the crystal, producing a voltage across the crystal proportional to the magnitude of the applied pressure. The voltage produced appears across the surface at right angles to the force. This type of transducer has a good high-frequency response and is fre-

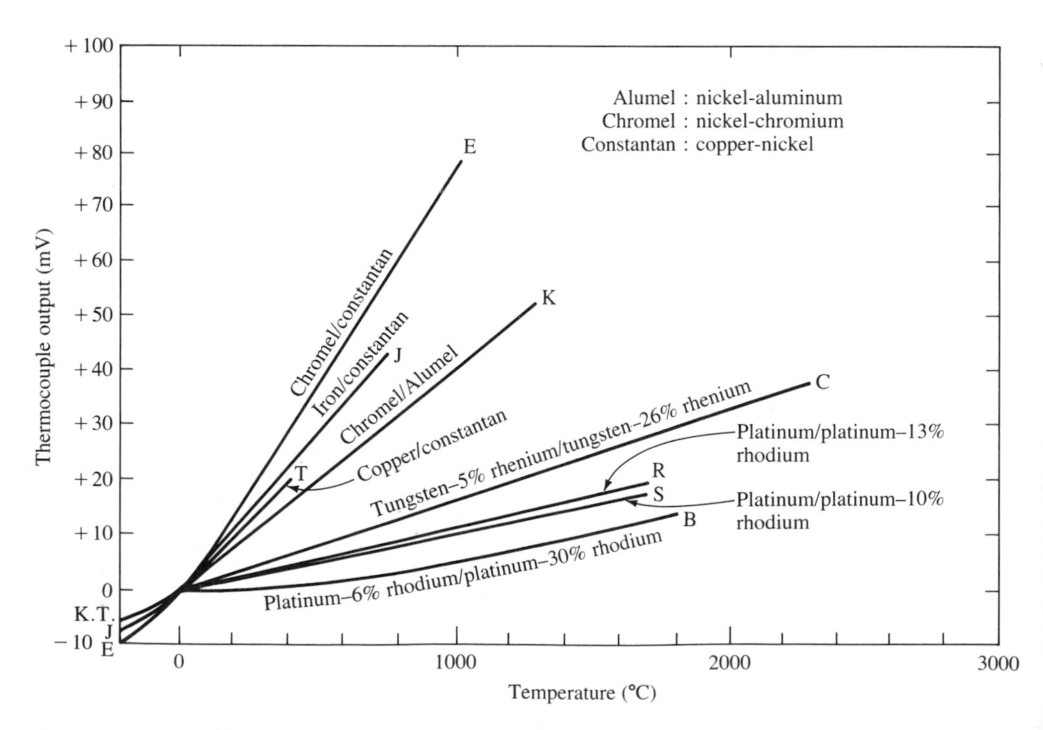

Figure 10–13 Output characteristics of thermocouples.

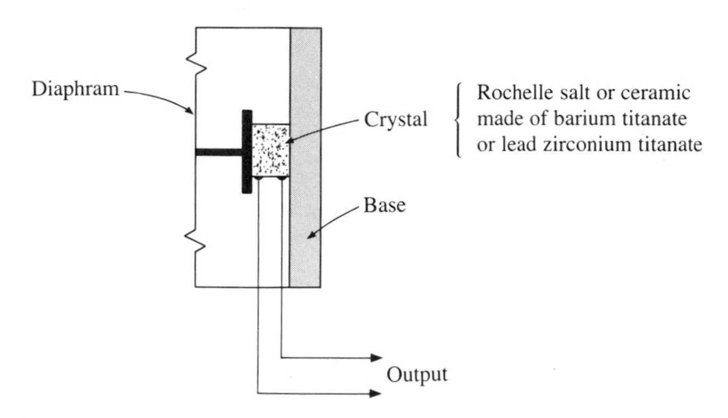

Figure 10–14 Piezoelectric transducer.

quently used both in the construction of microphones and in performing vibration measurements.

The piezoelectric transducer is not suitable for measuring static forces both because a dynamic change in the crystal structure is required, and because its output voltage is affected by temperature variations. Though it has a fairly high output impedance and can produce output voltages of up to several volts, for most applications it requires considerable amplification.

Like the thermocouple, the piezoelectric device can work in reverse. That is, when an AC voltage is applied to the crystal structure, the crystal vibrates in proportion to the signal, as does a piezoelectric speaker.

Magnetic-Induction Transducers

When a permanent magnet is mounted within a coil of wire, as shown in Figure 10–15, any movement of the magnet with respect to the coil induces a voltage across the coil. This is the principle used in the magnetic-induction transducer.

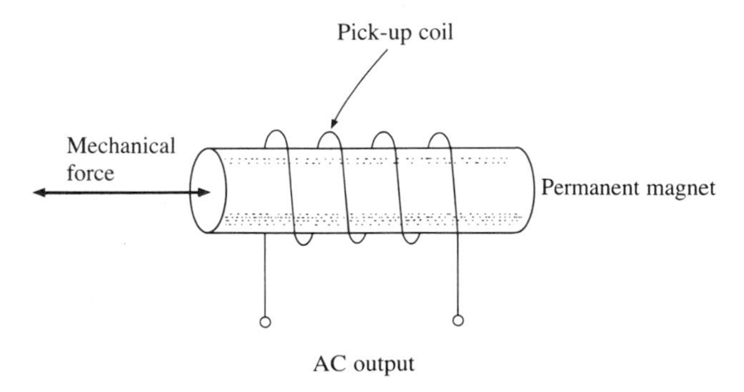

Figure 10–15 Magnetic induction-type transducer formed by moving a permanent magnet within a coil of wire.

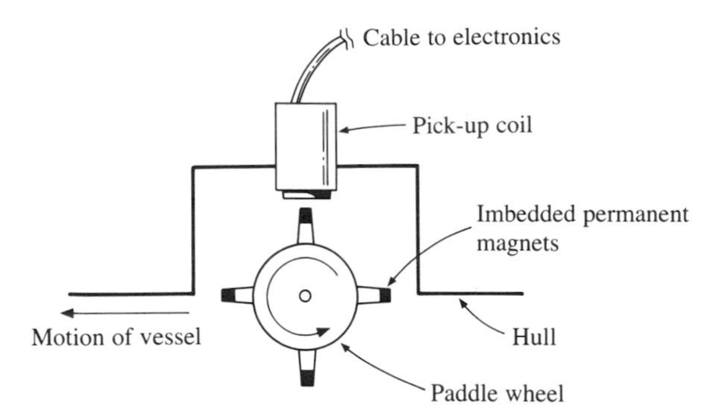

Figure 10–16 Paddle wheel magnetic transducer that measures fluid flow and the speed of boats.

In practice, the coil is usually mounted on a fixed support and the magnet driven by the mechanical force to be measured.

The output voltage is a function of both the rate and the amplitude of the relative motion. Figure 10–16 shows a variation of this principle, often used to measure both the flow in pipes and the speed of boats, in which application small magnets are mounted in the tips of nonmagnetic either turbine blades or paddle wheels. The magnets induce a voltage pulse in the coil as they are swept by it. These pulses are shaped and fed to either a frequency-to-voltage converter, driving a meter, or directly to a frequency counter circuit.

Photovoltaic Transducers

The **photovoltaic cell,** often called a **solar cell,** produces a voltage that is a function of both the color and intensity of the light falling on its surface. It is used in light meters, in the measurement of the presence of nuclear radiation, and the measurement of gas density (Figure 10–17).

The basic silicon *p–n* junction solar cell consists of a thin layer of *p*-material deposited over a thicker layer of *n*-material, forming a *p–n* junction, as illustrated in Figure 10–18. The *p*-material is so thin as to be transparent to light; this light energy on the junction region creates electron-hole pairs. The electrons are pulled to the *n*-material, and the holes to the *p*-material. This action creates a potential difference between the two sides. When an external load is connected across the device, a current, which is a function of the light intensity, flows. Thus, the photovoltaic cell is a self-generating source of EMF. As summarized in Table 10–2, the response speed of photovoltaic cells is relatively slow; they are not suitable for detecting high-speed events.

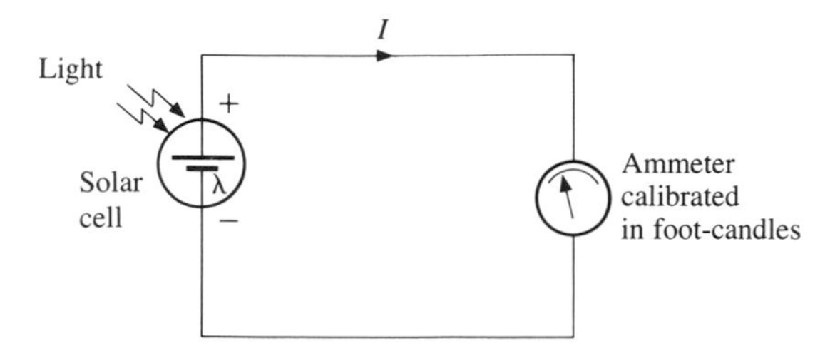

Figure 10–17 The solar (photovoltaic) cell used as a light meter.

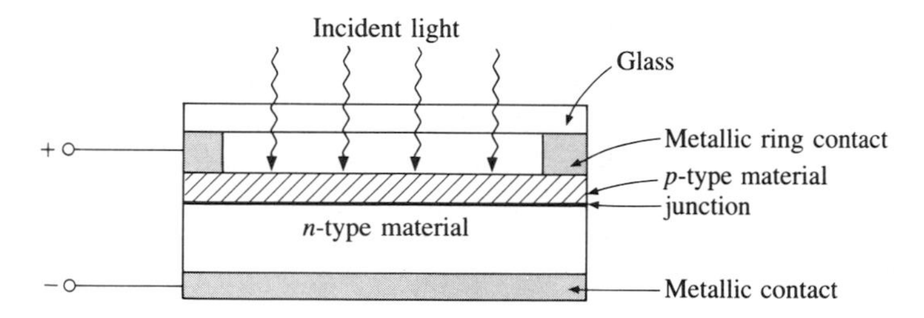

Figure 10–18 Cross-section construction of a semiconductor photovoltaic cell.

Table 10–2 Characteristics of Commonly Used Photocells

Cell material	Time constant	Peak wavelength
Selenium (Se)	2 ms	0.57 μm
Germanium (Ge)	50 μs	0.8 μm
Silicon (Si)	20 μs	1.55 μm
Indium antimonide (InSb)	10 μs	6.8 μm*
Indium arsenide (InAs)	1 μs	3.2 μm*

*Artificially cooled to increase its sensitivity in the infrared region.

10–6 ELECTROMAGNETIC FLOW METER

If a conductor were moved through a magnetic field in such a manner as to cut the lines of magnetic force, a voltage would be developed between the ends of the conductor. If the moving conductor were to cut the lines of magnetic force at right angles, the voltage induced in the conductor would be given by

$$E = BLv \tag{10–14}$$

where

E = induced EMF (V)

B = field strength (Wb/m^2)

L = actual conductor length cutting the field (m)

v = velocity of the conductor (m/s)

This principle can be applied to a conductive liquid flowing through a nonconductive pipe. Figure 10–19 shows the basic elements of a flow meter using this principle. Electrodes are mounted on the pipe perpendicular to the magnetic field, and the segment of liquid flowing between the electrodes at any given instant acts as the moving conductor. The voltage developed between the electrodes is proportional to the rate of flow.

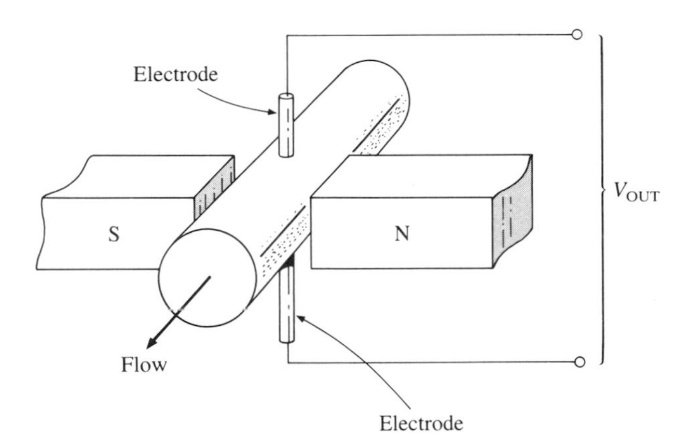

Figure 10–19 Simple electromagnetic flow meter using no moving parts.

This type of flow meter is not suitable for flow rates below about 1 ft/s or for liquids with a conductivity less than 0.1 mS/m (resistances greater than 10 kΩ/m). It does, however, offer the advantages of creating no pressure drop and requiring no direct connection with the liquid.

10–7 INDUCTIVE TRANSDUCERS

In a simple **inductance-changing transducer,** the magnetic path of an inductor is modified by the movement of an armature located in the magnetic path. An external force causes a displacement or rotation of the armature, varying the length of the air gap in the flux path and thereby causing a change in inductance. Figure 10–20 shows the elements of a simple inductive transducer.

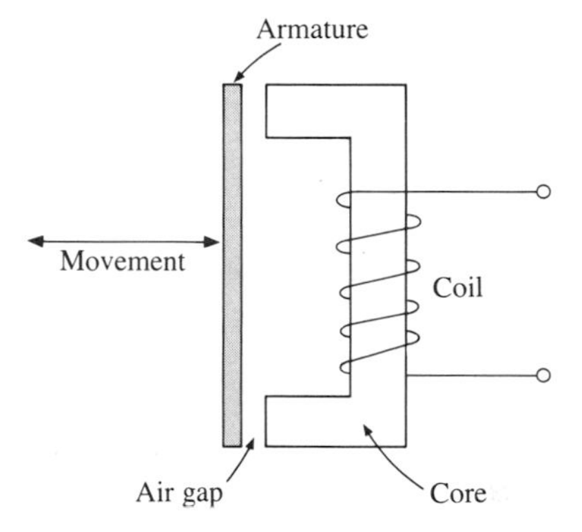

Figure 10–20 An inductive transducer that uses an armature to change the magnetic path.

A variation of this principle allowing the teeth on a rotating gear to pass through the magnetic path is frequently used as a pickup for a tachometer.

An example of the inductance-changing transducer is the **linear transformer,** which is often used to measure mechanical position. As shown in Figure 10–21, the transducer is a transformer with a secondary that can be rotated with respect to the primary. The primary is excited by a constant AC source, and the secondary is rotated by the mechanical member the position of which is to be measured. When the windings are parallel as in Figure 10–21*b*, the sec-

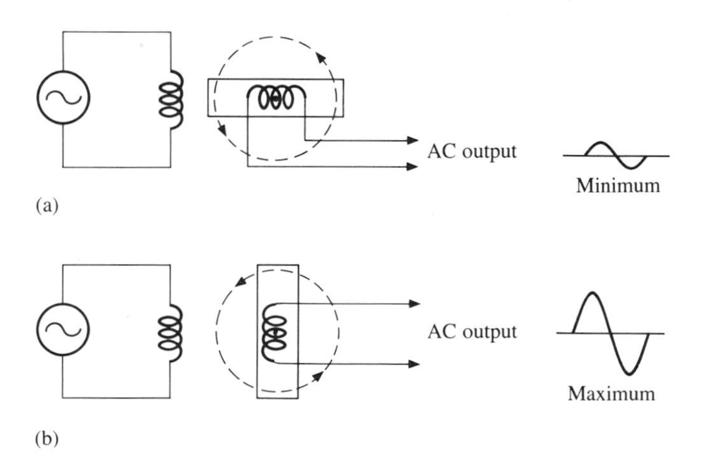

Figure 10–21 Linear transformer whose secondary winding rotates in relation to its primary winding. (a) Minimum output when windings are perpendicular. (b) Maximum output when windings are parallel.

ondary voltage is at a maximum. As the secondary rotates 90° away from the primary, the secondary voltage decreases until reaching a minimum value at 90° (Figure 10–21*a*). If rotation continues beyond 90°, the secondary voltage again starts to increase, but the relative phase between primary and secondary is reversed.

The inductance changes can be measured in two ways. The inductive transducer can be part of an unbalanced inductance bridge, like the Maxwell, Hay, or Owen bridges, so the output voltage of the bridge is proportional to the change in inductance. Otherwise, the transducer can be made part of the *LC* tank circuit of an oscillator (e.g., Colpitts oscillator) so that any inductance change modulates the frequency of the oscillator. The oscillator output must be demodulated to output useful information.

The **linear variable differential transformer,** or **LVDT,** is an inductance-changing transducer frequently used to measure *linear motion* from as little as ±0.05 in. to ±20 in. As shown in Figure 10–22*a,* the LVDT

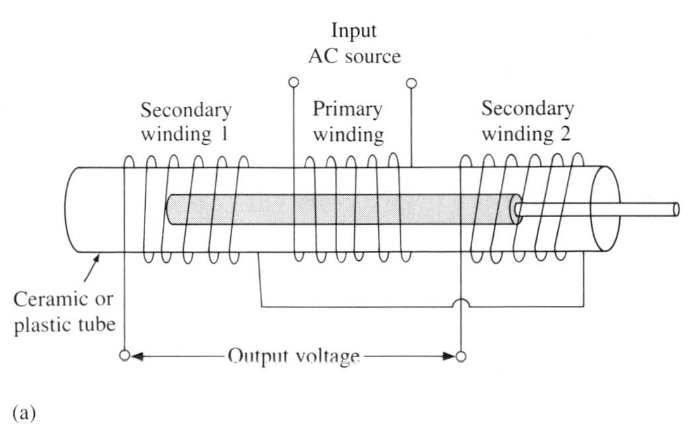

(a)

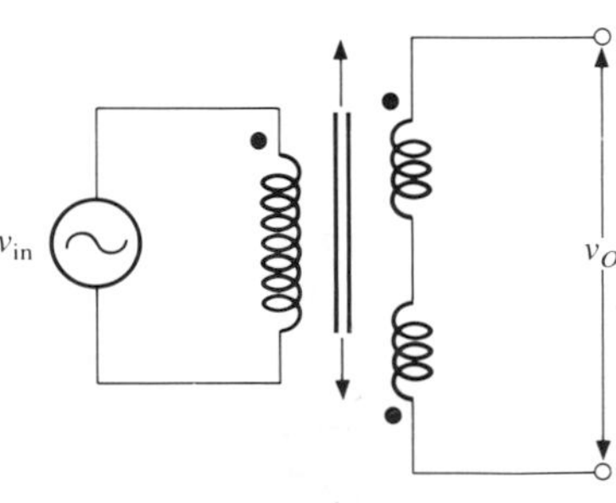

(b)

Figure 10–22 Linear variable differential transformer. (a) Windings. (b) Schematic showing secondary windings connected in series opposition.

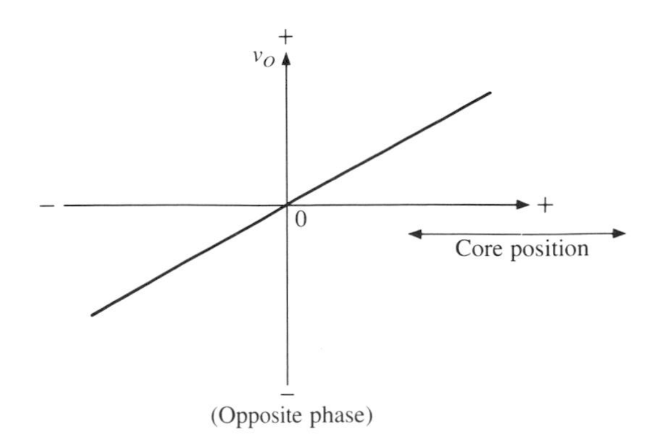

Figure 10–23 LVDT output voltage versus core displacement.

consists of a single primary and two secondaries wound over a hollow tube; the primary is located between the two secondaries. A movable cylindrical magnetic core slides inside the tube, which is mechanically coupled to the device whose position is to be measured.

Figure 10–22*b* shows that the two secondary windings are connected in series but opposing in phase. As a result, when the core is positioned exactly in the center of the tube, the voltages induced in the two secondaries are equal and out of phase by 180°. The voltage induced in one secondary winding increases and the voltage in the other decreases linearly with movement from the central position, even as little as ±0.001 inch. As graphed in Figure 10–23, the phase of the output signal indicates the direction of movement of the core, and the magnitude of the secondary voltage indicates the distance the core has moved from the center position.

Compared to other variable inductance transducers, the LVDT requires a minimum of external equipment while remaining capable of producing relatively higher output voltages (typically 50 to 250 mV per millimeter of change in core position).

10–8 CAPACITIVE TRANSDUCERS

Most capacitance-changing transducers have one movable and one fixed capacitor plate separated by a nominal distance d. The capacitance of two such parallel plates is

$$C = k\varepsilon_o\frac{A}{d} \tag{10–15}$$

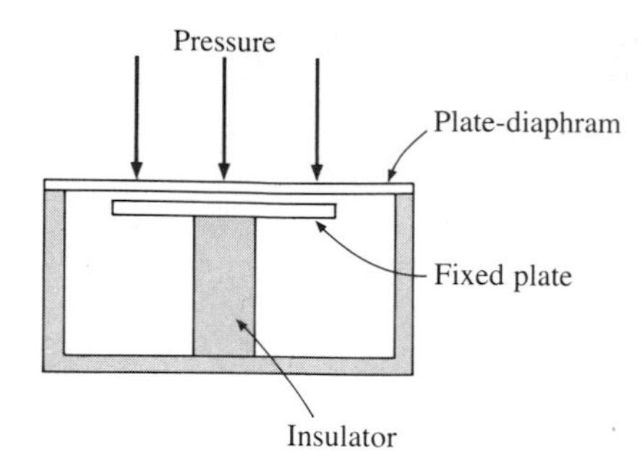

Figure 10–24 Capacitive pressure transducer construction.

where

A = area of each plate (m^2)

k = dielectric constant of the medium separating the plates ($k_{air} = 1$)

$\varepsilon_o = 8.854 \times 10^{-12}$ (F/m)

d = distance between both plates (m)

The movable plate is connected to a force-summing member. In pressure-measuring systems, a diaphragm often doubles as the movable plate. As increased pressure moves the diaphragm towards the fixed plate, the distance between the plates decreases, causing an increase in capacitance. If the pressure is decreased, the distance between the plates increases, causing a decrease in capacitance.

Figure 10–24 shows a simple capacitive pressure transducer in which the capacitance increases with pressure. This same principle is used in some microphones, since sound waves are pressure waves.

Like inductive transducers, capacitive transducers can be part of an unbalanced inductance bridge, like the series resistance, parallel resistance, or Schering bridges. The output voltage of the bridge is proportional to the change in inductance, or the transducer can be made part of either the *LC* or *RC* circuit of an oscillator, so any capacitance change modulates the frequency of the oscillator.

EXAMPLE 10–4

A capacitive pressure transducer has two parallel plates, each 15 mm square. If the dielectric is air, determine the capacitance change if the spacing between the plates changes by 1.7 mm.

Solution:

The area of each plate is

$$A = (15\text{ mm}) \times (15\text{ mm})$$
$$= 225\text{ mm}^2,\text{ or } 2.25 \times 10^{-4}\text{ m}^2$$

Therefore, the change in capacitance is

$$\Delta C = \frac{(1)(8.854 \times 10^{-12}\text{ F/m})(2.25 \times 10^{-4}\text{ m}^2)}{0.0017\text{ m}}$$
$$= 1.17\text{ pF}$$

10–9 PHOTOTRANSISTORS AND PHOTODIODES

Electromagnetic radiation, such as light, can affect the p–n junction characteristics of a semiconductor device; thus, both diodes and transistors can be made sensitive to light. When used as light-sensitive transducers, they are generally referred to as **photodiodes** and **phototransistors,** and are used in fiber optic receivers, isolators, and light-sensitive relays.

The photodiode, as shown in the circuit and load line of Figure 10–25, is normally reverse-biased. The diode is constructed with a transparent window placed over the p–n junction, allowing light to fall on the junction. The reverse current flowing in the diode is directly proportional to the light intensity striking the junction. In effect, it is similar in operation to a photoconductive cell. However, the response time of the photodiode is much faster. At maximum light intensity, the current through the diode is a maximum, equal to V/R. When no light falls on the photodiode, the current is zero and the voltage across the diode equals V.

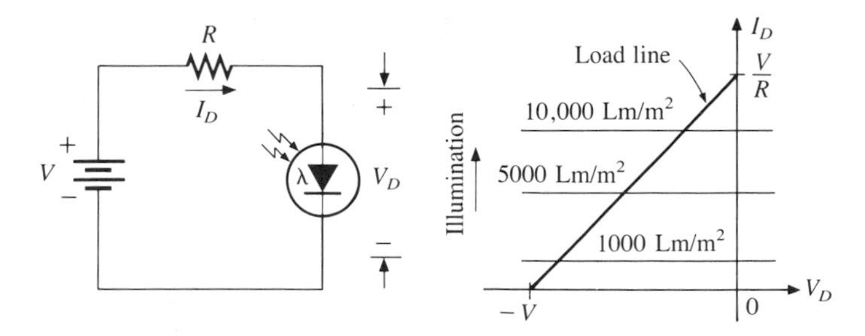

Figure 10–25 Photodiode.

By the addition of an additional p–n junction, a phototransistor has a sensitivity to light many times greater than that of a photodiode. A transparent window is placed over the phototransistor's base area (the n–p–n junctions), allowing light to fall there. Figure 10–26 shows a family of collector curves for a typical phototransistor. As the light intensity increases (similar to an increase in base current in a conventional bipolar transistor), the collector current (I_C) increases, causing the transistor's collector-to-emitter voltage (V_{CE}) to decrease, and vice versa. Consequently, I_C is directly and V_{CE} is inversely proportional to light intensity.

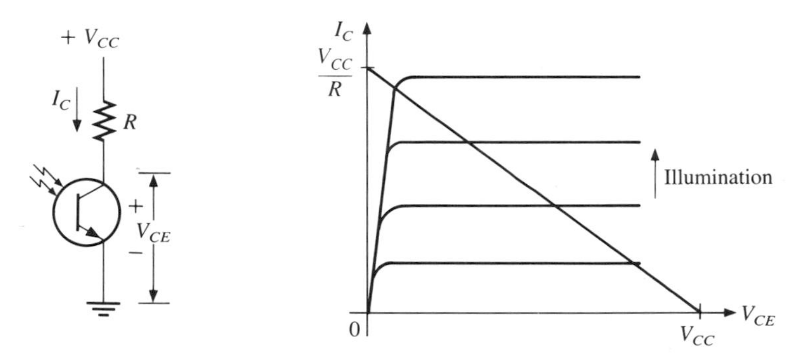

Figure 10–26 Phototransistor collector characteristics.

10–10 SUMMARY

This chapter discusses transducers that convert energy from one form to another and whose output is in the form of a useful electrical voltage or current. These transducers can be either externally powered or self-generating.

Table 10–3 Summary of Pressure Transducer Types

Transduction	Sensing element	Type variation	Excitation	Output
capacitive	diaphragm	unbalanced bridge	AC	AC
		LC oscillator	AC	frequency
		RC oscillator	AC	frequency
inductive	bellows or	*LC* oscillator	AC	frequency
	diaphragm	unbalanced bridge	AC	AC
piezoelectric	diaphragm	crystal structure	none	AC
strain gage	diaphragm	wire stretching	AC	AC
			DC	DC
		semiconductor	DC	DC

Although there are twelve different types of electrical phenomena that can be the basis for any given transducer, most transducers can be classified as being resistance-changing, self-generating, inductance-changing, or capacitance-changing. A single physical parameter can be measured by transducers from all these four categories, as illustrated by the pressure transducers listed in Table 10–3.

In operation, transducers can be used to measure a particular parameter, such as temperature, or to generate the necessary signal level controlling another device, such as a temperature controller or regulator.

10–11 GLOSSARY

alumel (AL) An alloy of 2% aluminum, 94% nickel, 3% manganese, and 1% silicon, used for thermocouples.

chromel (CR) An alloy of 10% chromium and 90% nickel, used for thermocouples.

cold junction Also called a **reference junction,** a thermoelectric junction that is maintained at a known constant temperature, so the voltage produced by a thermocouple may be measured.

constantan (CN) An alloy of 60% copper and 40% nickel, used for thermocouples.

gage factor A dimensionless measure of the ability of a given strain gage to produce a given amount of resistance change for a given amount of strain.

inductance-changing transducer A transducer whose output is proportional to the change of its self-inductance.

linear transformer An inductance-changing transducer used to measure mechanical position, whose transformer secondary can be rotated with respect to its AC excited primary.

linear variable differential transformer (LVDT) An inductance-changing transducer used to measure linear motion. It consists of two secondary windings connected in a series-opposing manner over a hollow tube. A magnetic core, coupled to the device whose position is to be measured, slides inside the tube.

Peltier effect When a voltage is applied in a thermocouple circuit, one junction becomes warmer and the other cooler. This is the opposite of the Seebeck effect.

photoconductive cell A transducer whose internal resistance varies inversely with the amount of incident light that strikes the cell's active material.

photovoltaic cell A self-generating semiconductor device which converts light intensity into electrical energy.

photodiode A semiconductor diode with a transparent window over its p–n junction, so that the diode current varies with the amount of light striking the junction.

photoresistor A photoconductive cell, the resistance of which varies inversely with light intensity.

phototransistor A bipolar transistor with a transparent window over its base region, so that the collector current varies with the amount of light falling on this region.

piezoelectric effect The generation of a voltage across a crystal of barium titanate, quartz, tourmaline, or Rochelle salt when a mechanical pressure or vibration is applied to the crystal.

reference junction See **cold junction.**

relative position transducer A transducer whose output is proportional to either a linear or angular displacement from a known reference point.

resistance-temperature-detector (RTD) A resistive material used to measure temperature by measuring the change in its resistance.

Seebeck effect When two dissimilar metals are joined to produce two junctions, an EMF is produced causing a current to flow in the loop if one junction is heated. The EMF and current vary with the temperature difference between the two junctions.

sensitivity factor A parameter of a transducer expressed in terms of voltage output per voltage of excitation per unit of the physical quantity being measured.

solar cell A semiconductor photovoltaic device that generates a voltage proportional to the intensity and color of the light falling on its surface.

strain gage A measuring device that converts strain due to pressure, force, or tension into a proportional electrical signal.

thermistor A semiconductor device the resistance of which varies with temperature.

thermocouple A device used to measure temperature by measuring the voltage generated across the junction of two dissimilar metals at that temperature.

transducer A device that converts energy from one form to another. Energy can be either thermal, electrical, chemical, mechanical, or electromagnetic radiation.

10–12 PROBLEMS

1. A 3-terminal potentiometer is used as a simple transducer to measure angular rotation using the voltage divider circuit of Figure 10–1. If the total potentiometer resistance varies from 0 to 5 kΩ when the potentiometer's shaft is rotated from 0 to 300° and the output voltage is 8.4 V when the circuit is excited with a voltage of 15 V, determine the angular rotation in degrees.
2. A slide-wire resistance displacement transducer with a maximum displacement of 1.25 inches is used in the circuit of Figure 10–1 with a 15-V DC source. What is the measured displacement if the measured output voltage is 4.295 V?
3. For the circuit of Figure 10–1, if the total potentiometer resistance is 634 Ω and the wiper arm of the slide wire is positioned so that resistance R_2 is 215 Ω, what is the output voltage if the DC excitation is 6 V?
4. The resistance of a copper resistance thermometer (i.e., an RTD) at 25 °C is 872.15 Ω. Determine the thermometer's resistance at 37 °C when its temperature coefficient is 0.038%/°C.
5. For the RTD of Problem 4, what is the temperature if the corresponding resistance is 1045.8 Ω?
6. Determine the resistance of a strain gage mounted on a steel beam if the nominal gage resistance is 176 Ω, the gage factor is 2.4, and the applied strain is 1.9×10^{-4}.
7. If the strain gage of Problem 6 is used in an unbalanced Wheatstone bridge, like that of Figure 10–7, determine the output voltage if the excitation voltage is 9 V.
8. A 145-Ω strain gage with a gage factor of 2.9 is mounted on a 10.019-in long steel rod to measure its change in length as a function of an applied strain. If the gage is used as one arm of a Wheatstone bridge with a DC excitation of 12 V, determine the length of the rod when the bridge output voltage is 8 mV.
9. A strain gage with a resistance of 154 Ω changes by 0.132 Ω when subjected to a strain of 3×10^{-4}. Determine the gage factor.
10. A transducer having a nominal resistance of 225 Ω is excited by a 5-V source and produces an output voltage of 50 mV when measuring a known pressure of 25 mm Hg. Determine the sensitivity factor of the transducer.
11. A pressure transducer has a sensitivity factor of 157 μV/V/in H_2O. Determine the output voltage when the transducer is excited by a 10-V source when measuring a known pressure of 15.7 in H_2O.

12. If the distance between the plates of a capacitive transducer increases from 0.59 to 0.82 mm, determine the corresponding percentage change in capacitance.

13. A copper-constantan (CU–CN) thermocouple, using the measurement scheme of Figure 10–12, measures an unknown temperature using an ice bath reference for the cold junction. If the thermocouple constants k_1 and k_2 are 0.042 mV/°C and 3.7×10^{-5} mV/°C^2 respectively, determine the unknown temperature when the potentiometer measures an EMF of 3.5 mV.

14. If the ice bath temperature of Problem 13 is actually 4 °C instead of 0 °C, what is the unknown temperature?

11

OSCILLOSCOPES AND PROBES

11–1 INSTRUCTIONAL OBJECTIVES

At the completion of this chapter, you will be able to

- Describe the function of the basic components of a cathode-ray tube.
- Describe the operation of both the vertical and horizontal deflection systems of a single-trace oscilloscope.
- Explain the difference between the "alternate" and "chopped" modes in a dual-trace oscilloscope.
- Describe the function and characteristics of 1X, 10X, high-voltage, current, and demodulator probes.
- Describe how an oscilloscope is used to make measurements of voltage, current, and time.
- Explain how frequency is measured using both the time-base and Lissajous-pattern methods.
- Explain how both the sweep method and Lissajous pattern can be used to measure the phase shift between two waveforms of the same frequency.
- Describe how rise- and fall-time measurements are made using an oscilloscope.

11–2 INTRODUCTION

The **cathode-ray oscilloscope (CRO)** is a universal instrument that displays waveforms on the phosphor-coated screen of a **cathode-ray tube (CRT).** Because its full name is so cumbersome, the abbreviations "oscilloscope" and "scope" are preferred by most users. The oscilloscope presents a two-dimensional graph of a signal versus either time or a second signal. The most common application by far of an oscilloscope is the display of a signal's amplitude on the vertical axis against time on the horizontal. Since the oscilloscope uses no moving parts, it can handle signals thousands of times faster than can mechanical plotters, recorders, or meters.

Although the general-purpose oscilloscope accepts voltage signals directly, other electrical and nonelectrical parameters can also be readily displayed using transducers, which convert the physical phenomenon of interest to a proportional voltage signal. There are transducers for current, power, pressure, acceleration, temperature, displacement, strain, light, and other physical quantities. This makes the oscilloscope a valuable tool in fields other than electronics. This versatility, combined with a relatively broad frequency response, makes the oscilloscope one of the most useful single items of test equipment available. It can be thought of as a "personalized" instrument, because many of its controls can be adjusted to suit the individual preferences of the operator.

Oscilloscopes come in many shapes and sizes; this chapter discusses those characteristics and features that are included in most available oscilloscopes.

11–3 OSCILLOSCOPE BASICS

Figure 11–1 shows a simplified block diagram of the basic oscilloscope. Besides a DC power supply, the basic oscilloscope has six subsystems:

(a) cathode-ray tube (CRT)
(b) vertical amplifier
(c) horizontal amplifier
(d) time base
(e) trigger circuit
(f) calibrated attenuator

The actual display is created by moving a focused beam, or a stream, of high velocity electrons across the phosphor-coated screen of the CRT. A small spot of light forms where the electron beam strikes the screen; this leaves a

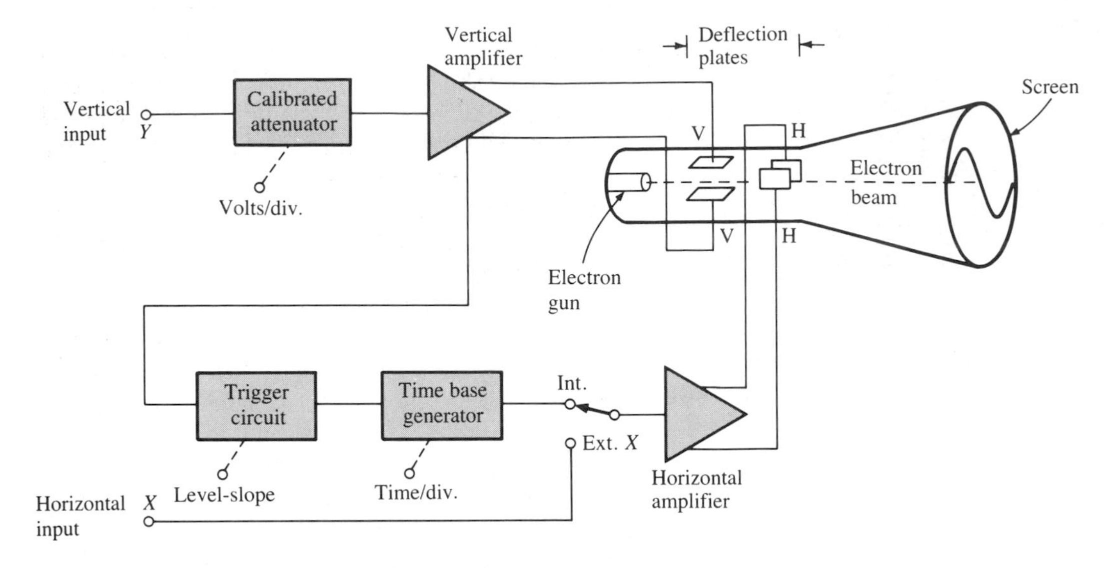

Figure 11–1 Cathode-ray oscilloscope block diagram.

glowing trail as the beam moves across the screen. The intensity of the trail, commonly called a **trace,** quickly fades unless traced over at regular intervals, as when repetitive waveforms are displayed.

Before coming to the screen, the electron beam passes first between a pair of vertical and then horizontal deflection plates. A voltage applied to the vertical deflection plates produces an electric field between them, which in turn deflects the electron beam either up or down in the vertical plane. The direction in which the beam is deflected depends on the relative polarity of the plates. Likewise, a voltage applied to the horizontal deflection plates deflects the electron beam either right or left in the horizontal plane. By simultaneously applying the proper horizontal and vertical deflection voltages, the electron beam can be directed anywhere on the face of the CRT.

Normally, the observed signal is displayed on the oscilloscope's vertical input. It first passes through a calibrated input attenuator that permits adjustment of the vertical gain. This attenuator usually has a front-panel multiposition switch calibrated in terms of *volts/division*. Then, the signal is magnified by the vertical amplifier, which has a fixed gain and a push-pull output stage that drives the vertical deflection plates with the required deflection voltage.

The **horizontal amplifier** also has a push-pull output stage, which drives the horizontal deflection plates. The input to the horizontal amplifier can be switched between two possible types of input signals. An external signal can be chosen as the input to the horizontal amplifier in a manner analogous to that of the vertical. Or, the output of an internal **time-base generator** can be selected, providing a horizontal sweep signal that is a linear function of time.

The oscilloscope's internal time-base generator, also called a **sweep generator,** provides a sawtooth waveform for the horizontal deflection of the trace. The positive-going ramp of the sawtooth is linear, and its slope is determined by a *TIME/DIVISION* front-panel control. At the beginning of the ramp, the trace is automatically positioned at the left of the screen. As the ramp voltage increases, the trace moves across the screen from left to right at a constant rate. At the end of the ramp, the trace is blanked, and the beam rapidly returns to the left side of the screen, ready for the next sweep to begin.

Figure 11–2 shows how the time-base waveform combines with the vertical input signal to produce the trace. If the input signal is periodic and the sweep always begins at the same point in the cycle, the electron beam continues over the same path during each successive sweep. This action maintains a bright, stable display.

The circuit initiating the sweep at a particular point in the waveform is called the **trigger circuit** and has two principal controls. One control is a variable *TRIGGER LEVEL*. It selects the voltage at which the input signal

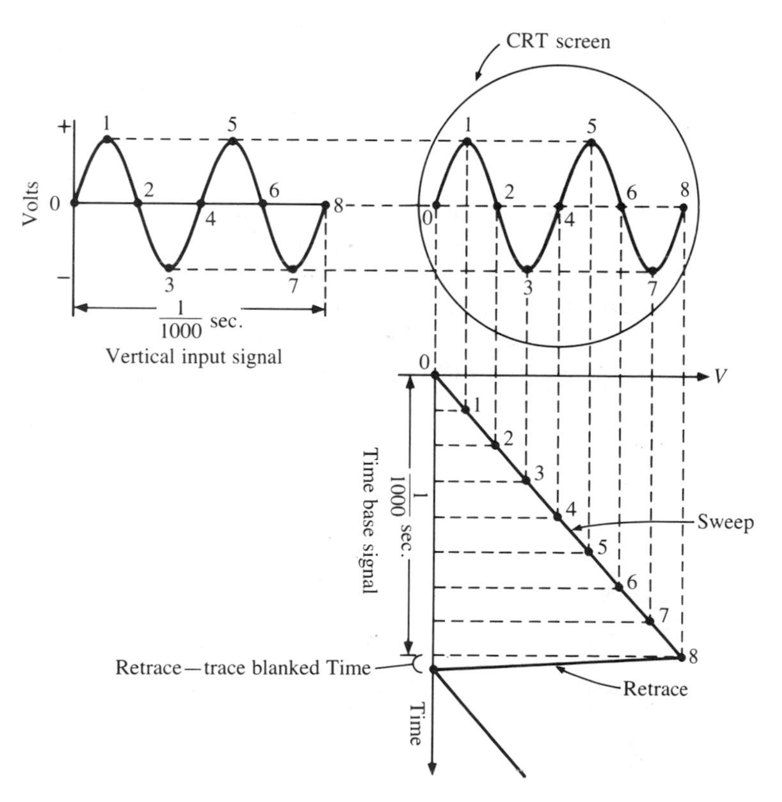

Figure 11–2 Signal versus sweep voltage.

initiates a sweep. The second control is called *SLOPE*. It determines whether the sweep begins on the positive- or negative-going slope of the waveform. The combined effects of these two controls permit the sweep to be triggered at any point in the waveform. As an example, the sweep shown in Figure 11–2 is triggered at zero volts on a positive-going slope. Faster sweep speeds cause fewer cycles of the input wave to be displayed. Obviously, slower sweep speeds display more cycles of the input waveform.

The Cathode-Ray Tube (CRT)

The construction of a typical CRT is illustrated in Figure 11–3. As shown, the CRT uses *electrostatic focus* and *electrostatic deflection,* an arrangement found in most oscilloscopes.

The directly heated cathode (*K*) releases free electrons when heated by the enclosed filament. The cathode is surrounded by the control grid (*G*), a cylinder with a small hole in its end for the passage of electrons. The purpose of the control grid is adjusting the magnitude of the electron stream that passes through it on the way from the cathode to the screen. A front panel *INTENSITY* (or brightness) control varies the negative bias voltage of the control grid with respect to the cathode. The more negative to the control grid voltage with respect to the cathode, the fewer electrons get through and the less intense the trace appears on the screen. This same principle applies to the control grid of a conventional vacuum tube.

Next, the electrons pass by the first accelerator anode (*H*), which is another disc or cylinder with a small hole in its center. The electrode is kept at a relatively high positive voltage with respect to the cathode in order to accelerate the electrons that have passed through the control grid. The focusing electrode (*F*) is a sleeve usually containing two small discs, each with a small hole. After the focusing electrode, the electrons pass through a second accelerating anode (*A*) that is biased at a high positive voltage with respect to the cathode.

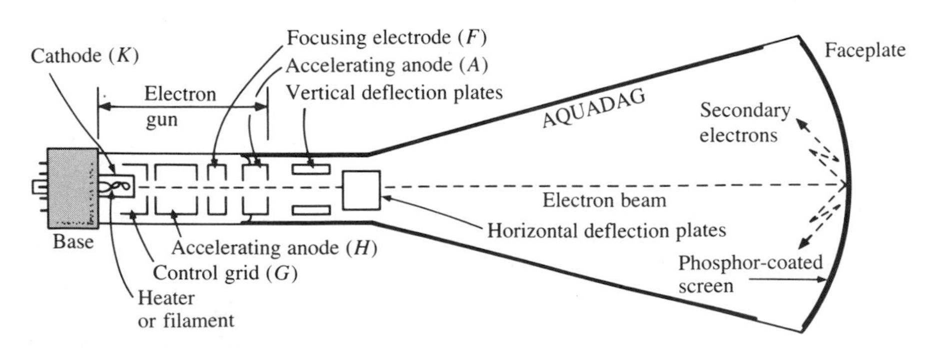

Figure 11–3 Cathode-ray tube construction.

The combination of the two accelerating anodes and the focusing electrode shapes the electric fields in the intervening gaps, forming the electrostatic equivalent of a double concave lens system. This lens system focuses the electron beam at a spot on the CRT screen. The focal point of this system is controlled by the front panel *FOCUS* control that adjusts the voltage on the focus electrode with respect to the accelerating anodes. After leaving the second accelerating anode, the electron beam passes between the deflection plates and strikes the phosphor-coated screen, where the kinetic energy of the electrons is converted to light and heat.

The inside of the conical part of the CRT is coated with a conductive material called **"Aquadag,"** which provides shielding from stray electromagnetic fields, prevents light from striking the back of the screen, and, most of all, gathers any secondary electrons emitted when the high velocity electrons of the beam strike the phosphor. An external magnetic shield made of "mu-metal" often surrounds the conical portion of the CRT, to prevent stray magnetic fields from bending or distorting the electron beam.

Electrostatic Deflection

The CRT used in modern oscilloscopes uses two pairs of electrostatic deflection plates to steer the electron beam to the desired location on the screen. One pair of plates exerts a force on the electron beam in the vertical plane, while the other pair exerts a force in the horizontal.

As shown in Figure 11–4, electrons passing between a pair of deflection plates are attracted by the positive plate and repulsed by the negative. Since electrons enter the region of the plates at a high velocity, the effect of the plates is only to bend the electron beam towards the more positive plate. The amount of deflection is directly proportional to the voltage difference between the plates, and the vertical and horizontal amplifiers provide these deflection voltages.

The deflection plates are normally driven in a *differential mode* by a push-pull amplifier. This means that one plate is driven positive, and the opposite driven negative. The positive rather than negative side of the high voltage

Figure 11–4 Electrostatic deflection system in a CRT.

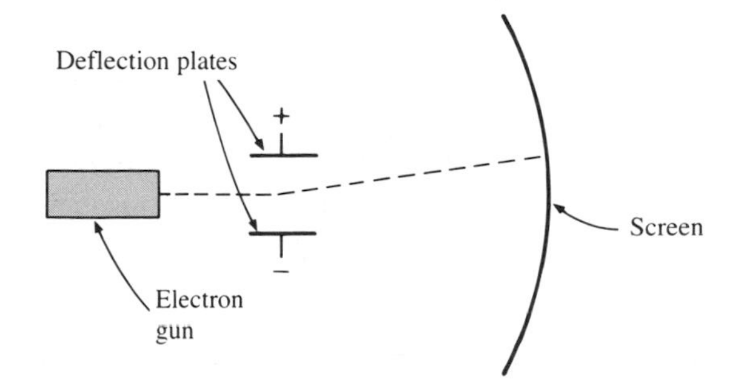

power supply is grounded, as is common practice in amplifiers, in order to permit operation of the deflection plates at a DC potential at or near ground. This greatly reduces the potential difference between the deflection plates and the second accelerating anode, which is also operated near ground potential, though at a high positive potential with respect to the cathode.

Push-pull deflection has two advantages over a single-ended system, where one plate is grounded and the opposite plate driven. First, the amount of deflection voltage needed by the deflection amplifiers is reduced by one-half. Second, a certain amount of defocusing of the electron beam is present near maximum deflection when voltage is applied only to one deflection plate. When the deflection plates are driven in push-pull, there is no defocusing, the average deflection voltage acting on the electron stream is zero, even though the net deflection voltage between the plates is twice that on either plate.

Screen Materials

The inner surface of the CRT screen is coated with a phosphor-based material that produces a spot of light when struck by a focused electron beam. The **phosphor** coating converts the kinetic energy of the beam electrons into radiant energy, which includes both light and heat. If the energy of the beam electrons becomes too great, the heat developed in the phosphor can alter its chemical characteristics and cause a permanently burned spot on the screen. It is important to keep the beam moving across the screen, never using an intensity greater than necessary for proper viewing.

The property of a phosphor to emit light when its atoms are excited is called **fluorescence.** The property of a phosphor to continue to emit light after the source of excitation is removed is called **phosphorescence. Persistence** is the term used to describe the period during which phosphorescence occurs. The human eye retains an image for about 1/16 second after seeing it. In a CRT, the spot on the screen can be moved so quickly that the path followed by the spot seems a solid line. If the beam traces over the same path at least 16 times a second, the trace appears to the eye as a stable line with very little flicker. Further, if the lighted path produced by the phosphor lingers after the beam has passed, the sweep speed and repetition rate necessary for a bright, stable display are both reduced.

Five types of phosphor coatings are commonly available for CRT applications. As summarized in Table 11–1, each type of phosphor is best suited to a particular application. The *P-1* phosphor, which is green and has a medium persistence, is used in general-purpose oscilloscopes. The *P-4* phosphor, which is white and has medium persistence, is used for television tubes, also called *kinescopes*. The *P-5* and *P-11* phosphors, both of which are blue, have a very short persistence and are used in oscilloscopes intended for photographic recording of the trace. The *P-7* phosphor produces a greenish-yellow color with a long persistence. It is used in displays for radar and sonar systems where retention of the image is required for several seconds after the electron beam has passed.

Table 11–1 Phosphor characteristics

Type	Color*		Persistence**	Use
	Fluorescent	Phosphorescent		
P1	YG	YG	M	oscilloscope, radar
P2	YG	YG	M	oscilloscope
P3	YO	YO	M	
P4	W	W	MS	television
P5	B	B	MS	photographic
P6	W	W	S	
P7	B	Y	MS(B), L(Y)	radar
P11	B	B	MS	photographic
P12	O	O	L	radar
P13	RO	RO	M	radar
P14	B	YO	MS(B), M(YO)	radar
P15	UV	G	M(YO)	radar
P15	UV	G	UV(VS), G(S)	flying-spot scanner
P16	UV	UV	VS	flying-spot scanner, photographic
P17	B	Y	S(B), L(Y)	oscilloscope, radar
P18	W	W	M-MS	projection TV
P19	O	O	L	radar
P20	YG	YG	M-MS	storage tubes
P21	RO	RO	M	radar
P22	W(R,B,G)	W(R,B,G)	MS	tricolor TV
P23	W	W	MS	television
P24	G	G	S	flying-spot scanner
P25	O	O	M	radar
P26	O	O	VL	radar
P27	RO	RO	M	color TV monitor
P28	YG	YG	L	radar
P31	G	G	MS	oscilloscope, TV
P32	PB	YG	L	radar
P33	O	O	VL	radar
P34	BG	YG	VL	oscilloscope, radar
P35	G	B	MS	oscilloscope
P36	YG	YG	VS	flying-spot scanner
P37	B	B	VS	flying-spot scanner, photographic
P38	O	O	VL	radar
P39	YG	YG	L	radar
P40	B	YG	MS(B), L(YG)	low repetition rate
P41	UV	O	VS(UV), L(O)	radar

*colors: B = blue, G = green, O = orange, P = purple, R = red, UV = ultraviolet, W = white, Y = yellow.

**persistence (to 10% level): VS = <1 μs, S = 1–10 μs, MS = 10 μs–1 ms, M = 1–100 ms, L = 100 ms–1 s, VL = >1 s.

The Graticule

A rectangular grid, called a **graticule,** as shown in Figure 11–5, is placed on or over the screen of a CRT to facilitate accurate measurements of signal voltages and time periods. The graticule can be in the form of either an inscribed plastic plate placed over the outside surface of the CRT screen or a grid either etched or silk-screened on the inside of the CRT faceplate.

Putting the graticule inside, on the same plane as the trace, eliminates measurement inaccuracies due to **parallax error,** which occurs when the trace and the graticule are on different planes and the observer is shifted from a direct line of sight. CRTs with internal graticules are more expensive to manufacture and must be provided with a means of *trace alignment*. External graticules are more easily aligned with the CRT trace and are less costly; the necessary concession is parallax error.

Graticule grids for general-purpose oscilloscopes are usually arranged in a pattern eight squares high and ten wide (Figure 11–5). Each of the eight vertical divisions represents a fixed number of voltage units, set by the attenuator *VOLTS/DIVISION* control. Each of the ten horizontal divisions can represent either time or voltage, depending on whether the internal time base or an external signal is selected as the source of horizontal deflection. When the internal time base is used, each horizontal division represents a fixed number of units of time; the horizontal scale is set by the *TIME/DIVISION* control. As a general rule, each major division both vertical and horizontal is 1 cm long.

Figure 11–5 The graticule.

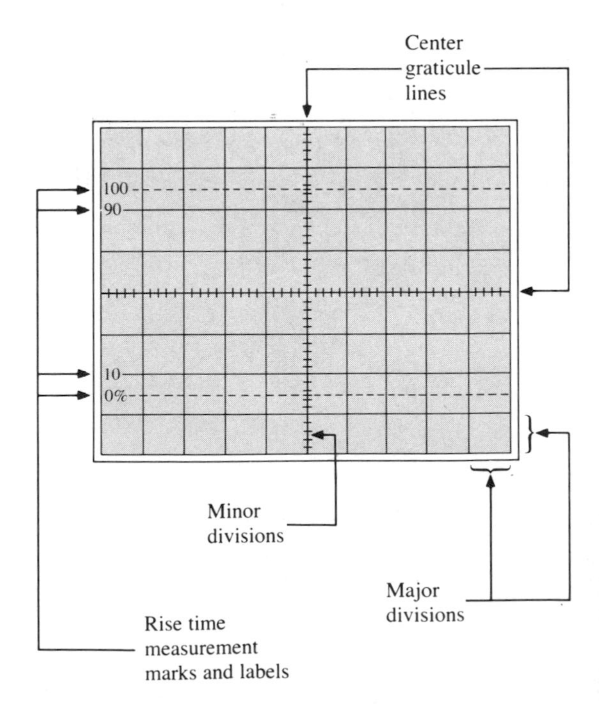

The vertical control can be also calibrated in terms of *volts/cm* instead of volts/division.

The labeling of the vertical and horizontal controls always refers to the *major divisions* on the graticule. The center vertical and horizontal graticule axes are further divided by "tick marks" into *minor divisions,* or subdivisions. There are five minor divisions to each major division; each minor division represents 0.2 times the value of a major division. For example, if a voltage has a vertical deflection of six major divisions and two minor divisions, then the total deflection is 6.4 divisions. If the vertical control is set to 0.5 V/DIV, then the voltage is

$$\begin{aligned}\text{voltage} &= (6.4 \text{ divisions}) \times (0.5 \text{ volts/divisions})\\ &= 3.2 \text{ volts}\end{aligned}$$

Accurate voltage and time measurements usually require that the trace be positioned so that measurements can be made using the minor divisions to obtain maximum resolution. Most oscilloscopes include additional graticule markings to facilitate **rise** and **fall time** measurements (Section 11–8). These can consist of two dashed lines for 0% and 100% points, in addition to a pair of labeled grid lines for the 10% and 90% levels. However, some graticules interchange the dashed and solid lines.

11–4 THE VERTICAL DEFLECTION SYSTEM

The vertical deflection system consists of a stepped input attenuator followed by several stages of amplification. It drives the vertical deflection plates of the CRT. The vertical section, or **channel,** must be capable of handling a wide range of signal amplitudes and frequencies, while introducing little or no distortion. It also must provide the same gain and phase shift for all signals within its frequency range, while providing a linear response throughout its amplitude range.

The highest frequency signal an oscilloscope can display is called its **oscilloscope bandwidth,** and is a major factor in determining its cost. Typical bandwidths range from 10 to 20 MHz, though they can be as high as 100 MHz in high-quality laboratory oscilloscopes. Closely related to bandwidth is the **rise time,** which is a measure of how quickly an oscilloscope can respond to an instantaneous change in voltage level at the input. Bandwidth (BW) and rise time (t_r) are related by a single equation, so that

$$\boxed{BW = \frac{0.35}{t_r}} \qquad (11\text{–}1)$$

Both are determined by the vertical deflection system.

EXAMPLE 11–1

If the vertical amplifier of an oscilloscope is rated at 20 MHz, determine the fastest rise time a square wave can have, displaying the waveform on the oscilloscope's screen without distortion.

Solution:

From Equation 11–1,

$$t_r = \frac{0.35}{20 \text{ MHz}}$$

$$= 17.5 \text{ ns}$$

As shown in Figure 11–6, the vertical deflection system can be subdivided into the following four functional units:

(a) input selector switch
(b) input attenuator
(c) vertical amplifier
(d) probe

Probes come in a variety of forms and are intended for specific applications; they are discussed separately in Section 11–7.

Input Selector Switch

The input selector switch permits the operator to select how the input signal is to be coupled to the vertical amplifier. It has three positions marked: AC, DC, and GND. The DC, or direct current, position couples the input signal directly

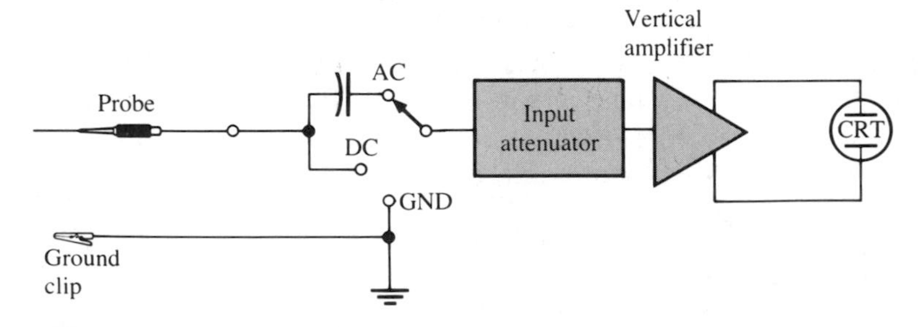

Figure 11–6 Vertical deflection system.

to the input attenuator and permits viewing of all components of a waveform simultaneously, including any DC component that may be present. The AC, or alternating current, position inserts a capacitor between the input connector and the input attenuator. The capacitor effectively blocks any DC component of the input signal from reaching the attenuator, permitting only the time-varying portion of the waveform to be seen. This position is particularly useful when a signal has a small AC component in relation to a large DC component, and it is desired to view the signal's AC component at a higher amplification than the large DC component would permit were it present. The *GND,* or *ground*, position disconnects the input connector completely, grounding the input of the attenuator. This feature is used to establish a ground or **baseline** reference level for the displayed trace in the form of a straight horizontal line. It does not, however, ground the input signal.

Input Attenuator

The **input attenuator** provides a calibrated means of controlling the gain of the vertical channel. Consequently, the input attenuator permits the oscilloscope to handle signals ranging from levels of millivolts to hundreds of volts by reducing the amplitude of an input signal, causing it to fall within the input range of the vertical amplifier. The input impedance the vertical attenuator presents to the signal connected to the vertical input is typically about 1 MΩ, shunted by a capacitance ranging from 30 to 40 pF. This loading effect can be further minimized by the use of special probes that increase input impedance but usually reduces gain.

The input attenuator is usually a number of frequency-compensated voltage dividers manually selected by a front panel *VOLTS/DIVISION* switch. The steps are normally in multiples of a "1–2–5 scale" arrangement. This means that if the lowest setting (i.e., highest sensitivity) of the input attenuator is 10 mV/division, higher settings (i.e., lower sensitivities) are 20 mV/division, 50 mV/division, 0.1 V/division, 0.2 V/division, and so on.

Figure 11–7 shows one section of a frequency-compensated voltage divider. C_i represents the input capacitance of the vertical amplifier, and C_v is an

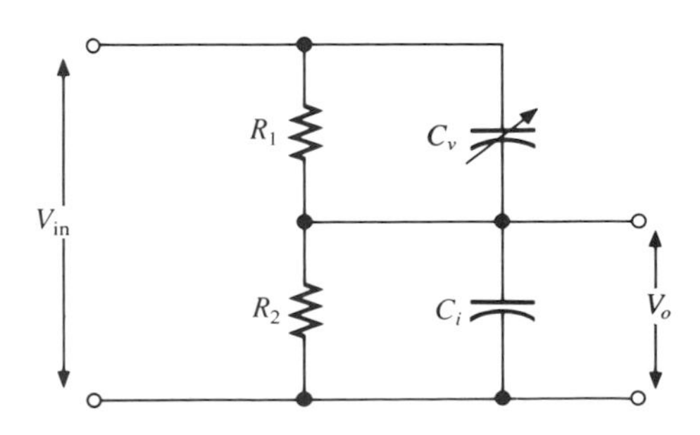

Figure 11–7 Frequency compensated voltage divider (attenuator).

adjustable capacitor that forms one leg of a balanced bridge. For AC signals, the output voltage v_o is related to the impedances Z_1 and Z_2:

$$v_o = v_{\text{in}} \frac{Z_2}{Z_1 + Z_2} \tag{11–2}$$

where

$$Z_1 = R_1 \parallel X_{C_v}$$
$$= \frac{1}{1 + 2\pi f R_1 C_v}$$

and

$$Z_2 = R_2 \parallel X_{C_i}$$
$$= \frac{1}{1 + 2\pi f R_2 C_i}$$

For DC signals however, the capacitive reactance is infinite, so that Equation 11–2 simplifies to

$$v_o = v_{\text{in}} \frac{R_2}{R_1 + R_2} \tag{11–3}$$

If the capacitance values are not in proportion to the resistance values, the output of the voltage divider is not the same at all frequencies. For the frequency-compensated voltage divider to have a constant output voltage over a wide frequency range, the relationships are

$$R_2 = \frac{R_1}{\alpha - 1} \tag{11–4}$$

and

$$C_v = \frac{C_i}{\alpha - 1} \tag{11–5}$$

where

$$\alpha = \frac{v_{\text{in}}}{v_o}$$

Bridge balance exists when $R_1C_v = R_2C_i$. The proper setting of C_v is usually determined by applying a square wave to the input and observing the output on the CRT screen. When C_v is adjusted properly, the output waveform is identical to the square-wave input signal. Otherwise, the square wave is distorted. The technique for compensating the input attenuator's frequency response is the same for oscilloscope probes. The effects of overcompensation and undercompensation are further discussed in Section 11–7.

EXAMPLE 11–2

For the frequency-compensated voltage divider of Figure 11–7, with

$$R_1 = 990\ \text{k}\Omega$$
$$R_2 = 10\ \text{k}\Omega$$
$$C_i = 7\ \text{pF}$$
$$C_v = 100\ \text{pF}$$

determine the output voltage when the input voltage is

(a) a DC voltage of 1 V,
(b) an AC voltage of 1 V peak at a frequency of 250 kHz.

Solution:

(a) For the DC signal, the output voltage is determined from Equation 11–3,

$$v_o = \frac{10\ \text{k}\Omega}{990\ \text{k}\Omega + 10\ \text{k}\Omega}(1\ \text{V})$$
$$= 0.01\ \text{V}$$

(b) For the AC signal, capacitive reactances X_{Ci} and X_{Cv} are 90,992 Ω and 6369 Ω respectively. Using equation 11–2,

$$v_o = \frac{3891\ \Omega}{83{,}333\ \Omega + 3891\ \Omega}(1\ \text{V})$$
$$= 0.045\ \text{V peak}$$

Since this voltage divider design does not respond equally to DC and AC signals, C_v would have to be adjusted to 1.01 pF (from Equation 11–5), so that the output voltage would be 1/100 times the input voltage at all frequencies, as is the case with DC.

The Vertical Amplifier

The **vertical amplifier,** or ***Y*-axis amplifier,** is a fixed-gain system that takes the output of the vertical attenuator and brings it up to the level required to drive the vertical deflection plates of the CRT. Like the vertical attenuator, the vertical amplifier must have a very flat, or constant, gain and phase response across the entire specified bandwidth of the oscilloscope.

A typical general-purpose oscilloscope CRT can require between 10 and 100 volts for each centimeter, or major division, of deflection. Therefore, the vertical amplifier would have to provide a gain of between 10,000 and 100,000 if the amplifier were to provide an overall maximum sensitivity of 1 mV/division. This means that, for an oscilloscope with a maximum vertical amplifier sensitivity of 1 mV/division (and, since there are eight vertical divisions on the graticule, four divisions above zero and four below), the vertical attenuator must be set to reduce the observed signal to a range of at most 4 mV lest the trace be deflected beyond the limits of the screen.

To achieve a high-gain, high-amplitude signal output, the vertical amplifier is often divided into two sections: a *preamplifier* and a *main amplifier.* The preamplifier, which usually has an FET input stage, provides high input impedance and most of the overall gain. The main amplifier provides the high-amplitude voltage swings necessary to drive the deflection plates and has a push-pull output stage to maintain deflection linearity. In some laboratory oscilloscopes, the preamplifier is a modular plug-in unit that is easily interchanged to permit altering of the vertical section for specific applications.

Changing the setting of the input attenuator switch changes in turn the vertical amplifier gain, or *sensitivity,* and, as a result, the number of volts per each major vertical division of the graticule. An additional vertical amplifier control, usually marked *VARIABLE,* provides a means by which the gain of the vertical channel can be varied smoothly between the calibrated steps of the input attenuator. This is used externally to adjust the calibration of the vertical amplifier. However, the calibration of the input attenuator switch is accurate only when the VARIABLE control is turned to the *CAL,* or *calibrated,* position.

Also associated with the vertical amplifier is a POSITION control, which allows the user to position the trace at any convenient vertical position on the screen, or to adjust the baseline at a specific graticule mark.

11–5 THE HORIZONTAL DEFLECTION SYSTEM

The horizontal deflection system requires the generation of a sweep signal, its synchronization or triggering, and its magnification. Each of these functions is under a control independent from the others.

The Sweep Generator

The *sweep generator,* or *time-base,* generator, as it is often called, permits the oscilloscope to display a vertical input signal as a function of time. To accomplish this, it is necessary to sweep the trace across the screen from left to right at a constant velocity. This requires a deflection voltage that increases linearly during the sweep, and, at the end of the sweep, drops back rapidly to its initial value, returning the trace to its starting position. The process of returning the trace to the left side of the screen is called **retrace,** or **flyback.** A *blanking* circuit turns off the electron beam during this time to prevent the retrace path from being displayed.

Figure 11–8 shows the waveform of a triggered sweep generator. Initially, the trace is at the extreme left side of the CRT screen, and the sweep voltage is at its minimum value, V_{min}. When a trigger signal is received, at $t = 0$, the generator produces voltage ramp increasing linearly to a maximum value (V_{max}), which causes the trace to move across the screen from left to right. At the maximum sweep voltage, the trace is at the extreme right side of the CRT screen. The beam is "blanked" while the sweep voltage rapidly returns to the minimum value, the flyback or retrace period, to await the next trigger signal.

Increasing the sweep rate increases the slope of the ramp voltage. Similarly, decreasing the sweep rate decreases the slope of the ramp. In either case, the maximum and minimum values of the ramp remain constant. The *sweep rate* is controlled by a *TIME/DIVISION* control switch like that of the vertical input attenuator but calibrated in the units of time required for the trace to

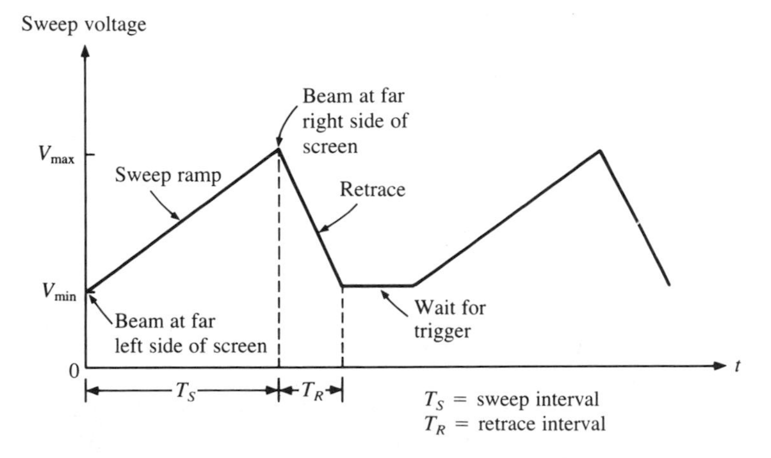

Figure 11–8 Triggered sweep generator waveform.

travel one major horizontal graticule division, such as 0.5 μs/DIV or 2 ms/DIV. Typical sweep accuracies are 3 to 5%.

Sweep generators usually produce a sawtooth waveform by taking the voltage across a capacitor as it is charged from a constant current source. By definition, the voltage across a capacitor is equal to the charge (Q) on the capacitor (in coulombs) divided by the capacitance (C, in farads):

$$V = \frac{Q}{C} \tag{11-6}$$

A constant-charging current causes Q to increase at a linear rate; and since C is a constant, the voltage across the capacitor also increases at a linear rate.

Figure 11–9 shows such a simple sawtooth generator scheme. When the transistor switch is turned on (i.e., saturated), the capacitor is discharged, and current is diverted around the capacitor. When the transistor switch is turned off (i.e., at cutoff), the capacitor is charged from the constant current source, and the voltage across the capacitor increases linearly.

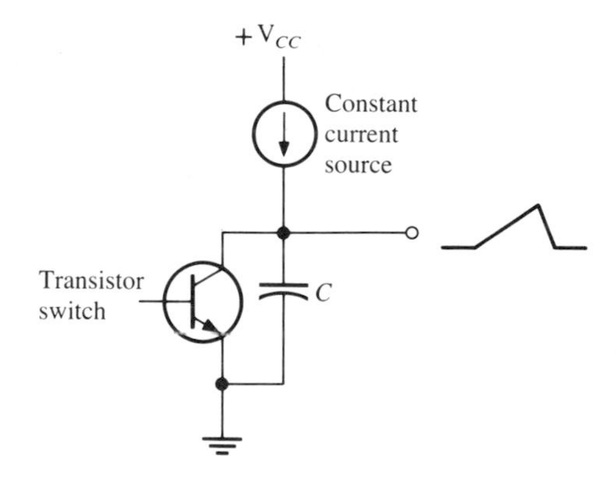

Figure 11–9 Circuit for a simple sawtooth generator.

Sweep Synchronization

To present a stationary display on the CRT screen, the sweep signal frequency must either be equal to or a submultiple of the frequency of the vertical input signal. If the sweep signal is free-running and not synchronized with the signal being viewed, the trace drifts across the screen, making viewing the waveform difficult or impossible.

All but the simplest oscilloscopes use a trigger circuit to initiate the sweep at a particular point on the trigger source signal (usually the signal being observed). The point selected corresponds to a certain value of voltage and slope. As long as the signal being observed is repetitive, the trace starts at the same point in a cycle each time and traces over the same path, resulting in a bright, stable display of one or more cycles.

There are four basic controls associated with the trigger circuit:

(a) level control
(b) slope selection
(c) source selection
(d) mode selection

The *TRIGGER LEVEL* control sets the reference level of a voltage comparator. The other input to the comparator is the signal used as a trigger source. This control determines the voltage at which the sweep begins. Figure 11–10 illustrates the relationship between the trigger and sweep signals.

The *SLOPE* control determines whether the level circuit initiates the sweep, as the trigger source signal passes through the selected level, in a positive-going or negative-going direction, as shown in Figure 11–11.

The *SOURCE SELECTION* switch is usually marked *TRIG SOURCE*. As shown in Figure 11–12, it permits the sweep to be synchronized to one of several different sources, which include the signal being observed, an external signal, and the 60-Hz power-line frequency. The corresponding positions of the source-selector switch are generally labeled *INT* (internal); *EXT* (external); and *LINE* (60-Hz power line). For most applications, the sweep is synchronized to the signal being observed.

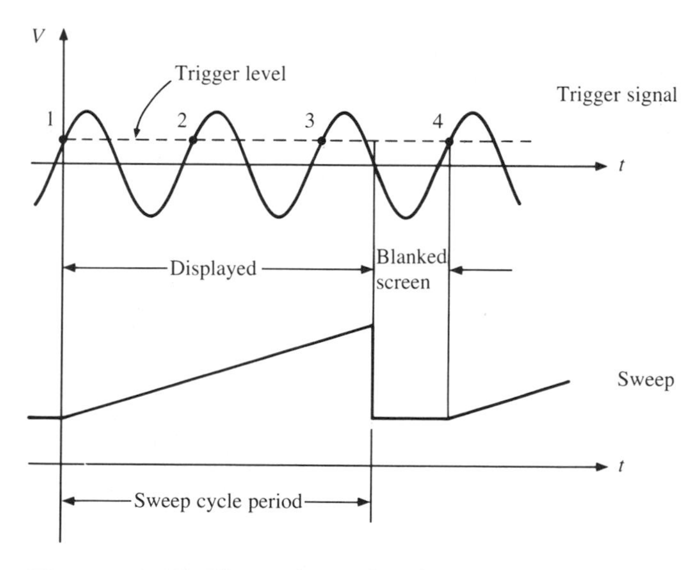

Figure 11–10 Time relationship for a sweep rate that is one-third the trigger signal frequency. Trigger points 1 and 4 initiate the sweep period while points 2 and 3 are ignored.

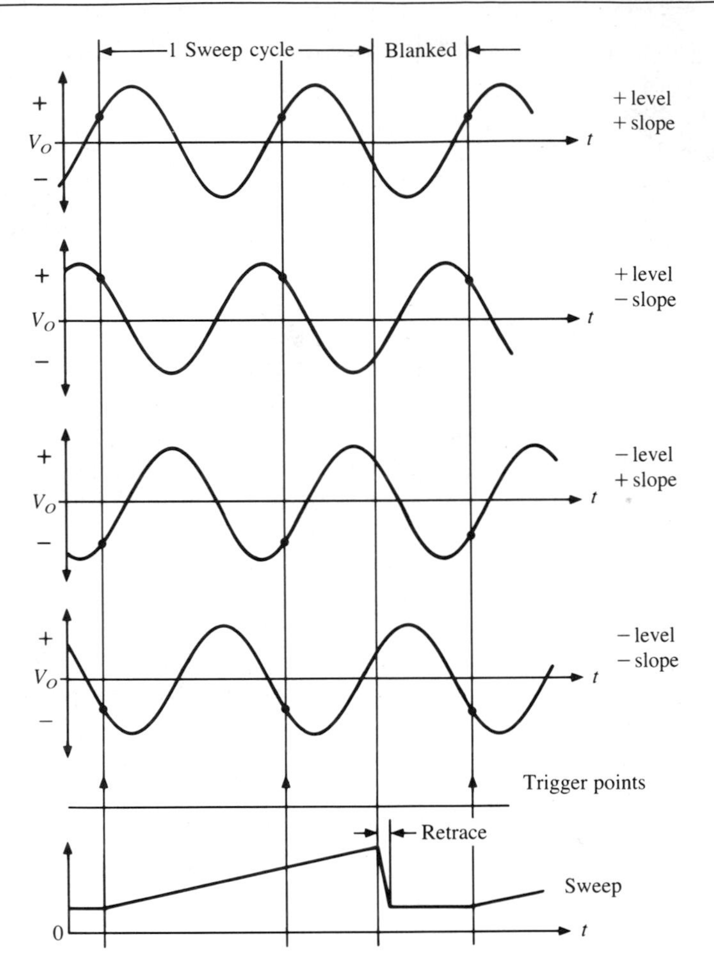

Figure 11–11 Combined effects of slope and level settings on an oscilloscope display.

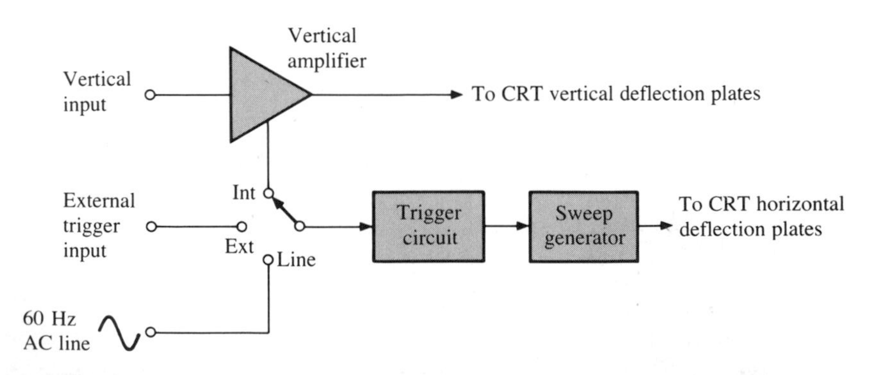

Figure 11–12 Trigger source selection.

Figure 11–13 Effect of 5X sweep magnification.

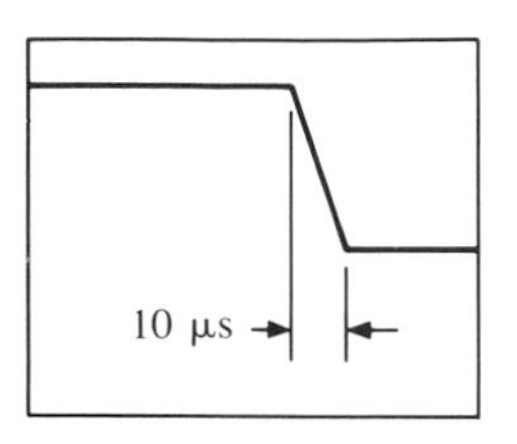

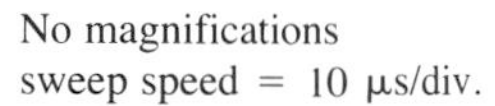
No magnifications
sweep speed = 10 μs/div.

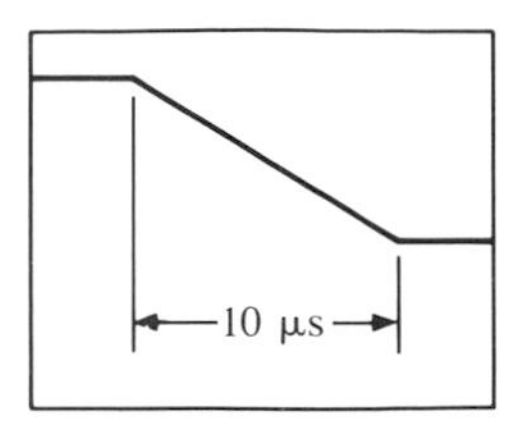

Magnified sweep = ×5
sweep speed = 2 μs/div.

The **MODE** selection switch has two positions: **NORMAL** and **AUTOMATIC.** In the *NORMAL* mode, the trigger point is determined by both the level and slope controls. Absence of a trigger signal source stops the sweep, resulting in a blank display. In the *AUTOMATIC* mode, the trigger generator automatically provides a trigger signal to the sweep generator when no vertical input signal is present. This provides a bright base line reference in the absence of a trigger signal. Most oscilloscopes in the *AUTOMATIC* mode also give a stable display if a signal is present, but the range of level adjustment and range of frequencies over which reliable triggering can be maintained is reduced. Normally, the trigger system does not respond to a triggering signal during either the sweep or retrace periods, and for a short time afterwards, called the **holdoff period.**

Sweep Magnification

Most oscilloscopes provide a means of magnifying the horizontal component of their display, expanding a small portion of the normal display along the time axis for a more detailed observation of a segment of the waveform. In effect, sweep magnification multiplies the sweep speed by increasing the gain of the horizontal amplifier. This feature is accomplished by the SWEEP MAGNIFICATION control, which provides an increase in the gain of the horizontal amplifier by a factor of five or ten. The sensitivity of the horizontal amplifier must be divided by this factor to give the correct reading. For example, if the time base is set at 10 ms/division and the sweep magnification expands the display by a factor of five, then each horizontal division is now 10 ms/5, or 2 ms. The effect of 5X sweep magnification is illustrated in Figure 11–13.

11–6 ADVANCED OSCILLOSCOPE FEATURES

Besides the basic features, many oscilloscopes have advanced features that enhance their flexibility. Delayed sweep, dual-trace display, input signal sampling, and delay lines are discussed in the following sections. Storage oscilloscopes are discussed in Section 11–9.

Delayed Sweep

Delayed sweep is a feature available on some laboratory oscilloscopes that facilitates the measurement of the rise time of a portion of a signal that does not occur at the beginning of a cycle. As a result, it adds a precise time delay between the trigger point and the beginning of the sweep. With delayed sweep, it is possible to trigger the sweep at any convenient point on the waveform being observed, using the *DELAY TIME* control to start the sweep at some later time closer to the point of interest.

Dual-Trace Oscilloscopes

A dual-trace oscilloscope has the capability of accepting two vertical input signals and displaying them simultaneously as two *separate* traces with respect to the same time axis on the CRT screen. This allows waveforms to be compared in terms of frequency, phase, and timing. A dual-trace oscilloscope is said to have two *channels,* often referred to as *CH–1* and *CH–2*. Each channel has its own vertical gain and vertical position controls and is independent of the other channel.

Two basic techniques are used in the construction of dual-trace oscilloscopes. The first, called a **dual-beam CRT,** uses a special CRT that actually produces two electron beams in a single CRT. The oscilloscope using a dual-beam CRT has two completely separate vertical attenuators, vertical amplifiers, and deflection systems, and has two full sets of vertical deflection plates. A single horizontal deflection system is common to both beams, to give them a common time base.

The more common, and less expensive, type of dual-trace oscilloscope is one that time-shares a single electron beam between its two trace paths. In this method, two separate input attenuators and vertical preamplifiers are electronically switched to feed alternately a common vertical amplifier, which in turn drives a pair of vertical deflection plates. The electronic switching circuit also provides a DC offset to each trace to provide vertical separation between the two traces.

The electronic switch has four modes of operation. It can display only Channel 1, only Channel 2, or display both input signals simultaneously when operated in either the **chopped** or **alternate modes.** In the chopped mode, the single beam is rapidly switched back and forth between the two traces in a vertical fashion, tracing each channel's signal out as a dashed line. The chopped mode is best suited for medium- and low-frequency signals, so that the short breaks in the traces become invisible on the display.

In the alternate mode, the beam first traces out one complete sweep for channel 1, say, in the upper half of the screen. The vertical amplifier is electronically switched to the other input channel, and the beam traces out one complete sweep for Channel 2 in the lower half of the screen. The two channels are repeatedly traced in this alternating pattern. The repetition rate is normally so high that both traces appear to be on the screen simultaneously. The alternate mode is usually used for viewing *high-frequency signals* where the sweep speed is much faster than phosphor decay time. However, were two high-frequency signals displayed in the chopped mode, the short breaks in the two waveforms resulting from the switching action would be visible.

As a general practice, since the controls for channel 1 are physically located above those for channel 2 on the front panel of the oscilloscope, the channel 1 trace should be positioned *above* the channel 2 , in order to avoid confusion as to which trace corresponds to which channel. Also, the vertical spacing between the two traces in either the chopped or alternate mode can be reduced to zero, using the POSITION control of either channel, to superimpose one trace on top of the other for direct comparison of amplitude or phase, when both are referenced to a common baseline.

Sampling Oscilloscopes

Standard oscilloscopes are available that have vertical amplifier bandwidths up to 100 MHz. At very high frequencies, however, the writing speed of the electron beam becomes so fast that it does not transfer sufficient energy to the phosphor coating to create a bright trace as it moves. In addition, the bandwith requirement of the vertical amplifier becomes extremely large, and thus difficult to obtain.

In order to view very high-frequency, repetitive waveforms, oscilloscopes have been developed that make use of a *sampling technique*. A low-frequency representation of the observed signal is synthesized from samples of the input signal taken over many hundreds of cycles. As shown in Figure 11–14, discrete samples of signal amplitude are taken during recurrent cycles, each succeeding sample being taken *slightly later* in the present cycle than in the one before. As each sample is taken, the beam is moved slightly to the right and is deflected vertically to the value of the new sample voltage. In this fashion, a dotted representation of the original signal is created, but at a much lower frequency. If the number of dots displayed is sufficiently large, the trace appears to be continuous.

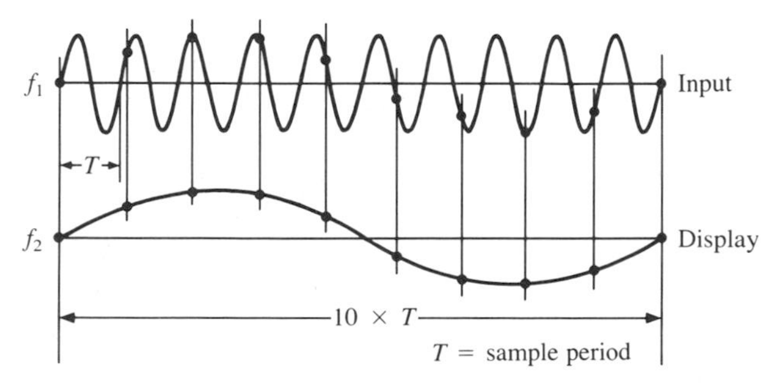

Figure 11–14 One cycle of display synthesized from ter. successive samples of a very-high frequency waveform (f_1).

For example, if a screen display of one cycle is created by sampling 500 cycles of a given input signal, the frequency that must be handled by the vertical amplifiers and the CRT is only 1/500 of the signal frequency. The only part of the oscilloscope that must operate at very high frequency is the sampling circuitry. The disadvantage of this technique is that the signal must be repeated *exactly* over a relatively long period of time; cycle-to-cycle fluctuations are not displayed faithfully. To ensure that the resultant signal can be displayed at lower frequencies by the vertical amplifier, sampling oscilloscopes are available that operate up into the gigahertz range.

Vertical Display Lines

High-quality oscilloscopes often provide a **delay line** in the vertical channel signal path. The purpose of a delay is to provide time for the trigger circuit to react and initiate the sweep before the signal arrives at the vertical deflection plates. This permits the operator to view the leading edge of the waveform being observed. The design and adjustment of this delay line are critical lest distortion be introduced into the delayed signal.

11–7 OSCILLOSCOPE PROBES

The signal to be observed is normally connected to the vertical input connector of the oscilloscope via a probe and a coaxial cable. Ideally, the connection of the oscilloscope probe to a circuit does nothing to upset the normal function of the circuit under test, and, further, the probe and cable system do not distort the signal to be seen on the oscilloscope. To fulfill these requirements, a variety of active and passive probes have been developed for many general and specialized oscilloscope applications.

1X Probes

The **1X probe,** or *times-one probe,* is the simplest oscilloscope probe in use. It consists of a length of shielded coaxial cable with a convenient probe tip at one end and a connector compatible with the oscilloscope input (usually BNC) at the other end. This simple probe is satisfactory for most DC and low-frequency AC measurements, but at higher frequencies the shunting effect of the cable capacitance becomes a problem. As shown by the equivalent circuit of the oscilloscope-probe-signal source connection in Figure 11–15, the input impedance of a typical general-purpose oscilloscope is approximately 1 MΩ, shunted by 35 pF. A three-foot length of cable might add another 90 to 100 pF of capacitance in parallel with this input impedance.

When the source resistance is small compared to the oscilloscope's input resistance, and at frequencies where the total capacitive reactance of the cable and oscilloscope ($C_c + C_i$) is very large compared to the total resistance (R_S +

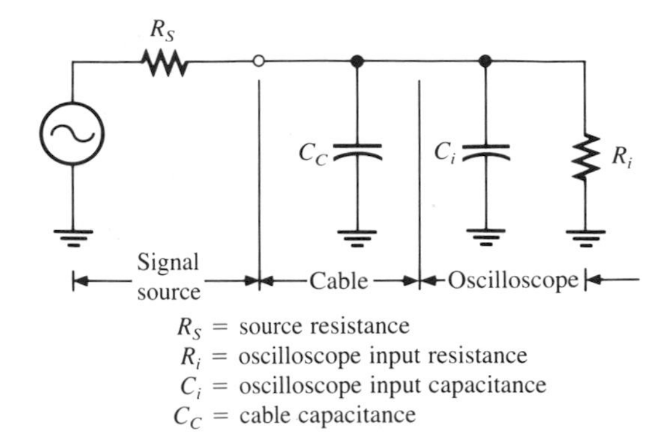

Figure 11–15 Equivalent circuit of a signal source and a 1X probe, forming a low-pass filter network.

R_i), the amplitude attenuation and phase shift due to this simple probe are negligible. The frequency of the input signal must satisfy the condition

$$f_{\text{in}} < \frac{1}{2\pi(R_S \parallel R_i)(C_c + C_i)} \tag{11–7}$$

However, the amplitude and phase shift errors that occur at higher source impedance and frequency levels become significant; also, the probe seriously loads down the circuit under test.

Since the shield side of the coaxial cable is grounded at the oscilloscope, all measurements with this probe are *made with respect to ground*. Therefore, the shield connection must never be connected to a circuit point that can not be grounded without affecting normal circuit operation. Particular caution must be exercised when working on or near a circuit that is not isolated from the AC-power line. Connecting the probe shield to the "hot" side of the power line can result in both serious operator injury and damage to the oscilloscope.

10X Probes

A more versatile probe is the **10X,** or *times-ten,* **probe,** whose equivalent circuit is shown in Figure 11–16*a*. The 10X probe *attenuates* the signal by a factor of ten, but is frequency compensated to present a constant high impedance to the test circuit over a large frequency range. This is similar in function to the input attenuator associated with the vertical amplifier. Using the 10X probe, the input impedance is ten times higher than the input impedance of the oscilloscope itself.

With the probe-oscilloscope input circuit redrawn as shown in Figure 11–16*b,* the probe body contains a parallel resistor-capacitor network that, together with the cable capacitance and the oscilloscope input resistance and capacitance, forms a balanced bridge. When $R_i(C_c + C_i) = R_1C_1$, the bridge is

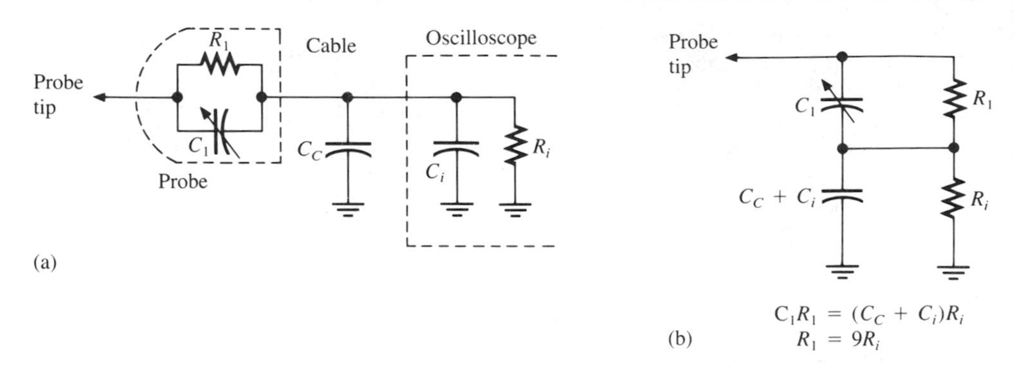

Figure 11–16 A 10X probe. (a) Equivalent circuit. (b) Probe circuit redrawn as a balanced bridge.

balanced. The circuit behaves as a 10:1 voltage divider for all frequencies from DC to the upper frequency limit of the oscilloscope. The capacitor in the probe body is made adjustable, so precise compensation can be obtained by applying a square wave and adjusting the capacitor for the best waveform (Figure 11–17). Most oscilloscopes provide a calibrated square-wave signal on the front panel suitable for this purpose. It is important to check 10X probes periodically for proper adjustment, especially when used with a new or different oscilloscope.

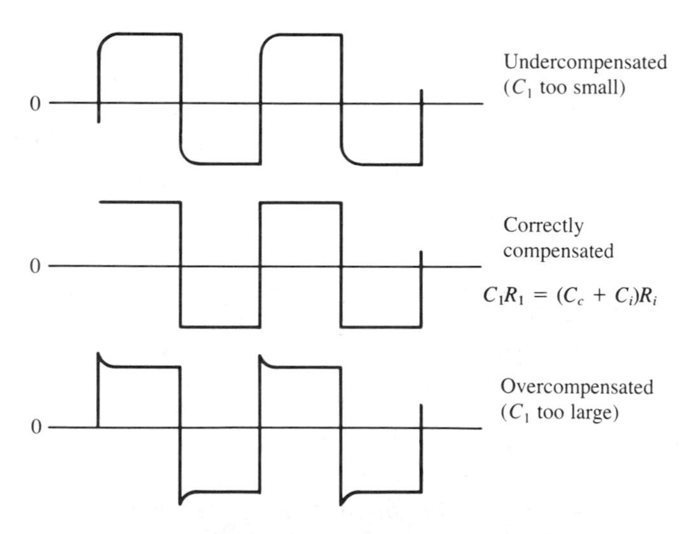

Figure 11–17 Effects of under, correct, and overcompensation of 10X probe on the shape of a square-wave input signal.

High-Voltage Probes

Most general-purpose oscilloscopes have a maximum vertical input voltage rating of about 600 V. When measuring voltages above 600 V, a special attenuator probe, called a **high-voltage probe,** must be used. This type of probe is essentially a frequency-compensated voltage divider designed to withstand voltages in the kilovolt range, and having a probe body designed to protect the operator from flashover. **Flashover** is a disruptive discharge or electric arc through air, around the surface of the probe body.

Current Probes

The most common type of current probe is constructed around a small split-core current transformer, which terminates in a resistive network. The core is clamped around the current-carrying conductor. The induced current in the secondary of the current transformer develops a voltage in the resistive network that has the same waveshape and is proportional to the amplitude of the current in the test circuit. The current waveform is displayed on the CRT screen in the normal fashion, but each unit of vertical deflection is read in units of current, rather than voltage. The probe body is usually marked with a conversion factor indicating the number of units of current per volt of deflection.

Although the split-core current probe is rugged, it is limited to measuring AC currents over a limited frequency range; the presence of high DC current components can cause saturation of the transformer core and a resultant loss in accuracy.

Active Probes

Active probes contain active elements (such as transistors, FETs, and vacuum tubes) either to provide better isolation or to buffer between the oscilloscope and the circuit under test. The active device is usually connected as a voltage follower (unity voltage gain amplifier). As such, when taking measurements in high-impedance circuits, it provides both less circuit loading and a broader frequency response than can be obtained with a passive probe. However, it has the disadvantages of requiring a power source and of tending to be more fragile and expensive than a passive probe.

Demodulator Probes

Demodulator probes are used for separating and displaying the audio component of an amplitude-modulated (AM) radio-frequency (RF) signal. The probe contains a simple AM diode detector circuit, like that of Figure 11–18, which rectifies the modulated RF signal and filters out the RF component, passing only the audio components to the oscilloscope input.

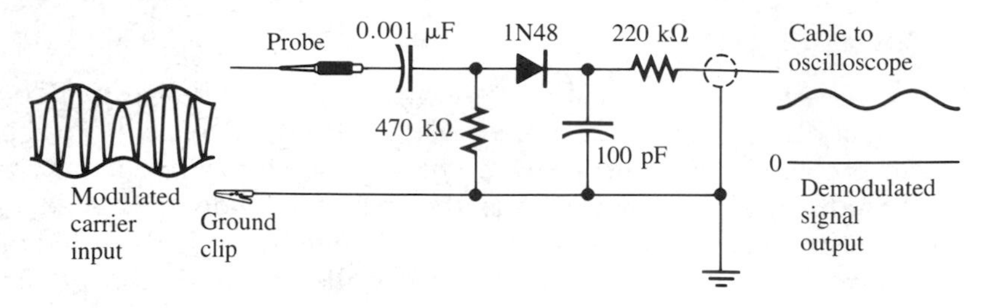

Figure 11–18 AM diode demodulator probe.

11–8 MEASUREMENTS WITH THE OSCILLOSCOPE

Because the oscilloscope is an extremely flexible and versatile instrument, it can be used to measure a number of parameters associated with DC and AC signals. Using a single-channel oscilloscope, it is capable of making measurements of voltage, current, time, frequency, and rise/fall times. If a dual-trace oscilloscope is used, the phase shift between two synchronous signals can be measured. It also provides alternate methods of measuring frequency.

Voltage Measurements

Oscilloscopes are best suited for the measurement of peak and peak-to-peak values of AC-voltage waveforms, although DC-coupled oscilloscopes also permit the display and measurement of an AC signal with a DC component. Before making a voltage measurement, it is important to be sure that (1) the probe is properly compensated, and (2) the fine adjustment control of the vertical attenuator is in the *CAL* or "calibrated" position. This control should not be disturbed during a measurement.

The signal to be measured is connected to the vertical input via the probe. The vertical SENSITIVITY, TIME BASE, COUPLING, and TRIGGER controls are set to provide a stable display that covers as many vertical divisions as possible without exceeding the limits of the screen. The vertical and horizontal POSITION controls can be used to reposition the trace slightly with respect to the graticule, in order to use the minor divisions of the graticule (i.e., "tick marks") to best advantage. If desired, a zero-voltage, or baseline, reference can be established by switching the vertical input selector switch to ground (GND) and adjusting the vertical position control to make the zero level coincide with a major horizontal grid line. The oscilloscope is a "personalized" instrument, which means that the control settings can be adjusted to suit the operator. As a general rule, two operators rarely adjust the oscilloscope's controls exactly the same way.

AC-voltage measurements are easiest when taken as *peak-to-peak* readings from the oscilloscope screen. A ground reference is established at the mid-horizontal graticule line, and then the AC waveform displayed. Peak-to-peak voltage is read as

$$V_{p\text{-}p} = \text{vertical divisions} \times \text{volts/divisions} \times \text{probe attenuation} \tag{11–8}$$

The number of divisions is the number of major vertical graticule divisions measured between the negative and positive peaks of the waveform. The "volts/division" is the setting of the vertical sensitivity control, while probe attenuation depends on the type of probe (such as 1X or 10X).

EXAMPLE 11–3

For the display of Figure 11–19, determine the peak-to-peak voltage of the triangle wave if the vertical sensitivity is set at 2 V/division.

Figure 11–19 Triangle waveform for Example 11–3.

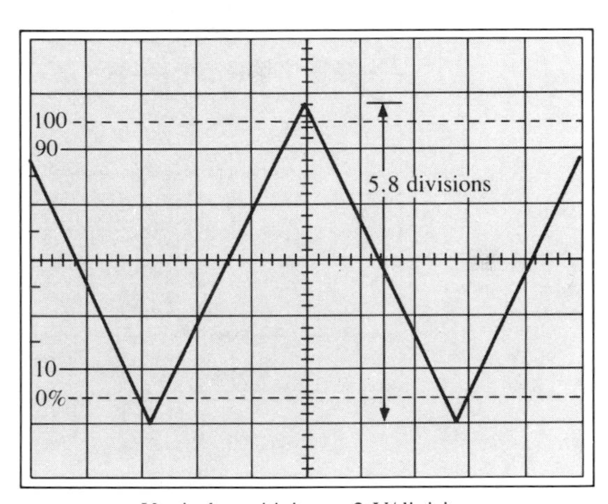

Solution:

As positioned on the graticule, the peak-to-peak vertical deflection spans 5 major divisions and 4 minor divisions. Consequently, the peak-to-peak deflection is expressed as 5.8 (major) divisions. For a 2-V/division vertical sensitivity, the peak-to-peak voltage is

$$V_{p\text{-}p} = (5.8 \text{ divisions}) \times (2 \text{ V/divisions}) = 11.6 \text{ V}$$

The voltage of an AC signal with a nonzero DC (average) level can be measured in several ways. If the peak value is being measured, the signal must be *DC coupled,* or the true peak value is lost. If only the peak-to-peak value is desired, the input can be either AC or DC coupled. However, if the peak-to-peak voltage is significantly smaller than the DC level, it is better if the input is AC coupled. This way, the sensitivity of the vertical amplifier can be sufficiently increased to display only the time-varying part of the signal.

EXAMPLE 11–4

For the display of Figure 11–20, determine the peak voltage of the sawtooth waveform if the vertical sensitivity is set at 2 V/division. The channel's zero reference is arbitrarily set to coincide with the second horizontal graticule line from the bottom of the display.

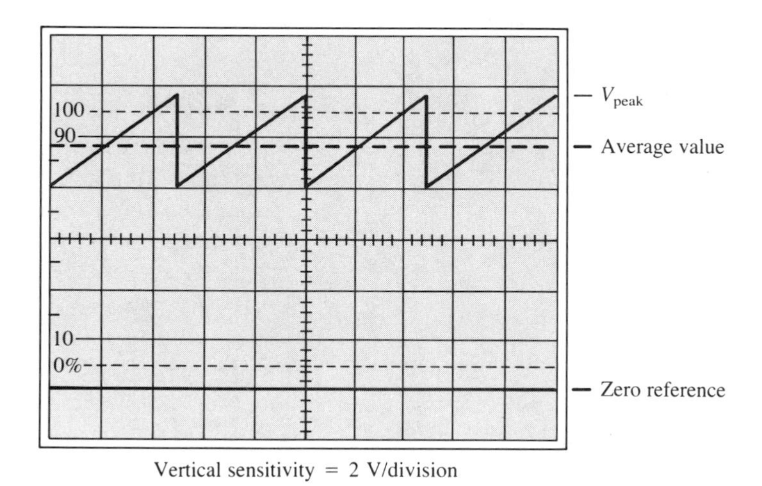

Figure 11–20 Sawtooth waveform for Example 11–4.

Solution:

With DC coupling, the peak voltage occurs at 5.8 divisions above the zero reference. At a vertical sensitivity of 2 V/division, the peak voltage is

$$V_p = (5.8 \text{ divisions}) \times (2 \text{ V/division})$$
$$= 11.6 \text{ V}$$

If the peak-to-peak voltage is required, the input could be *AC coupled* and the zero reference adjusted at the vertical midpoint of the display. The vertical sensitivity can be increased to, for example, 1 V/division or 0.5 V/division, so that the waveform occupies the maximum area possible without exceeding the limits of display. If the sensitivity were 0.5 V/division, the peak-to-peak deflection would probably be four times as great as shown in Figure 11–20. In this case, the span would take 5.6 divisions, offering greater resolution than at 1.8 divisions.

To obtain the RMS or average values of standard waveforms, it is best to measure the peak-to-peak value and convert mathematically using the factors given in Table 11–2.

DC-voltage measurements are made in a similar fashion to AC measurements, with the exceptions that (1) the reference line can be established near either the top or bottom of the screen depending upon the polarity of the signal and (2) DC coupling must be used.

Current Measurements

There are two ways to measure current with an oscilloscope. AC and DC currents can be measured by looking at the voltage across a known value of resistance and applying Ohm's Law to the observed trace. Application of this technique is limited by the need for one side of the resistor and oscilloscope to be at ground potential, although some oscilloscopes are equipped with differential preamplifiers that permit viewing the voltage drop across components that have both terminals "floating" above ground. Differential preamplifiers have provision for input leads, and the display represents the voltage difference between the two inputs, each of which can be above ground. The second method requires the use of a current probe (see Section 11–7), and is only applicable to the measurement of AC currents.

Table 11–2 Peak-to-peak conversion factors for average and RMS values

Waveform	V_{AV}	V_{RMS}
half-wave rectified sine wave	$0.318\ V_p$	$0.354\ V_p$
full-wave rectified sine wave	$0.636\ V_p$	$0.707\ V_p$
square wave	0	$0.5\ V_p$
pulse	$(PW/T)\ V_p$	$(PW/T)^{1/2}\ V_p$
triangular wave	$0.5\ V_p$	$0.577\ V_p$
sawtooth wave	$0.5\ V_p$	$0.577\ V_p$

PW = pulse width, T = period.

Time-Base Method of Time and Frequency Measurement

The time-base method of measuring frequency and time uses the internal time-base of the oscilloscope to measure the period of one cycle. The frequency (in hertz) is determined by the reciprocal of the period:

$$f = \frac{1}{T} \tag{11-9}$$

To determine the frequency of an AC waveform, the oscilloscope is adjusted to display one complete cycle of the waveform over as much of the horizontal axis as possible. It is important to check that the variable sweep adjust control is in the CAL (calibrate) position. The number of major horizontal graticule divisions required for one complete cycle is observed, and the period (T) is determined from

$$T = \text{horizontal divisions} \times \text{time/division} \tag{11-10}$$

The accuracy of this method is limited to the sweep accuracy of the time base and the resolution to which the graticule can be read. This usually produces an accuracy in the region of three to five percent.

EXAMPLE 11–5

For the display of Figure 11–21, determine the frequency of the square wave if the time base is set at 10 μs/division.

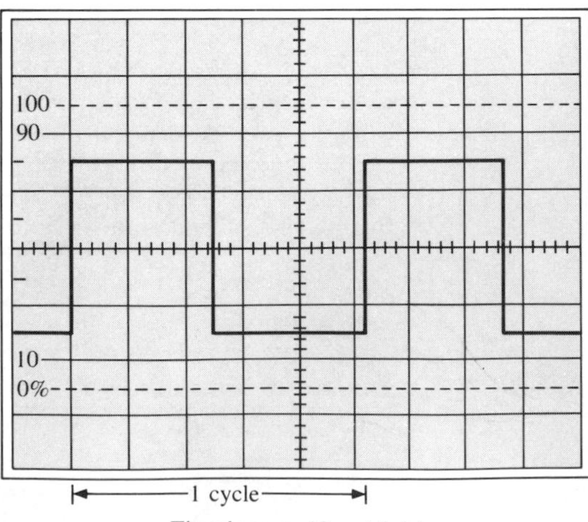

Figure 11–21 Square wave for Example 11–5.

Solution:

The start of one cycle, using the horizontal POSITION control, is adjusted so that it coincides with one of the major vertical graticule lines. One complete cycle takes 5.2 divisions. At 10 μs/division, the period of the square wave is

$$\begin{aligned} T &= (5.2 \text{ divisions}) \times (10\ \mu\text{s/division}) \\ &= 52\ \mu\text{s} \end{aligned}$$

The square-wave frequency is

$$\begin{aligned} f &= \frac{1}{52\ \mu\text{s}} \\ &= 19.2 \text{ kHz} \end{aligned}$$

Z–Axis Time Marks

Many oscilloscopes have a terminal on the rear panel called the ***Z*–axis input.** The *Z*–axis input allows the trace to be **intensity-modulated** by an external signal. If positive-going pulses from an accurate external frequency source are fed to the *Z*–axis input, bright timing marks appear on the trace that facilitate accurate time measurements on the displayed signal. The spacing of the timing marks is given by

$$\boxed{t = \frac{1}{f}} \qquad (11\text{–}11)$$

where f equals the frequency of the external frequency source connected to the *Z*–axis input.

Lissajous Method of Frequency Measurement

If two different signals are both sine waves, the **Lissajous method** can be used to determine the frequency ratio of the two signals. If the frequency of one of the signals is known, the frequency of the other can be easily determined from the resulting **Lissajous pattern** on the oscilloscope screen (Figure 11–22), providing that the frequencies are related by the ratio of two integers (such as 1:1, 3:1, 3:5). The oscilloscope's internal time base generator is disabled, so the time base control probably reads "EXT", "*X–Y*", "CH–2", or some other mode indicative that the sweep speed no longer applies. Consequently, one signal is applied to the oscilloscope's horizontal amplifier, and the other is connected to the vertical input.

The convention is to connect the known signal to the *horizontal* channel and the unknown to the *vertical* channel. A laboratory audio or RF oscillator with good frequency calibration and stability is often used as the known fre-

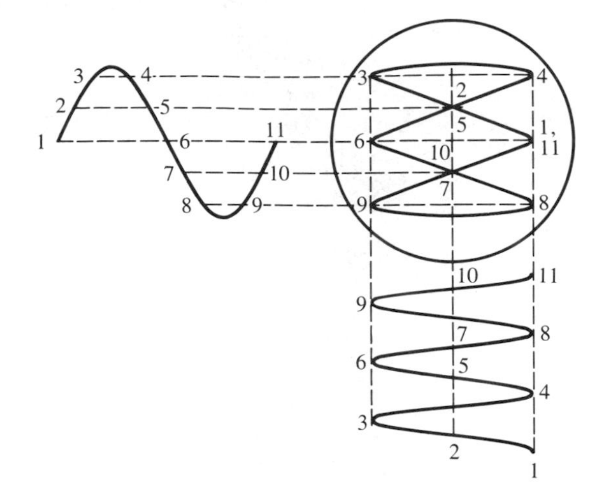

Figure 11–22 Lissajous pattern when a sine wave is applied to the horizontal amplifier is three times the frequency of the sine wave signal applied to the vertical amplifier.

quency. Furthermore, the upper practical limit of this method is for frequency ratios greater than 10:1.

To determine a frequency ratio

1. Connect the known signal to the horizontal input and the unknown signal to the vertical input.
2. Adjust the oscilloscope gain and generator amplitude controls so that the resulting pattern covers as much of the graticule area as possible (Figure 11–23*a*).
3. Slowly and carefully adjust the oscillator frequency to establish a stationary pattern on the screen.
4. Using the oscilloscope vertical and horizontal POSITION controls, position the resulting Lissajous pattern so that its upper edge is tangent to the top horizontal graticule line and its left edge is tangent to the left-most vertical graticule line.
5. Count the number of loops touching the top horizontal graticule line (N_H). This number represents the part of the ratio corresponding to the *vertical* (unknown) input signal. In Figure 11–23*a*, there are three such loops.
6. Count the number of loops touching the left-most vertical graticule line (N_V). This number represents the part of the ratio corresponding to the *horizontal* (known) input signal. In Figure 11–23*a*, there are five such loops.
7. The ratio of the number of horizontal to vertical loops (N_H/N_V) is the ratio of the unknown (vertical input) frequency to the known (horizontal input) frequency, so that

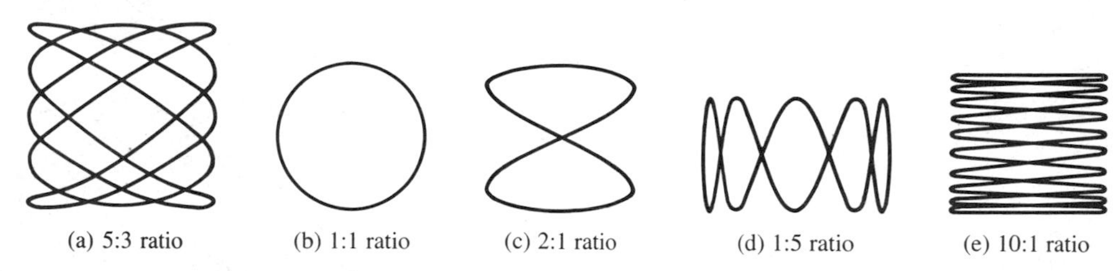

Figure 11–23 Frequency measurement using Lissajous patterns. (a) 5:3 ratio. (b) 1:1 ratio. (c) 2:1 ratio. (d) 1:5 ratio. (e) 10:1 ratio.

$$\frac{N_H}{N_V} = \frac{f_V \text{ [unknown]}}{f_H \text{ [known]}} \tag{11–12}$$

For Figure 11–23*a*, the ratio is 3:5. If the known frequency is measured as 1200 Hz, then the unknown frequency is 3/5 of 1200 Hz, or 720 Hz.

The Lissajous pattern method of measuring an unknown frequency may seem simple. In theory this is so, but in practice it requires much skill and patience. In many cases, it is difficult to obtain a stationary pattern, rather than one that appears to rotate in a "cartwheel" or "merry-go-round" fashion. This is especially true if the frequency of either of the input signals is not stable.

Phase-Shift Measurements with Lissajous Patterns

The phase shift of two sine-wave signals of the same frequency can be measured with the Lissajous method. Because both signals have the same frequency, the resulting Lissajous pattern is in the form of an "eccentric" ellipse. The degree of eccentricity is a function of the phase shift between the two signals. Figure 11–24 shows the patterns formed for phase shifts ranging from 0° to 315°. For phase shifts of either 0° or 180°, the Lissajous pattern is a straight line; the ellipse is essentially "closed up." At either 90° or 270°, a full circle is displayed; the ellipse's major and minor axes are equal.

To make a phase difference measurement using the Lissajous ellipse method

1. Disable the oscilloscope's internal sweep.
2. With no signals applied to either the vertical or horizontal amplifiers, center the resulting spot on the screen.
3. Connect one signal to the vertical input and adjust the vertical gain so the

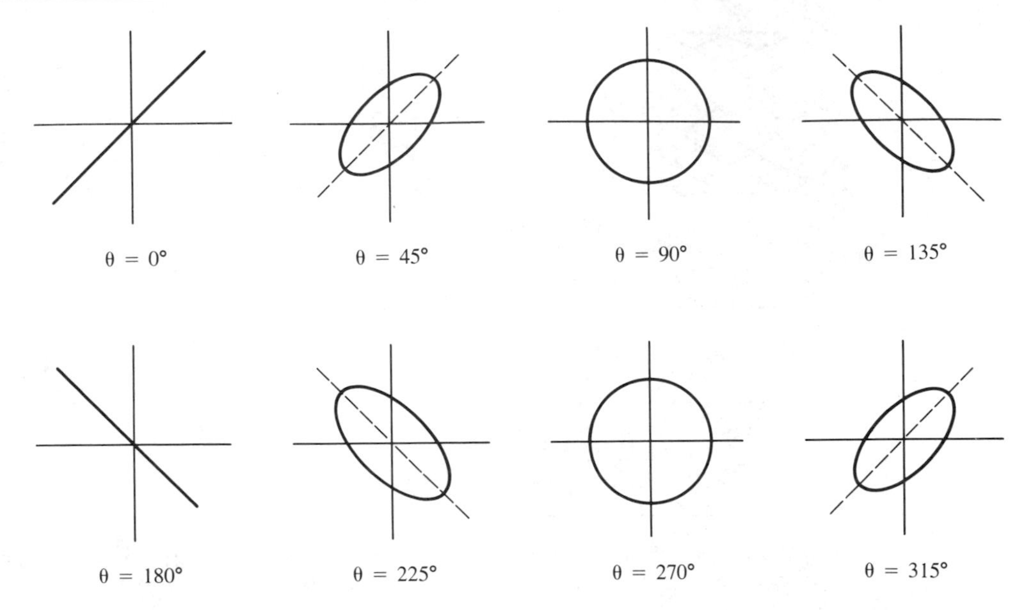

Figure 11–24 Lissajous elipses for phase shifts from 0° to 315°.

resulting vertical line spans the full height of the display (usually eight divisions high).

4. Disconnect the vertical input and apply the other signal to the horizontal input. Adjust the horizontal gain so the resulting horizontal line is equal in length to the vertical line of Step 3 (i.e., eight divisions).
5. Reconnect the vertical input signal. An ellipse with equal horizontal and vertical deflections should appear.
6. As in Figure 11–25, the phase shift in degrees is determined from

$$\theta = \sin^{-1}\left(\frac{Y_1}{Y_{\max}}\right) \tag{11–13}$$

Figure 11–25 Lissajous (eccentric) elipse method for measuring the phase difference between two signals of the same frequency.

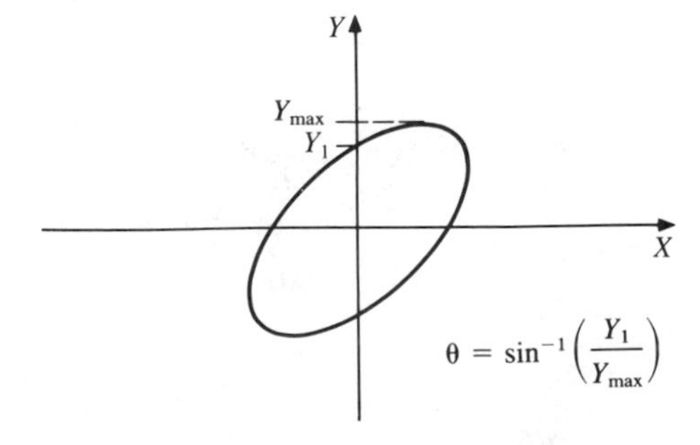

where

$$Y_1 = \text{positive } Y\text{-intercept value}$$
$$Y_{max} = \text{maximum } Y\text{-value of the ellipse}$$

EXAMPLE 11–6

Determine the phase shifts of the four patterns of Figure 11–26 using the intercept formula of Equation 11–13. For convenience, the gain of both the vertical and horizontal amplifiers has been adjusted so that the maximum vertical and horizontal deflections are each 2.0 divisions.

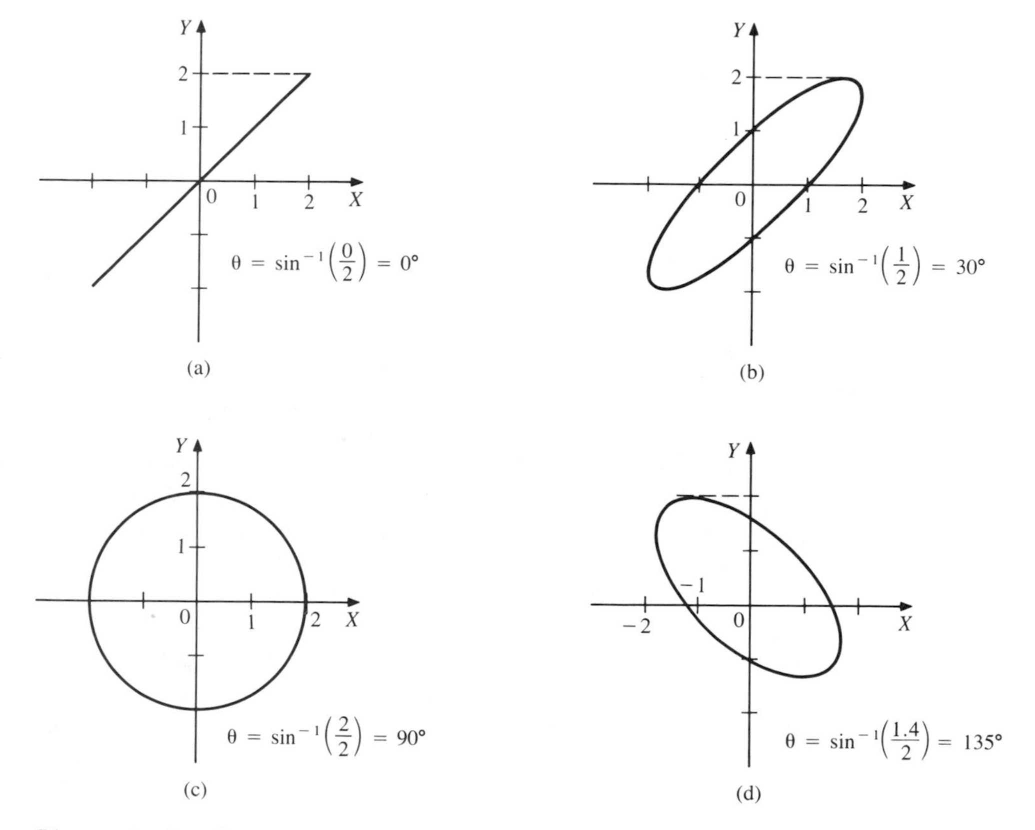

Figure 11–26 Lissajous patterns for Example 11–6.

Solution:

(a) In Figure 11–26*a*, a straight line at 45° automatically indicates a phase shift of 0° (or 360°). This can be verified using Equation 11–13:

$$\theta = \sin^{-1}(0/2)$$
$$= \sin^{-1}(0)$$
$$= 0°$$

(b) In Figure 11–26*b*,

$$\theta = \sin^{-1}(1/2)$$
$$= \sin^{-1}(0.5)$$
$$= 30°$$

The signal connected to vertical input leads that of the horizontal input by 30°.

(c) In Figure 11–26*c*, a circle automatically indicates a phase shift of 90°. This can be verified using Equation 11–13:

$$\theta = \sin^{-1}(2/2)$$
$$= \sin^{-1}(1)$$
$$= 90°$$

(d) In Figure 11–26*d*,

$$\theta = \sin^{-1}(1.4/2)$$
$$= \sin^{-1}(0.7)$$
$$= 135°$$

This method can measure phase shifts as small as 1°. However, measurement becomes more difficult as the angle approaches 90° (a circle). Besides indicating the phase shift between the two input signals, the Lissajous method also includes any phase shifts attributed to the oscilloscope's vertical and horizontal amplifiers, leads, and probes. To determine if the oscilloscope measurement system introduces any additional shift in phase, a single sine-wave signal is connected simultaneously to both the horizontal and vertical inputs. If no phase shift is introduced by the amplifiers, cables, or probes, then the resultant Lissajous pattern is a straight, 45° line. If there is an ellipse, the phase shift should be measured and *subtracted* from the value obtained when measuring the phase shift of two different sine waves. A good oscilloscope should have no visible phase shift. Should this not be the case, variable delay lines can be connected in series with the input leads to compensate for the induced phase shift.

Sweep Method for Measuring Phase Shift

The phase shift between two waveforms can also be determined by displaying both signals on the screen of a dual-trace oscilloscope, directly measuring the time difference (t_1) between points of similar amplitude, and comparing it with the time required for one complete cycle (T). The easiest approach is to measure the time difference between the points at which the waveform crosses the "zero" axis. The period for one complete cycle is measured, and the phase shift in degrees determined as a fraction of the total cycle:

$$\theta = 360° \frac{t_1}{T} \tag{11–14}$$

EXAMPLE 11–7

Determine the phase shift between the two sine waves in the display of Figure 11–27 if the time base is set at 1 ms/division.

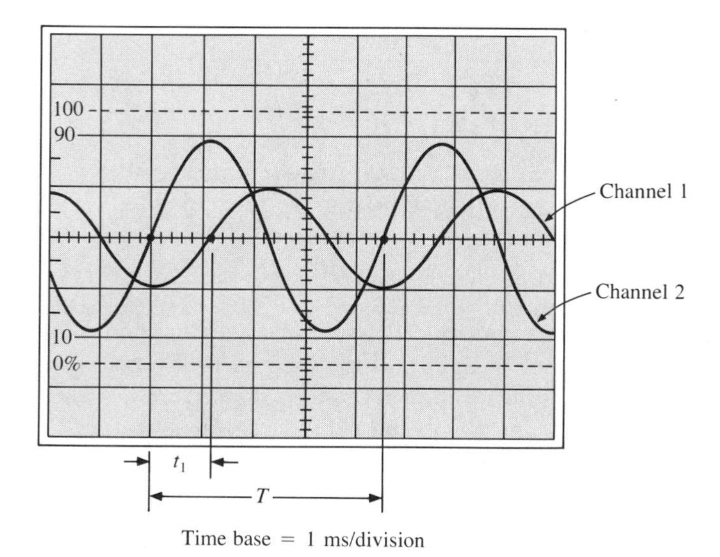

Figure 11–27 Measurement of phase difference using the oscilloscope's sweep.

Solution:

For a horizontal sweep speed of 1 ms/division, the period of either sine wave is

$$T = (4.6 \text{ divisions}) \times (1 \text{ ms/division})$$
$$= 4.6 \text{ ms}$$

The time delay between both waveforms, taken for convenience where both waveforms cross zero, is

$$t_1 = (1.2 \text{ divisions}) \times (1 \text{ ms/division})$$
$$= 1.2 \text{ ms}$$

so the phase shift is

$$\theta = 360° \left(\frac{1.2 \text{ ms}}{4.6 \text{ ms}}\right)$$
$$= 93.9°$$

If both the period and time delay measurements are made using the same sweep speed, the phase shift calculation can be done simply in terms of the number of horizontal divisions involved. However, if the sweep speed is changed to improve the resolution of the time delay measurement, all time measurements must be converted from terms of divisions to the actual time in terms of seconds.

Rise-Time Measurements

As mentioned earlier in this chapter, the display graticule is frequently marked with four horizontal lines (either solid or dashed) labeled 0%, 10%, 90%, and 100%. These markings are used specifically in making rise (and fall) time measurements. There are two 90 and 100% markings on some screens. If this is the case, the lower 100% and 90% markings are used to represent 0% and 10%, respectively, when making rise-time measurements. When measuring fall times, the upper 100% and 90% lines represent 0% and 10%, respectively, so the percentage of full amplitude increases *downward*. These markings are specifically used in making rise- (and fall-) time measurements.

The vertical amplifier calibration is adjusted so the pulse baseline coincides with the 0% graticule line and the maximum pulse amplitude coincides with the 100% line (Figure 11–28*a*). The time difference between the 10% and 90% points is the rise time. The same technique applies to the measurement of fall time (Figure 11–28*b*), except that the falling edge of the pulse is used. In either case, the 0–10–90–100% (or 100–90–90–100%) marking arrangement on the display graticule is independent of the sensitivity setting of the vertical amplifier, which eliminates the need to calculate and find the 10% and 90% voltage points on the waveform.

EXAMPLE 11–8

Determine the 10–90% trailing edge fall time of the pulse of Figure 11–28*b* if the sweep speed is 5 μs/division.

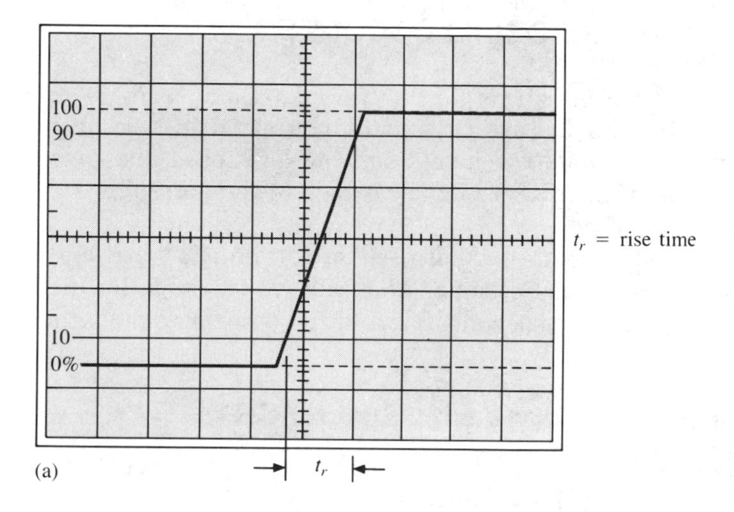

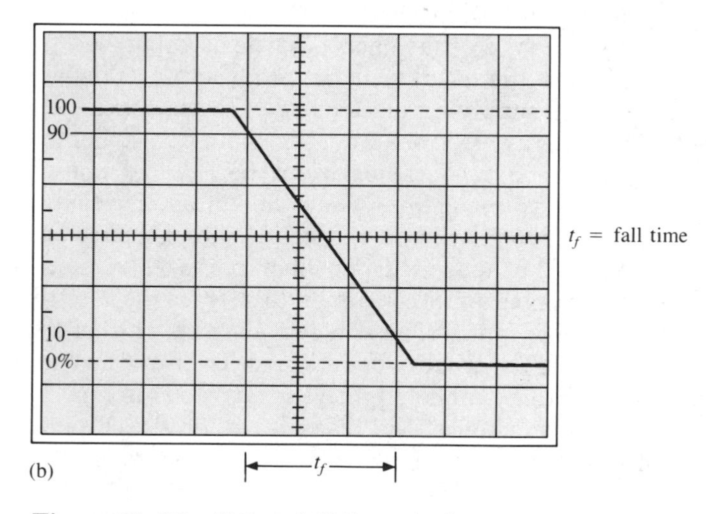

Figure 11–28 10% and 90% graticule markings for rise and fall time measurements. (a) Rise time. (b) Fall time.

Solution:

The time difference between the 10% and 90% points is 3.0 divisions. At 5 μs/division, the 10–90% time is

$$t_f = (3 \text{ divisions}) \times (5\ \mu\text{s/division})$$
$$= 15\ \mu\text{s}$$

11–9 SPECIAL PURPOSE OSCILLOSCOPES

A storage oscilloscope can retain the trace caused by a single sweep for a long period of time. This feature is particularly useful in studying nonrepetitive events such as turn-on transients or very low speed phenomena where the required sweep time is very long compared to the persistence of the standard oscilloscope phosphor.

There are two basic methods used for retaining a trace on the oscilloscope screen for a long period of time. As discussed in the following sections, one method makes use of a special CRT, and the other uses a digital storage technique.

Storage Oscilloscopes with Special CRTs

There are two types of special CRTs used for storage applications. One uses the phosphor coating on the screen as its storage medium. The phosphor does not have long persistence, but is deposited on a dielectric layer over a transparent metal film on the screen, so that each particle is separated from adjoining particles. In addition to the usual writing gun, there are low-energy electron guns called *flood guns* supplying electrons to the phosphor particles (Figure 11–29). When high-energy electrons from the write gun strike the screen, they cause secondary emission of electrons from the isolated phosphor particles, leaving a path of net positive charge when the voltage is greater than 150 V. The metal film is held positive with respect to the phosphor and collects the secondary electrons. The unexcited phosphor particles remain near ground potential, and the charged path continues to attract most of the low-energy electrons from the flood guns. These electrons pass through the positively charged region, causing the trace to be retained. To erase the stored image, the entire metal film is briefly made slightly negative. This type of CRT has no provision to control persistence; the trace is either stored or it is not. As a consequence, it is sometimes referred to as *bistable storage.*

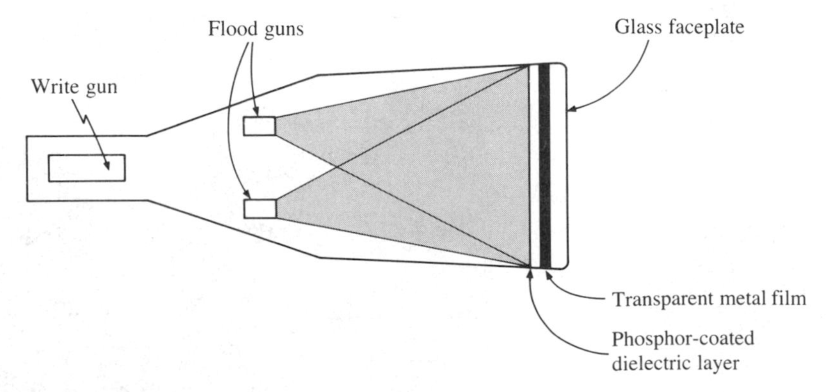

Figure 11–29 Construction of a CRT which uses the screen phosphor for direct bistable storage.

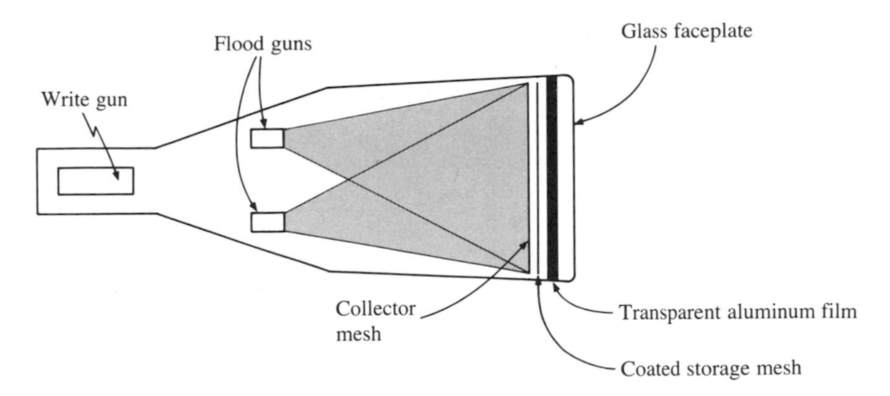

Figure 11–30 Location of a storage mesh to provide variable persistance storage in a CRT.

Another type of storage CRT uses a conventional phosphor screen backed by aluminum film (Figure 11–30). There are two wire meshes directly behind the screen. The mesh closest to the guns is the *collector mesh,* and the one closest to the screen the *storage mesh.* The collector mesh is maintained at a positive potential, and the storage mesh at a zero or slightly negative potential. The storage mesh has a layer of storage material deposited on its inside surface. The high-energy electrons from the write gun produce secondary emission at the storage layer, leaving a positively charged path where the beam has been. Electrons from the flood guns are diverted from the negatively charged areas of the storage mesh and pass through the positively charged areas, producing a continuous trace on the screen. Secondary and flood gun electrons that do not pass through the storage mesh are attracted back to the collector mesh.

Erasure is accomplished by momentarily bringing the storage mesh to the same positive potential as the collector mesh. Secondary emission caused by the flood guns at the storage material layer effectively erases the stored trace. Application of a variable width negative pulse train to the storage mesh can be used to control its discharge rate. This provides a persistence control; narrow pulses give long persistence, and wider pulses give progressively shorter persistence.

Digital Storage Oscilloscopes

The **digital storage oscilloscope** stores a signal by converting successive samples to binary numbers, which are stored in a digital memory and used to recreate a composite waveform in much the same manner as the sampling oscilloscope display is created. This type of oscilloscope samples the voltage of the waveform over many cycles, each time a little later in the cycle. The stored values are used to synthesize a dotted (discrete) representation of the signal.

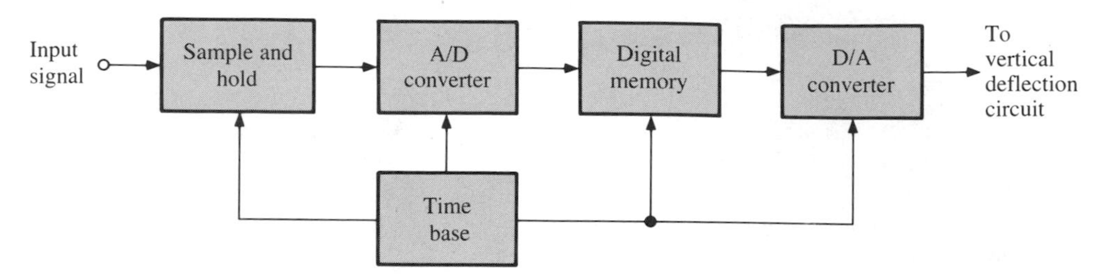

Figure 11–31 Block diagram of a basic digital storage oscilloscope.

As in the block diagram of Figure 11–31, the samples are taken by a sample-and-hold circuit that is connected to the input signal for a period of time very short compared to the length of one cycle. The sample-and-hold circuit effectively "snaps a picture of the voltage level." The output of the sample-and-hold circuit is connected to an analog-to-digital converter, where the analog voltage level is converted to a digital number and stored in memory. When enough samples have been taken, the storage digital numbers are successively converted back into analog values by a digital-to-analog converter, and are sent to the vertical deflection circuit as the trace is swept horizontally in synchronism. This digital-to-analog conversion process is repeated continually, maintaining the trace on the screen as long as desired.

Since the technique used is virtually the same, the digital storage oscilloscope has the same shortcoming as the sampling oscilloscope, in that the waveform displayed is the result of samples taken over many cycles, and it does not give a true representation of cycle-to-cycle changes.

11–10 SUMMARY

This chapter is perhaps the longest and most extensive in this book, and rightfully so. Next to the versatile multimeter, the (cathode-ray tube) oscilloscope is perhaps the most useful and frequently used instrument for a variety of measurements. This chapter discussed the fundamentals of the cathode-ray tube and how it interacts with vertical and horizontal amplifiers, a sweep generator, and triggering circuits to form an oscilloscope. In addition, various types of probes are discussed, as well as their features and limitations.

As a measurement tool, the oscilloscope is capable of measuring the basic electrical quantities of voltage, current, frequency, time, rise/fall times, and phase. Other quantities, such as vibration, temperature, velocity, pressure, and flow are possible to measure using transducers that express these quantities as calibrated functions of voltage, or frequency.

Oscilloscopes are available that have advanced features such as multi-trace displays, display storage, delayed sweep, and input signal sampling, which can meet virtually any situation.

11–11 GLOSSARY

active probe A probe, usually with an FET amplifier circuit, used as a buffer to minimize interaction between the circuit under test and the input circuitry of the oscilloscope.

Aquadag A conductive material used on the inside of the conical part of a cathode-ray tube to shield from stray magnetic fields.

alternate mode A method of simultaneously displaying two high-frequency traces on the screen of a dual-trace oscilloscope. The two channels are repeatedly traced in an alternating pattern on the screen.

automatic mode A triggering mode where the oscilloscope automatically provides a trigger signal to the sweep generator when no vertical input signal is present.

baseline The resulting line on an oscilloscope display that the sweep produces in the absence of a vertical input signal.

cathode-ray tube (CRT) Also called a *kinescope,* an electron-beam tube in which electrons emitted by a hot cathode are formed by an electron gun into a narrow beam that is focused onto a phosphor-coated screen.

channel In an oscilloscope, input to either one or more vertical amplifiers or to the horizontal amplifier.

chopped mode A method of simultaneously displaying two low-frequency traces on the screen of a dual-trace oscilloscope. A single beam is rapidly switched back and forth vertically between the two traces. As a result, each channel signal is represented by a finely dashed or dotted line.

delay line An electronic circuit or transmission line used to produce a time delay of a signal.

digital storage oscilloscope An oscilloscope which displays a signal by converting successive samples to binary numbers, which are then stored in a digital memory.

dual-beam CRT An oscilloscope in which the cathode-ray tube produces two separate electron beams that may be individually or jointly controlled.

fall time The amount of time for a pulse waveform to decrease from 90 to 10 percent of its maximum amplitude.

flashover An electric discharge or arc around an insulated surface, such as a probe.

flyback See **retrace.**

fluorescence The property of a phosphor material to emit light when its atoms are excited.

graticule A calibrated screen placed in front of an oscilloscope display to measure beam deflection.

high-voltage probe A probe with a high internal resistance used for the measurement of extremely high voltages.

holdoff period The time period, during either the sweep or retrace periods, when the oscilloscope's trigger circuit will not respond to a triggering signal.

horizontal amplifier Also called the ***X*-axis amplifier,** an amplifier in an oscilloscope for signals intended for horizontal deflection on the screen.

input attenuator A frequency-compensated resistive voltage divider that reduces the input voltage by a given amount without distortion.

intensity modulation Also called *Z*-axis modulation, the modulation of beam intensity in a cathode-ray oscilloscope.

Lissajous method A measurement technique using an oscilloscope to measure either the phase shaft between two sine waves having the same frequency or the measurement of a frequency of an unknown sine wave in comparison with the frequency of a known sine wave.

Lissajous pattern A pattern produced when sine waves of given amplitude, phase, and frequency relationships are applied simultaneously to the vertical and horizontal amplifiers of an oscilloscope.

normal mode A triggering mode in which the trigger point is determined by both the level and slope controls of an oscilloscope.

1X probe A probe having no voltage attenuation.

oscilloscope (CRO) A test instrument using a cathode-ray tube to make visible on a phosphor-coated screen instantaneous values and waveforms as functions of either time or some other quantity.

oscilloscope bandwidth The maximum frequency of the input signal that can be displayed without distortion or attenuation by the vertical amplifier.

parallax error An error that occurs when the trace and the graticule of an oscilloscope are on different planes and the observer is shifted from a direct line of sight. Can also be applied to analog meters.

persistence The period during which phosphorescence occurs.

phosphor A material emitting light when struck by a focused electron beam by converting the kinetic energy of the beam electrons into radiant energy.

phosphorescence The property of a phosphor material of continuing to emit light after its excitation source has been removed.

retrace Also called **flyback,** the return of the electron beam in a cathode-ray tube to its starting point after a sweep across the screen.

rise time The amount of time for a pulse waveform to increase from 10 to 90 percent of its maximum value.

sweep The horizontal movement of the electron beam across the cathode-ray tube's screen.

time base A voltage generated by the sweep generator circuit of an oscilloscope so that its trace is linear with respect to time.

10X probe A probe that reduces the input voltage by a factor of 10.

trace The pattern, usually a single line, on the screen of an oscilloscope.

triggering The action of starting the sweep in an oscilloscope display.

vertical amplifier Also called the ***Y*-axis amplifier,** an amplifier in an oscilloscope for signals intended for vertical deflection on the screen.

***Y*-axis amplifier** See **vertical amplifier.**

***Z*-axis input** The input circuit that permits an external voltage to modulate the intensity of the oscilloscope's trace.

***Z*-axis modulation** See **intensity modulation.**

11–12 PROBLEMS

1. If a rectangular pulse has a rise time of 27.2 ns, what is the maximum bandwidth that the oscilloscope must have to display the pulse without distortion?
2. What is the maximum rise time a square wave or rectangular pulse can have if it is to be displayed on a 20-MHz oscilloscope without distortion?
3. For the 10X probe shown in Figure 11–16, $R_1 = 9\ M\Omega$, $R_i = 1\ M\Omega$, and $C_i = 12$ pF. If the capacitance of the probe's 3-foot cable is 20 pF/foot, determine the value for C_1 for which the probe will be properly compensated.
4. A 10X-high impedance probe, having a resistance of 9 MΩ and a capacitance of 6 pF, is connected to an oscilloscope with an input impedance of 1 MΩ. If the equivalent capacitance of the probe-oscilloscope connection is 2 pF, determine the input capacitance of the oscilloscope.
5. The vertical amplifier of an oscilloscope uses a frequency-compensated voltage divider like that of Figure 11–7. If $R_1 = 900$ kΩ, $R_2 = 100$ kΩ, $C_v = 5$ pF, and $C_i = 85$ pF, determine the output voltage when the input voltage is
   (a) a sine wave with a peak-to-peak voltage of 3.5 V at a frequency of 25 kHz,
   (b) a DC voltage of 1.8 V.
6. The waveform of Figure 11–32 is observed on an oscilloscope. If the vertical amplifier and time base sensitivities are set respectively at 0.02 V/division and 5 ms/division, determine both the RMS voltage and the frequency of the signal.

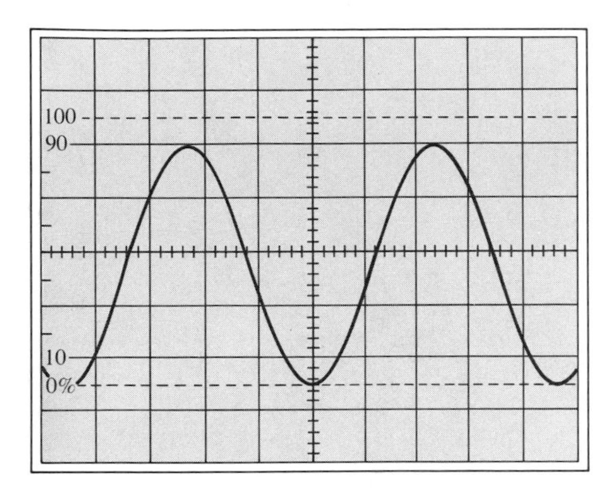

Figure 11–32 Waveform for Problem 6.

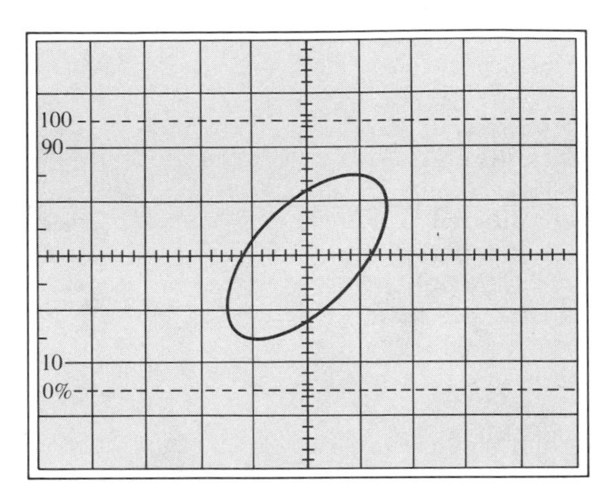

Figure 11–34 Lissajous pattern for Problem 9.

7. If an oscilloscope display graticule is 10 divisions wide, determine the time-base setting required to display exactly two cycles of a 100-kHz sine wave.
8. A 975-Hz sine wave is connected to the horizontal input of an oscilloscope. When another sine wave is connected to the vertical input, the Lissajous pattern of Figure 11–33 is observed. Determine the frequency of the signal connected to the vertical amplifier.

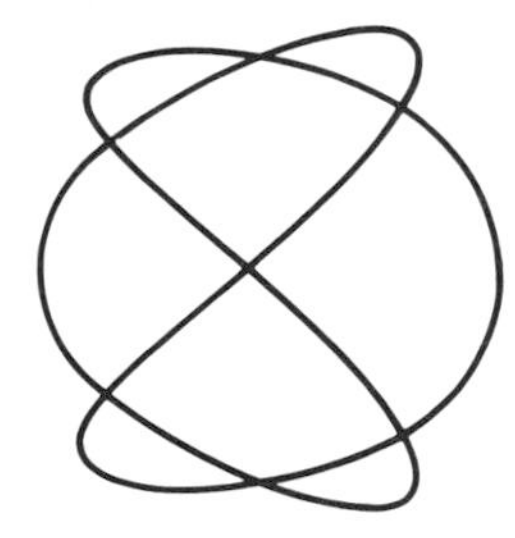

Figure 11–33 Waveform for Problem 8.

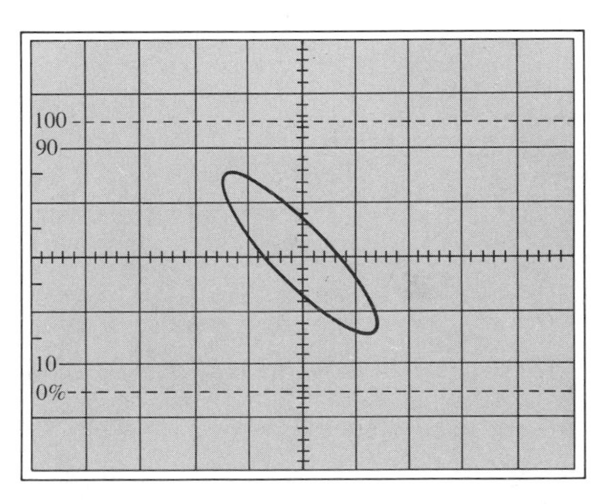

Figure 11–35 Lissajous pattern for Problem 10.

9. Determine the phase shift between two sine waves, using the Lissajous pattern of Figure 11–34.
10. Determine the phase shift between two sine waves, using the Lissajous pattern of Figure 11–35.
11. Prior to the measurement of the phase shift of two sine waves, one of the sine waves is connected simultaneously to both the vertical and horizontal inputs. The resulting Lissajous pattern is shown in Figure 11–36*a*. Then, the two sine waves are connected to the oscilloscope, giving the pattern of Figure 11–36*b*. Determine the phase shift between the two sine waves.

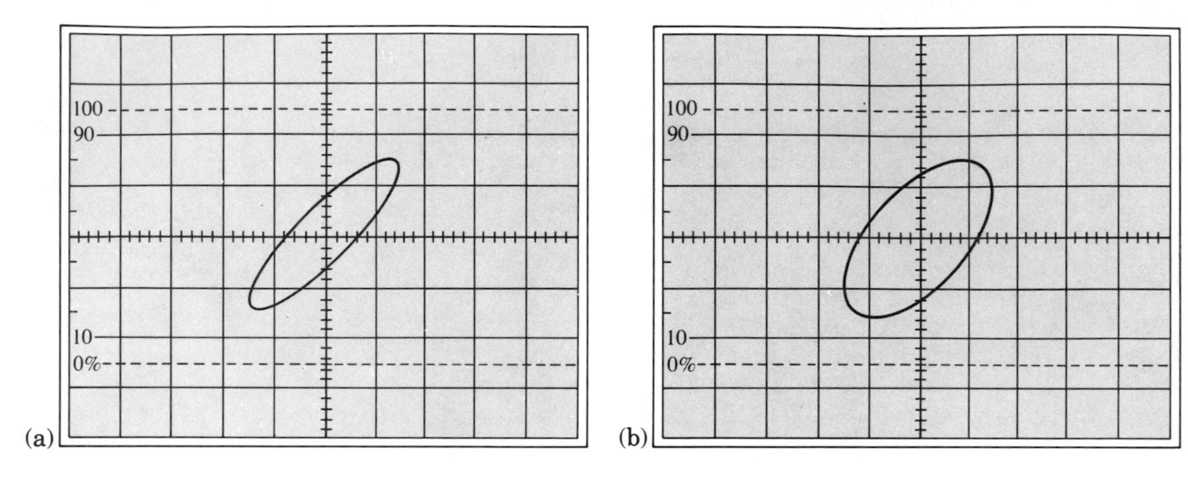

Figure 11–36 Lissajous patterns for Problem 11.

12. For the two sine waves of Figure 11–37, determine the phase shift if the time-base sensitivity is set at 0.1 μs/division.

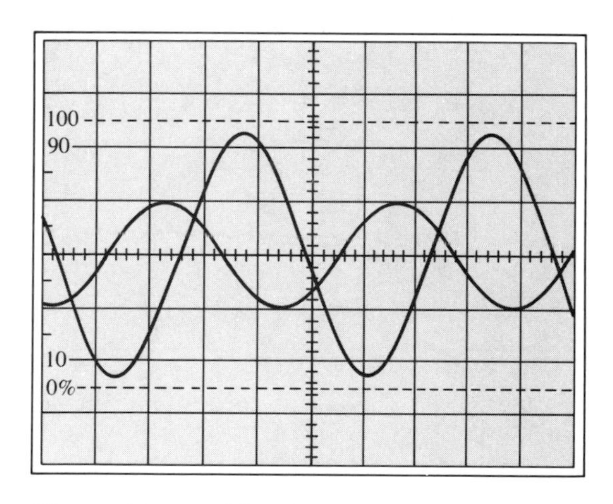

Figure 11–37 Waveforms for Problem 12.

13. Determine the fall time of the portion of the square wave shown in Figure 11–38 if the time-base sensitivity is set at 0.1 μs/division.

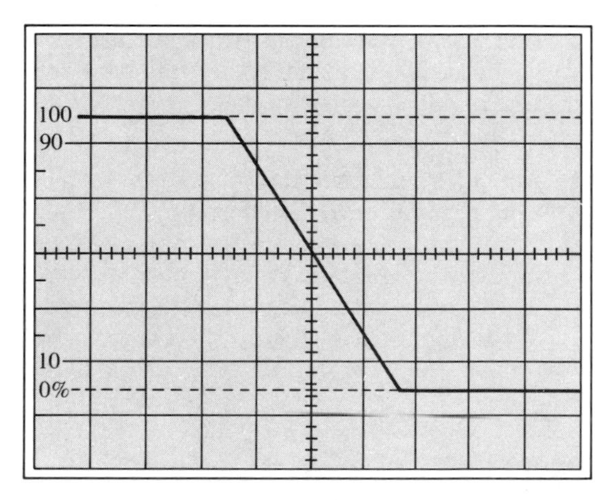

Figure 11–38 Waveform for Problem 13.

14. Repeat Problem 13 if the time base is magnified by using the 5X control.

12

MEASUREMENT OF FREQUENCY AND TIME INTERVALS

12–1 INSTRUCTIONAL OBJECTIVES

At the completion of this chapter, you will be able to

- Explain the operation of a basic frequency counter.
- Describe how to use a universal counter/timer to measure frequency, periods and time intervals, frequency ratios, and events.
- Describe the operation of a reciprocal counter.
- Describe how high-frequency measurements are made using prescalar, transfer oscillator, and heterodyne frequency converter techniques.
- Describe methods for making low-frequency, phase-shift, and rise-time measurements.
- Explain how calibrations are made and frequencies measured using direct comparison methods against both another time base and the National Bureau of Standards.
- Describe the operation and limitations of a resonant-reed meter measuring frequency.

12–2 INTRODUCTION

Over the years, many instruments have been developed to measure frequency and time, but relatively few are still in common use. This chapter concentrates on those most likely to be encountered in a modern service shop or engineering lab.

Instruments for the measurement of frequency can be broken down into two groups: active and passive. An *active instrument,* as its name suggests, contains active analog and digital circuits and requires a power source. On the other hand, a *passive instrument* does not contain active circuits, therefore not requiring a power source. The active instruments are generally more accurate and more expensive. Since the measurements of frequency and time are closely related, some of the instruments that measure frequency can also be used to measure time, and vice versa.

12–3 ACTIVE FREQUENCY MEASURING INSTRUMENTS

At present, most practical frequency and time measurements are made using either an oscilloscope or an electronic counter/timer. The oscilloscope is satisfactory for low-precision frequency and time measurements, using the oscilloscope's internal time base. However, more precise measurements can be made using a high-quality external oscillator in conjunction with either the Z-axis input, to provide timing marks; or the Y-axis input, to form Lissajous patterns. The techniques involved in measuring both frequency and time with the oscilloscope are discussed in Chapter 11.

12–4 THE FREQUENCY COUNTER

A **frequency counter** is a device that counts selected input signal transitions for a fixed period of time and displays the resultant frequency on a multidigit LED/LCD display. The block diagram of a basic counter is shown in Figure 12–1.

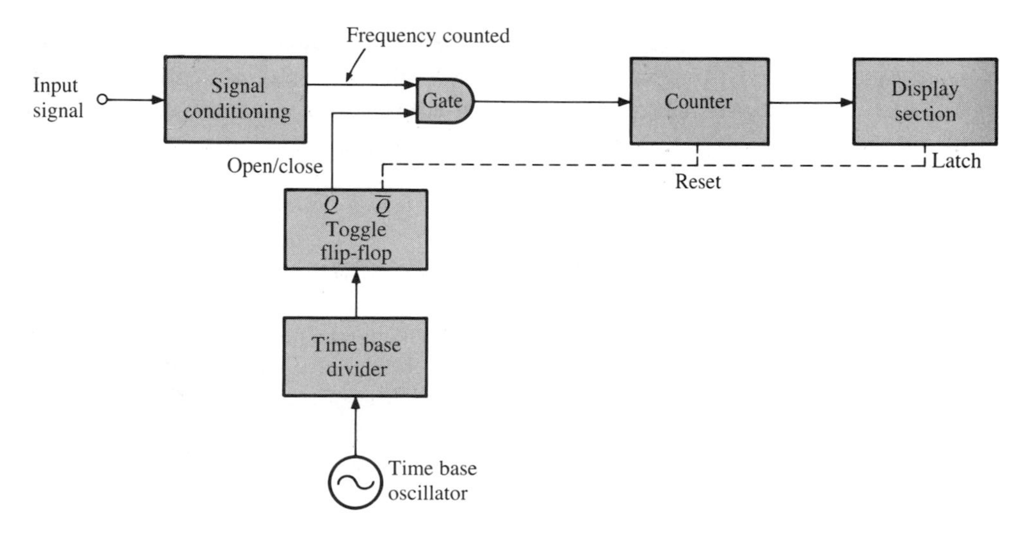

Figure 12–1 Frequency counter block diagram.

The input signal is applied to a high-impedance input circuit where it is both amplified and conditioned ("squared up") to make its waveshape and amplitude compatible with the digital circuitry that follows. The conditioned input signal is applied to one input of a "count" gate. The output frequency of the 1-MHz time base oscillator is divided down by the time-base divider chain and applied, as the clock signal, to a toggle flip-flop. The division ratio of the frequency divider is selectable in steps of factors of ten, providing timing pulses at frequencies of 1 kHz, 100 Hz, 10 Hz, and 1 Hz. This corresponds to gate times of 1 ms, 10 ms, 100 ms, and 1 s, respectively. When the time base selector switch is shown in the 1-kHz (1-ms) position, for example, it clocks the flip-flop once each millisecond. This means that the Q output of the flip-flop is alternately high and low for one 1 ms. The signal on the flip-flop's complementary output ($\overline{Q}$) is inverted with respect to Q.

While Q is high, the "count" gate is opened, so the input signal passes onto the counter section. Thus, the counter counts for 1 ms. At the end of the 1-ms count period, output Q goes low, disabling the gate. Since output $\overline{Q}$ goes high at this time, its leading edge triggers the first one-shot. Once triggered, a narrow output pulse of the one-shot signals the digital display section to be latched with the new count information. On the trailing edge of the update pulse, a second one-shot is triggered, and it produces a narrow pulse that resets (i.e., "zeros") the counter section. The counter is ready to begin the next count cycle. If, for example, we assume that 234 pulses arrived during the 1-ms period that the gate was opened, the counter would contain information (in digital code) equal to 234 when the gate was closed. When the display section was updated, the display would numerically read "234." Since the time base would be set to the 1-kHz position, the *least significant* digit of the display (i.e., the "4") represents a resolution of 1 kHz. Consequently, the measured frequency is 234 kHz; the three trailing zeros are understood.

If the time base were now switched to a gate time of 100 ms (10 Hz), the least significant digit of the display would represent a resolution of 10 Hz. If 234 pulses were counted during the 100 ms gate interval, the actual frequency would be 2340 Hz.

12–5 THE UNIVERSAL COUNTER/TIMER

The **universal counter/timer,** or **UCT,** combines the frequency- and time-measurement functions in one instrument at a relatively inexpensive price. Based on various combinations of slopes and triggering levels, the UCT is capable of operating in any one of five basic modes:

- Frequency measurement
- Period measurement
- Event counter

- Time-interval measurement
- Frequency ratio

The slope and triggering controls function in the same way as those of an oscilloscope (Chapter 11).

With the exception of frequency measurement, all modes of the UCT are discussed in the following sections. The UCT scheme used for frequency measurement is identical to that discussed in the previous section.

Period Measurements

Period measurements determine the time required for a signal to complete one full cycle of oscillation. Mathematically, it is the reciprocal of frequency. For the scheme shown in Figure 12–2, the counter section counts pulses from the time base section, and the input signal controls the gate flip-flop. The number of pulses counted is directly proportional to the length of one cycle of the input signal.

The gate is opened at a selected point on the input waveform and closed at the same point in the next cycle; this corresponds to a time period equal to the duration of one cycle. For example, if the time base output frequency were 100 kHz and the gate opened by the input signal for 1 ms, the resulting count would be 100. Since the clock pulses would be arriving at the rate of one every 10 μs, this count would represent 100 × 10 μs, and would be displayed as 1.00 ms.

For high-frequency signals in which both the gating period is relatively short and the count resolution low, many UCTs provide the option of measuring the time duration of several complete cycles and then displaying the *scaled*

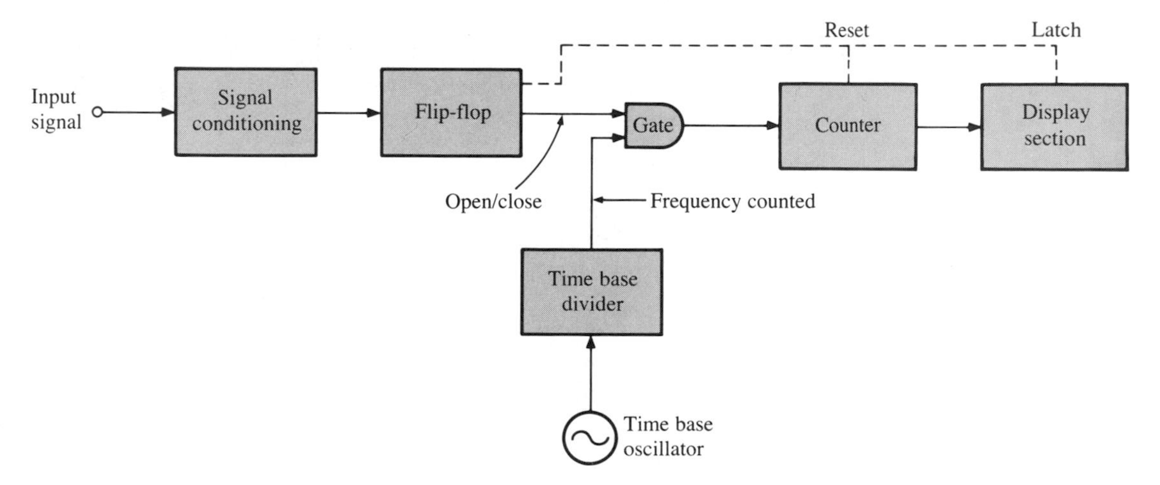

Figure 12–2 Period measurement.

average value of those measurements. However, measuring the frequency of low-frequency signals is best accomplished by measuring the period and computing the frequency by taking the reciprocal of the period. For example, it is necessary to count for 10 seconds to provide a ±0.1-Hz resolution on a 100-Hz signal. But, in the "period measurement" mode, a display of 10,000 μs can be arrived at in a fraction of a second, providing resolution of ±0.01 Hz.

Frequency Ratios

When measuring frequency ratios, the UCT time-base oscillator is disconnected. The lower-frequency input signal serves as the control for the gate flip-flop, and the higher-frequency signal is counted (Figure 12–3).

For example, were the lower frequency 1 kHz and the higher frequency 14 kHz, the gate would be open for 1 ms and 14 cycles of the higher frequency would be counted. The display would read "14," indicating a frequency ratio of 14:1.

Time Intervals

Time intervals are measured by using external "start" and "stop" signals to control the count gate (Figure 12–4). The counting of the pulses of the UCT's internal time base begins with the presence of a start trigger signal and ends with a stop trigger signal. The accumulated count held by the counter section is directly proportional to the time interval between start and stop trigger signals, so the final count is scaled and displayed according to the time-base frequency in use.

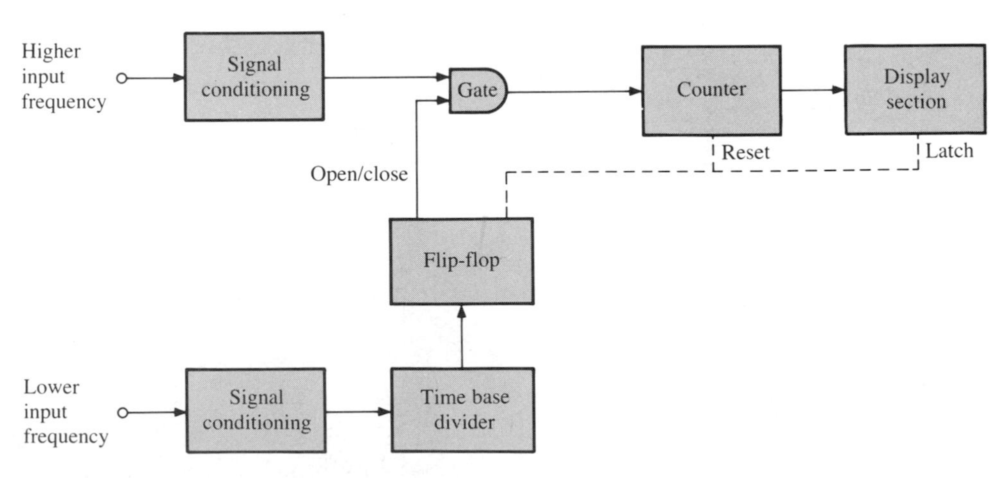

Figure 12–3 Measurement of frequency ratios using a UCT.

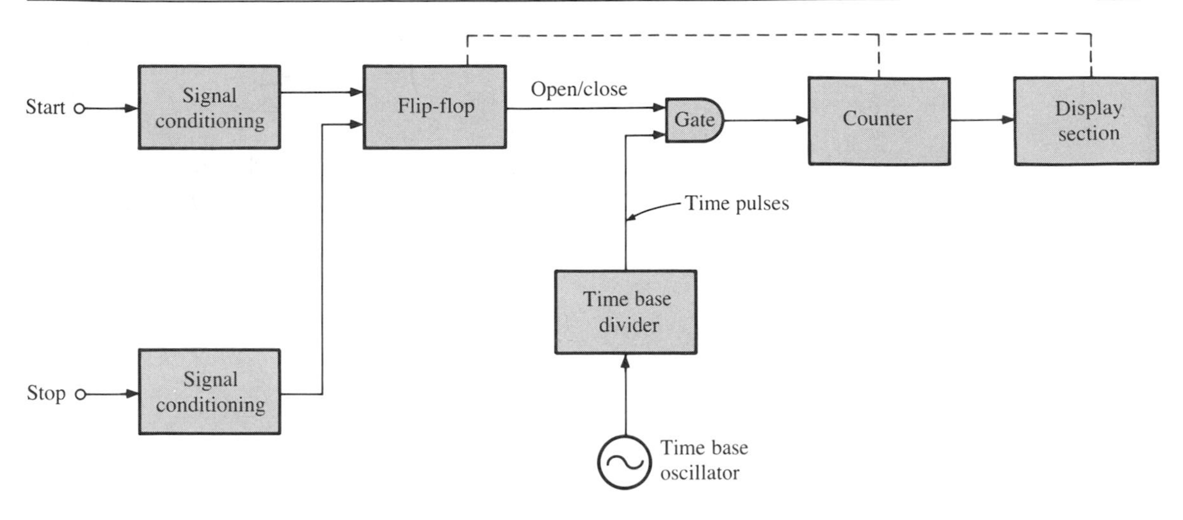

Figure 12–4 Measurement of time intervals using a UCT.

The time-interval mode is particularly convenient for measuring pulse width and spacing. The measured pulses are fed simultaneously to the "start" and "stop" inputs, and identical threshold levels are selected to reduce errors due to pulse rise and fall times. To measure pulse width, the leading edge is selected as the "start" and the trailing edge as the "stop" signal. To measure pulse spacing, the trailing edge is selected as the "start" signal and the leading edge the "stop" signal.

The basic time interval approach can be extended, and is only limited, by the ingenuity of the user. For example, time-interval measurements can be used to determine the velocity of moving objects if the objects break two light beams at a known distance apart. The first light beam provides the "start" signal, and the second provides the "stop" signal. The velocity is calculated by dividing the spacing by the measured time interval.

Event Counting

Event counting is similar to frequency measurement, except that an external signal is used to control the gate. The UCT counts selected signal transitions until it is stopped or reset. For example, it can be set to count the number of leading or falling edges or switch closures.

Input-Signal Conditioning

The input-signal conditioning section (Figure 12–5) gives the input signal a relatively noise-free waveshape and sufficient amplitude to be compatible with the digital circuitry of the counter section. Analogous to the input stage of a radio or television receiver, the input section is often referred to as the "front

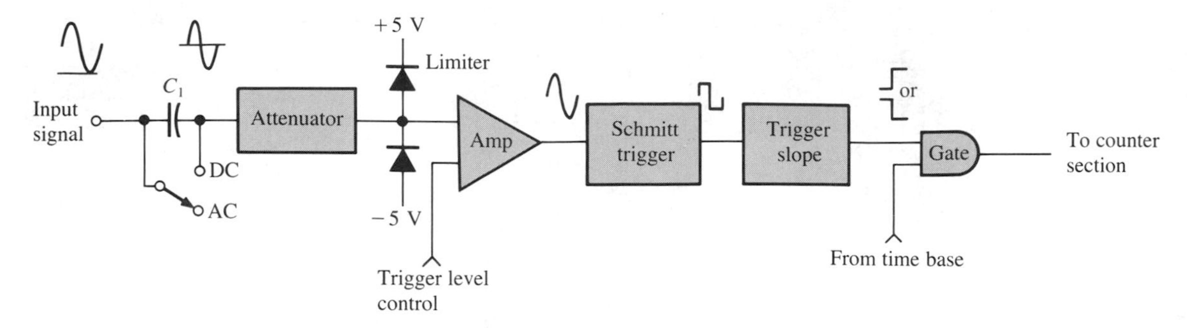

Figure 12–5 Frequency counter input signal conditioning.

end," frequently using an amplifier to increase the amplitude of low-level signals and an attenuator to decrease the amplitude of high-level signals.

The attenuator usually consists of a frequency-compensated voltage divider, or attenuator, and typically provides switch-selectable division ratios of 1, 10, and 100, or a continuously variable division control. The voltage divider section is often followed by a voltage limiter network, which ensures that the following amplifier stage is not overloaded by high-level input signals. The input signal can be either AC coupled (C_1 in the signal path) or DC coupled (direct coupled, C_1 shorted). The frequency-compensated attenuator that follows is similar to that used at the input of an oscilloscope, and its operation is described in Chapter 11. Basically, it provides a fixed value of voltage division for all frequencies over a broad range. The simple limiter circuit that follows limits the voltage excursions at the output of the attenuator to values between one diode drop above +5 V (approximately +5.7 V) and one diode drop below −5 volts (−5.7 V). As long as the signal voltage at the junction of the two diodes is within the −5.7- to 5.7-V range, both diodes are reverse biased. If the input signal forward-biases one of the diodes, that diode conducts heavily and prevents any further voltage increase. The ±5-V limits shown here assume that TTL devices are used for the counter section. If CMOS devices are used, different voltage limits may be appropriate.

Following the limiter is a broadband amplifier, which increases the amplitude of any low-level signals that may be present. Following the amplifier is a logic-compatible Schmitt trigger, which ensures that (1) the signal has been squared up and given fast rise and fall times before being applied to the counter section, and (2) the signal levels are compatible with the digital logic family used in the counter section. In the output of the Schmitt trigger shown in Figure 12–6, each cycle of the input signal is represented by a rectangular digital pulse the duty cycle of which is unimportant.

The Time-Base Oscillator

The accuracy of an electronic counter is almost entirely dependent on its time-base oscillator. Consequently, time-base error is the frequency counter's great-

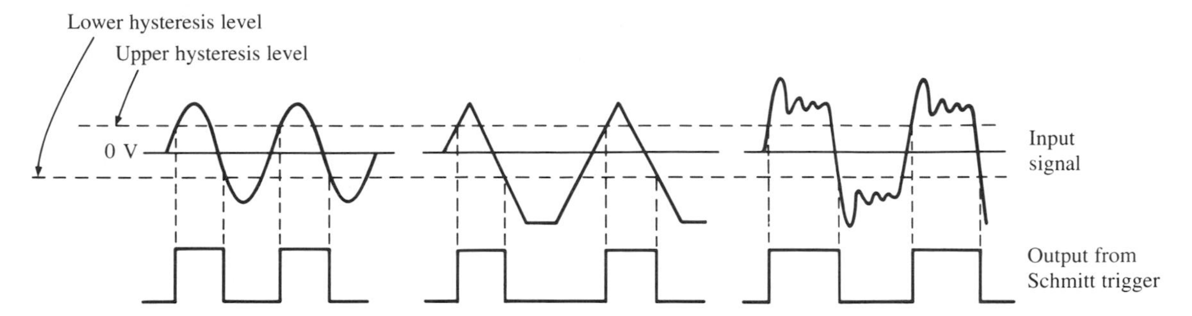

Figure 12–6 Effect of using a Schmitt trigger on various waveforms.

est error source. Most UCTs use a very stable crystal-controlled time base oscillator of a fixed frequency between 1 and 10 MHz. The frequency is divided down through a series of decade (divide-by-10) counters to provide the selected gating period. A front-panel mounted switch permits selection of the desired gating period by switching in the proper number of decade dividers.

Three types of time-base oscillators are used in UTCs:

(1) a room-temperature crystal oscillator without any form of temperature compensation,
(2) a temperature-compensated crystal oscillator, and
(3) an oven-controlled crystal oscillator.

The basic difference between these three types is the degree of temperature-induced frequency drift that each exhibits.

Oscillator Stability and Drift

The uncompensated room-temperature crystal oscillator is the least expensive of the three and has a typical frequency drift of ±5 ppm over a temperature range of 0 to 50 °C. The **temperature-compensated crystal oscillator (TCXO)** uses special components and circuits to, at least partially, compensate for the effects of varying temperature on the oscillator frequency. A TCXO can be expected to give a frequency drift in the range of ±2 ppm over a temperature range of 0 to 50 °C.

The **oven-controlled oscillator** houses its crystal, or frequently its entire oscillator circuit, in a thermostatically controlled, insulated container called an **oven** to maintain a nearly constant crystal temperature. The simpler ovens use a heating element controlled by a thermostat, while the more sophisticated ovens use electronic circuitry to *proportionally control* the oven heater, providing even tighter control of the oven temperature.

The most expensive counters normally use an oven-controlled oscillator to provide the minimum oscillator frequency drift due to ambient temperature change. Simple on/off oven control can provide both temperature stability and proportional oven control of better than ±0.2 ppm. However, the physical properties of the crystal change somewhat with time, resulting in a cumulative frequency drift called **aging.**

A less-significant source of oscillator frequency drift is random frequency and phase fluctuation within the crystal itself. **Short-term stability** is a measure of the effects of the noise generated internally within the oscillator on the average frequency measured over a short period of time (usually one second). Since the oscillator frequency is also affected by changes in power supply voltage, the supply voltage must be well regulated to achieve high-frequency stability. This drift is usually expressed in terms of a 10% change in line voltage.

Time-Base Frequency Divider Chain

The time-base divider section consists of a series of switch-selectable decade counter stages. For every ten input pulses, the decade counter stage produces one output pulse. In modern counters, the decade counters are in integrated circuit form. The time-base switch selects the submultiple of the oscillator frequency that determines the count period. This same switch also often controls the position of the decimal point in the display. For example, if the counter has an eight-digit display, a one-second count period provides a ±1-Hz resolution with a maximum reading of 99,999,999 Hz. Further, a 10-second count period provides a ±0.1 Hz resolution with a maximum reading of 9,999,999.9 Hz. The decimal point is moved one position to the left each time the count gate period is increased.

When a binary counter completes a count cycle, a ±1-count ambiguity exists in the display's least significant digit. As shown in Figure 12–7, depending upon the exact instant that the gate opens and permits the first pulse to be counted, a pulse can be missed. As a consequence, counter accuracy specifications are generally followed by a "±1-count, ±time-base error" qualifier.

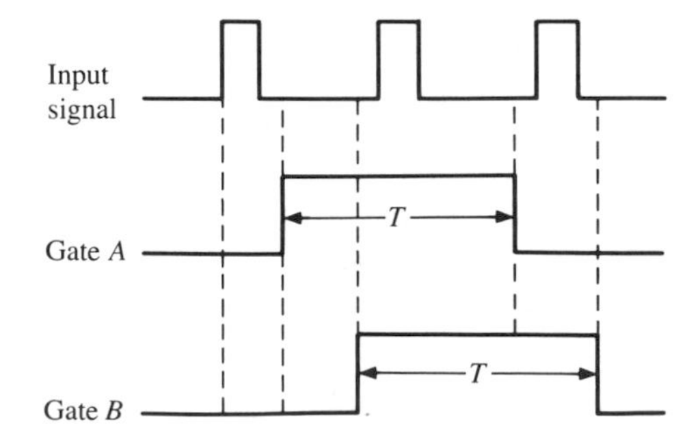

Figure 12–7 Example of a one-count error. Even though gates *A* and *B* are enabled for the same time, gate *A* passes only one pulse while gate *B* passes two pulses.

EXAMPLE 12–1

A frequency counter has an accuracy of ±1 count and a time-base error of 3 ppm. Determine the percent error that results when measuring the following frequencies:

(a) 60 Hz
(b) 1 kHz
(c) 25 MHz

Solution:

(a) At 60 Hz, the magnitude of the error is first determined

$$\begin{aligned}\text{error} &= \pm(1 \text{ count} + 60 \text{ Hz} \times 3 \text{ ppm})\\ &= \pm(1 \text{ count} + 1.8 \times 10^{-4} \text{ counts})\\ &\simeq \pm 1 \text{ count}\end{aligned}$$

so that the percent error is

$$\%\ \text{error} = \pm\frac{1 \text{ count}}{60 \text{ Hz}} \times 100\%$$

$$= \pm 1.7\%$$

(b) At 1 kHz in similar fashion,

$$\begin{aligned}\text{error} &= \pm(1 \text{ count} + 1 \text{ kHz} \times 3 \text{ ppm})\\ &= \pm(1 \text{ count} + 0.003)\\ &\simeq \pm 1 \text{ count}\end{aligned}$$

$$\%\ \text{error} = \pm\frac{1 \text{ count}}{1 \text{ kHz}} \times 100\%$$

$$= \pm 0.1\%$$

(c) At 25 MHz,

$$\begin{aligned}\text{error} &= \pm(1 \text{ count} + 25 \text{ MHz} \times 3 \text{ ppm})\\ &= \pm(1 \text{ count} + 75 \text{ counts})\\ &= \pm 76 \text{ counts}\end{aligned}$$

$$\%\ \text{error} = \pm\frac{76 \text{ counts}}{25 \text{ MHz}} \times 100\%$$

$$= \pm 0.000304\%$$

Display Section

The display section contains latches and decoder/drivers as well as the multidecade display elements themselves. The latches store the counter's running total at the end of the **gate period.** The stored count is applied to the input of the decoder/drivers, where it is converted into the appropriate segment code for each digit of the LED or LCD display. The driver portion supplies the proper voltage and current levels to the display segments. Thus, the last count is stored and continuously displayed until it is updated at the end of the next gate period.

Other Sources of Error

There are two notable sources of error other than inaccuracies. These are the *systemic error* and the *trigger error*.

Systemic error is caused by a slight mismatch between the rise times and propagation delays of the UCT's start and stop input channels. Mismatched probes and cable lengths can also cause this type of error. **Trigger error** is usually caused by the random noise that can accompany an input signal. The noise can cause the signal to cross the hysteresis level either too soon or too late. As a result, when making frequency measurements, the displayed frequency can be either high or low. When making period measurements, the signals themselves are used to control the opening and closing of the count gate. As a consequence, the time the gate is open can be either too long or too short.

A significant trigger error occurs only when an external signal is used to control the gate. This applies when period, time-interval, and ratio measurements are being made.

12–6 THE RECIPROCAL COUNTER

Example 12–1 demonstrates that the greatest percent measurement error occurs at the lowest frequency. The *reciprocal counting* technique is used to reduce the errors associated with low-frequency measurements. As illustrated by the block diagram of Figure 12–8, the **reciprocal counter** is a variation of the conventional frequency counter in that it uses *separate registers* to count elapsed time and events. The contents of these counters are mathematically processed to provide a display reading in terms of either the signal period or frequency.

When the count gate is open, the "events" counter accumulates pulses from the external input and the time counter accumulates pulses from the internal clock. The arithmetic circuits process this information and provide the appropriate data to the display. The advantage of reciprocal counting is that it can give significantly more accurate digits at lower frequencies.

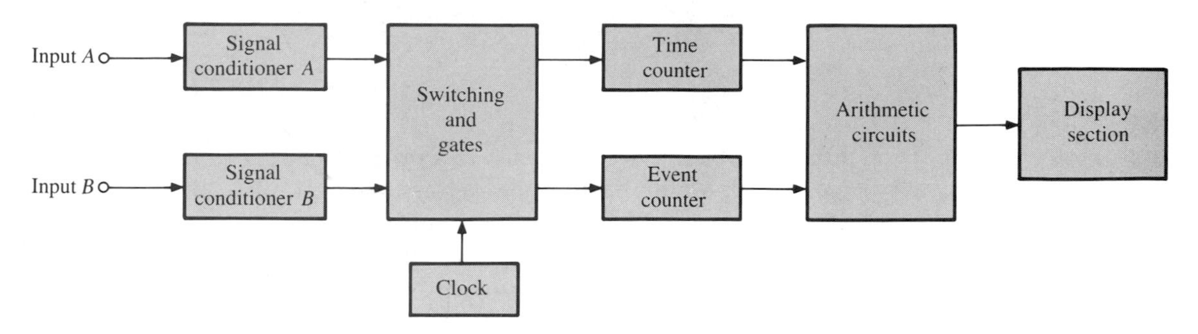

Figure 12–8 Reciprocal frequency counter.

EXAMPLE 12–2

Repeat Example 12–1 if the frequencies are measured with a reciprocal counter with a time base of 1 MHz.

Solution:

(a) At 60 Hz, the gate is opened for 1/60 Hz, or 16.7 ms. Consequently, the number of pulses of the 1-MHz time base counted during this 16.7-ms period is

$$\text{number of pulses} = (16.7\text{ ms}) \times (1\text{ MHz}) = 16{,}700$$

so that the accuracy of the measurement of the 60-Hz signal is ± 1 count in 16700, or 60 Hz/16700 = 0.0036 Hz. Expressed as a percentage of the 60-Hz measured frequency, the error is

$$\%\text{ error} = \pm\frac{0.0036\text{ Hz}}{60\text{ Hz}} \times 100\% = \pm 0.006\%$$

(b) At 1 kHz, the number of pulses counted is 1 ms/1 μs, or 1000. The accuracy of the measurement is ± 1 count in 1000, or 1 kHz/1000 = 1 Hz. Expressed as a percentage of the 1-kHz measured frequency,

$$\%\text{ error} = \pm\frac{1\text{ Hz}}{1\text{ kHz}} \times 100\% = \pm 0.1\%$$

(c) At 25 MHz, the number of pulses counted is 40 ns/1 μs, or 0.04. The accuracy of the measurement is ± 25 counts in 1, or 25 MHz/0.04 = 625 MHz. Expressed as a percentage of the 25-MHz measured frequency,

$$\% \text{ error} = \pm\frac{625 \text{ MHz}}{25 \text{ MHz}} \times 100\%$$

$$= \pm 2500\%$$

Even if the time base were increased to a practical upper limit of 10 MHz, the percent error would be 250%, which is unacceptable. While the measurement accuracy of the reciprocal counting technique is superior at lower frequencies (typically below 1 kHz), the conventional method should be used in high-frequency measurements.

12–7 HIGH-FREQUENCY MEASUREMENT

The measurement of frequencies greater than 50 MHz often presents problems, as these frequencies are often beyond the upper range of frequency counters. However, this limitation can be overcome by using either a prescaler, transfer oscillator, or a heterodyne frequency converter.

The Prescaler

The **prescaler** is a frequency divider that is inserted ahead of the frequency counter's input. It often uses a high-speed emitter-coupled logic (ECL) divider circuit to divide the measured frequency down to a range that can be handled directly by the counter. In this manner, frequencies as high as 2 or 3 GHz divided by 100 can be measured by low-cost counters. To compensate for the division factor of the prescaler, either the gate time can be expanded by the same factor or the displayed frequency mentally multiplied by it. For example, the range of an eight-digit, 50-MHz frequency counter can be extended to 500 MHz, using a divide-by-10 prescaler such as the circuit shown in Figure 12–9.

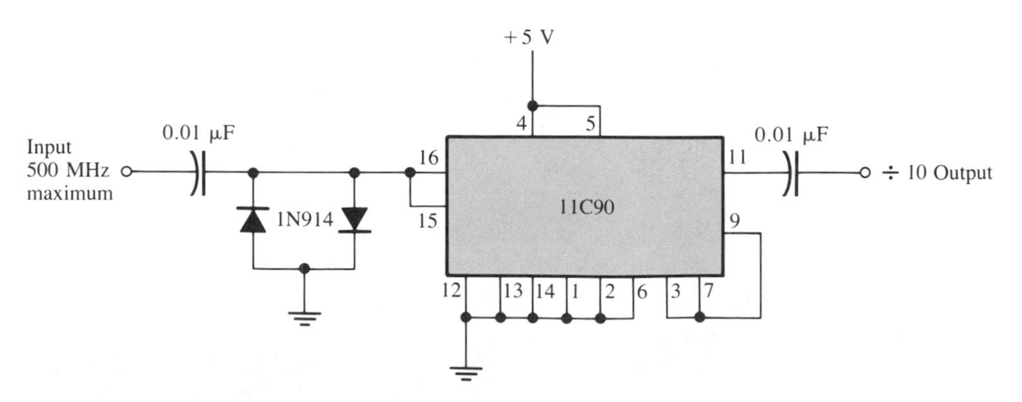

Figure 12–9 Divide-by-10 ECL prescaler.

If the unknown input frequency to be measured were actually 378.524 MHz, the frequency counter, assuming a 1-second gate time, would display a frequency of 37,852,400 Hz (37.8524 MHz). Since the prescaler divides the high-frequency input signal by a factor of 10, the actual input frequency is ten times the displayed reading, or 378.524 MHz.

The Transfer Oscillator

In the **transfer oscillator** scheme, shown in Figure 12–10, harmonics of a variable low-frequency oscillator (VFO) are mixed with the measured signal. The VFO is tuned until one of the harmonics is **zero-beat** with (made equal to) the measured frequency. At this point, the measured frequency is N times the VFO frequency displayed by the frequency counter, where N is the harmonic number.

If the value of the input frequency f_X is unknown, it is necessary to identify the harmonic value of the VFO in use. This is accomplished by slowly tuning the VFO either lower or higher in frequency to find an *adjacent* zero-beat. The harmonic number of the first VFO frequency is given by

$$H_1 = \frac{f_2}{f_1 - f_2} \text{ (round up answer)} \tag{12–1}$$

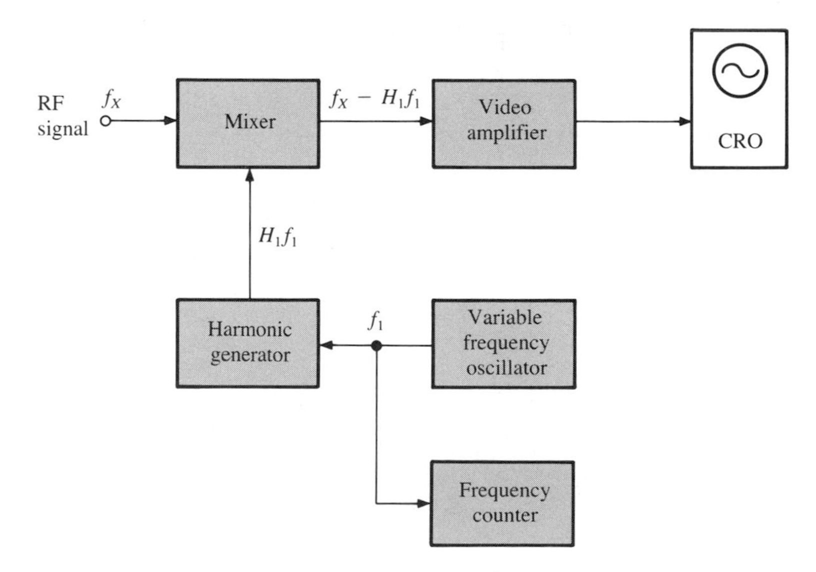

Figure 12–10 Using a transfer oscillator and frequency counter to measure an unknown frequency that is above the frequency counter's range.

where

$$f_1 = \text{VFO frequency at first zero-beat}$$
$$f_2 = \text{VFO frequency at second zero-beat}$$
$$H_1 = \text{harmonic number of } f_1$$

The unknown frequency is found from

$$f_X = H_1 \times f_1 \quad (12\text{–}2)$$

The unknown frequency can also be determined by first determining the harmonic number of the second VFO setting:

$$H_2 = \frac{f_1}{f_1 - f_2} \text{ (round up answer)} \quad (12\text{–}3)$$

and

$$f_X = H_2 \times f_2 \quad (12\text{–}4)$$

The greatest sources of error with this method are the precision with which the zero-beat is determined and the instability of the VFO used.

EXAMPLE 12–3

Using the transfer oscillator scheme of Figure 12–10, determine the unknown input frequency. The VFO frequency where the first zero beat occurs is measured as 2,471,429 Hz, and, adjusted for a second zero beat, the VFO frequency is measured as 2,544,118 Hz.

Solution:

Using Equation 12–1, the harmonic of the first zero beat at 2 MHz is

$$H_1 = \frac{2{,}544{,}118 \text{ Hz}}{2{,}544{,}118 \text{ Hz} - 2{,}471{,}429 \text{ Hz}}$$
$$= 35.0$$

so that the unknown frequency is

$$\begin{aligned} f_X &= 35 \times 2{,}471{,}429 \text{ Hz} \\ &= 86.5 \text{ MHz} \end{aligned}$$

The Heterodyne Frequency Converter

The heterodyne frequency converter technique (Figure 12–11) involves mixing the measured frequency (f_X) with a harmonic from a stable oscillator. Although this technique measures frequencies up to 20 GHz, the mixing process yields a difference frequency within the range of a frequency counter.

The time base serves as the stable oscillator, as well as performing its usual function. The oscillator output (f_C) is multiplied and fed to a harmonic generator. The output of the harmonic generator is passed through a calibrated turnable filter to select a particular harmonic (Nf_C) for application to the mixer. The frequency difference ($f_X - Nf_C$) between this harmonic and the measured frequency must be within the passband of the video amplifier.

In a simple system, the tunable filter can be manually operated. In more sophisticated systems, however, the filter is usually automatically tuned under the control of a microprocessor. In either case, if the filter starts at the low end of the converter's frequency range and searches upward in frequency, the first indication always occurs when the harmonic frequency is *below* the input fre-

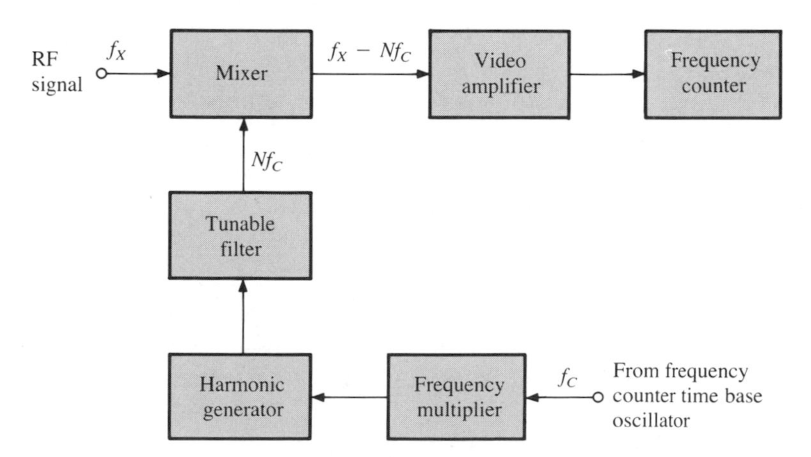

Figure 12–11 Heterodyne-down frequency converter to measure frequencies up to 20 GHz.

quency. The measured frequency is equal to the sum of the counted and the filter frequencies. There are heterodyne frequency converters for frequency ranges from 50 MHz to about 20 GHz.

12–8 LOW-FREQUENCY MEASUREMENT

One of the most accurate methods of determining the frequency of a low-frequency signal is measuring its period, then calculating its frequency by the reciprocal of the period. This is because many clock pulses are counted when the period is measured, as opposed to a few cycles of measured signal, so the resolution is improved.

As an alternate approach, some frequency counters include a **frequency multiplier,** which provides a result opposite that of a prescaler; the displayed frequency is an integer multiple of the measured frequency. The frequency multiplier is usually in the form of a **phase-locked loop (PLL)** synthesizer circuit, and produces an output that is an exact integer multiple of the input frequency, usually by a power of ten.

The PLL synthesizer is often used to produce a frequency that is an exact multiple of a stable reference frequency. When used to multiply a frequency, the PLL is basically an electronic feedback loop system that constantly compares a submultiple of its output frequency against its input frequency, thereby producing an output frequency that is exactly a given integer multiple times larger than the input.

As shown in the block diagram of Figure 12–12, the PLL synthesizer consists of four basic functions: a phase detector, a loop filter, a VCO, and a frequency divider. The **VCO, or voltage-controlled oscillator,** is an oscillator whose free-running frequency is usually determined by a resistor-capacitor network. The output frequency of the VCO (f_o) is reduced by a digital divide-by-N counter whose output is fed back to the phase detector, where it is com-

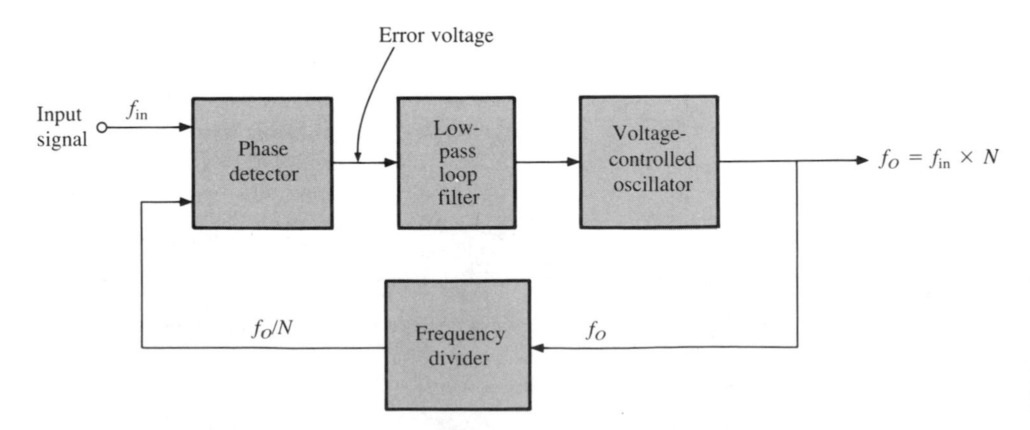

Figure 12–12 Phase-locked loop frequency multiplier (synthesizer) block diagram.

pared to the frequency of the input signal (f_i). The average (DC) output voltage from the phase detector, called the **error voltage,** is proportional to the difference in both the frequencies ($f_i - f_o/N$) and phases of its two inputs. The error voltage is sent through a low-pass filter to remove any high-frequency noise.

The filtered error voltage, in turn, controls the frequency of the VCO. The error voltage forces the VCO frequency to change in a direction that reduces the *frequency difference* between the inputs to the phase detector. Once the VCO starts to change frequency, the loop is said to be in the **capture** state. This process continues until the reference frequency (f_i) *and* the divided VCO frequency (f_o/N) are exactly the same. At this point, the loop is **phase-locked.** During phase-lock, the VCO frequency is exactly N times the frequency of the reference signal. For example, were the reference frequency 1 MHz and the divide-by-N counter to have a divisor or *modulus* (N) of 62, the output frequency would be 62 MHz.

Once phase-lock is established, the loop tracks, or follows, any change in input frequency over a limited range, called the **lock range.** Although the lock range is limited, it is still usually greater than the loop's capture range. Once either the input reference frequency exceeds the lock range or the modulus of the divide-by-N counter is changed, the loop becomes unlocked and the unlock/capture/phase-lock process starts all over again. When unlocked, the PLL VCO's output frequency equals the VCO free-running frequency. The divide-by-N counter usually has a division ratio of either 10 or 100. Figure 12–13 shows the circuit of a PLL synthesizer capable of multiplying the input frequency by either 10 or 100. Further use of the PLL as a frequency-synthesized generator is discussed in Chapter 13.

12–9 MEASURING PHASE DIFFERENCE

Frequency counters that have a "start/stop" interval mode and are capable of making period measurements can be used to measure the phase difference between two signals of the same frequency. This is usually accomplished by measuring the time between two successive positive or negative zero crossings and then using that time to calculate the phase difference (Figure 12–14).

First, the period (T) is measured in the conventional manner. To measure the time between zero crossings (t_ϕ), the frequency counter is set up to operate in the start/stop-interval mode. The start and stop trigger inputs are AC-coupled and set to operate at a level of zero volts DC. To obtain the greatest accuracy, the two input signals should have the maximum amplitude possible within the limits of the input circuits. Once T and t_ϕ have been measured, the phase difference in degrees is calculated from

$$\phi = \frac{t_\phi}{T} 360° \qquad (12–5)$$

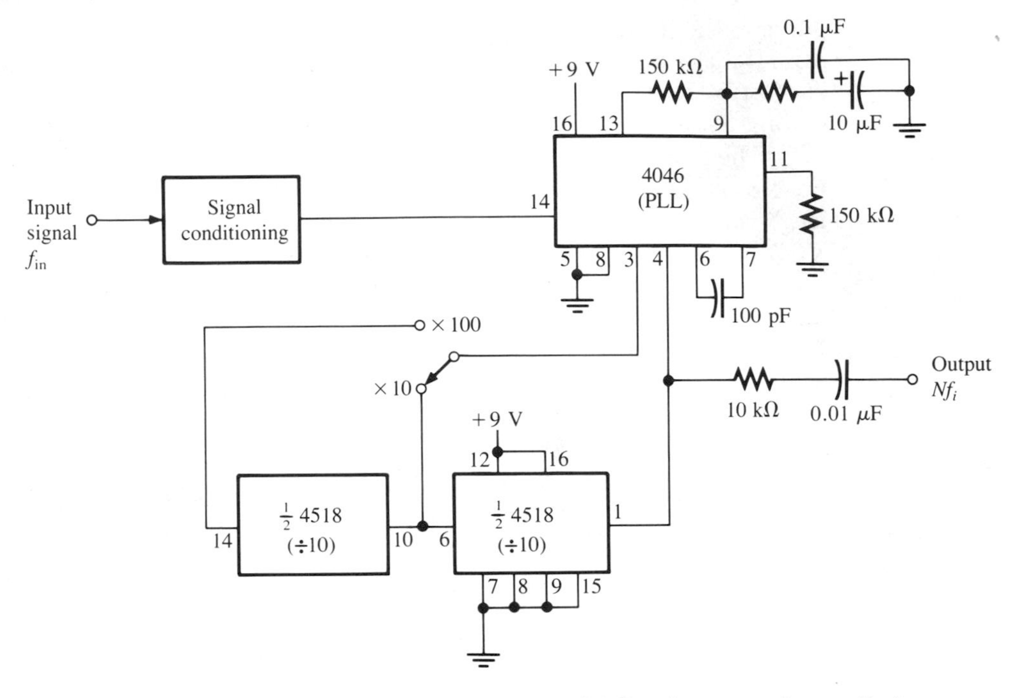

Figure 12–13 10X and 100X PLL frequency multiplier for measuring audio frequencies with greater resolution using a frequency counter.

12–10 MEASURING RISE TIME

By definition, the rise time of a pulse is the time taken to rise from 10% to 90% of its peak amplitude (see Chapter 13). A frequency counter may be used to measure rise time by using the period mode. The "start" control is set to trigger

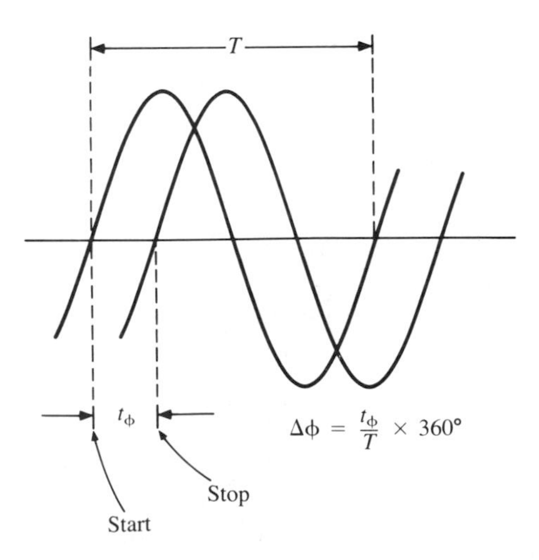

Figure 12–14 Phase difference measurement using a frequency counter having a "start/stop" interval mode.

Figure 12–15 Measurement of rise time using a frequency counter with a "start/stop" interval mode.

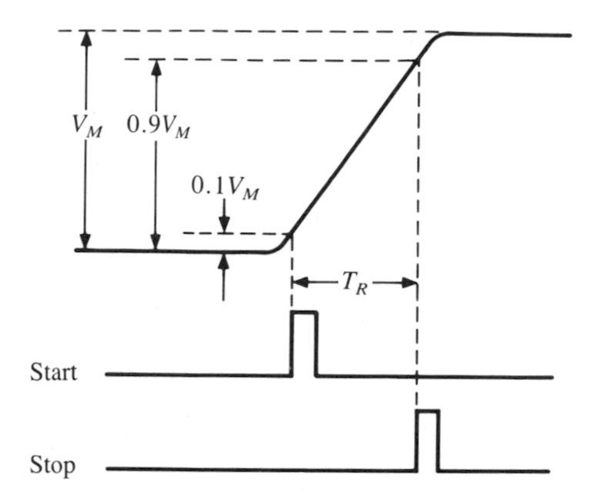

at the 10% point, and the "stop" control is set to trigger at the 90% point (Figure 12–15). For example, a certain digital signal rises from 0.1 V to 3.8 V. The "start" and "stop" trigger points would then be set at

$$V_{\text{start}}(\text{trigger}) = 0.1\text{ V} + (0.1)(3.7\text{ V}) = 0.47\text{ V}$$

$$V_{\text{stop}}(\text{trigger}) = 0.1\text{ V} + (0.9)(3.7\text{ V}) = 3.43\text{ V}$$

By reversing the "start" and "stop" levels, the same technique can be used to measure the fall time of a pulse (the time it takes to fall from 90% to 10% of its maximum amplitude).

12–11 CALIBRATION

The only calibration required for most counter/timers is the accurate setting of the frequency of the counter's internal time-base oscillator. Calibration of the time-base oscillator usually consists of comparing the time-base oscillator frequency to a standard frequency source. This should be done periodically, both according to the manufacturer's recommendations and before making any critical measurements that require an accuracy approaching the limits of the counter.

There are several methods used to calibrate time-base oscillators, and the choice usually depends on the available equipment as much as the accuracy and stability of the time-base oscillator.

Direct Comparison

If a frequency standard with an accuracy of better than five times that of the counter is available, its output can be applied directly to the counter input. The internal time-base oscillator of the counter is adjusted so the counter reads the standard frequency. Of course, the time-base oscillator and crystal should be at their normal operating termperature before any adjustment is made.

Using the Time-Base Output

The counter is provided with a time-base output, and, if the output and the frequency standard are harmonically related, a more accurate frequency adjustment can be made. The time-base output is connected to the vertical input of an oscilloscope, and the frequency standard is connected to the horizontal. The time base is adjusted to display a Lissajous pattern consisting of a single ellipse, which indicates a frequency ratio of 1:1 with a phase difference (see Section 11–8). Because it is extremely difficult to obtain a 1:1 match, the ellipse rotates slightly in a "cartwheel" fashion. If the number of rotations in a given time period is counted, the value of the measured frequency can be interpolated. As an example, were the ellipse to rotate 19 times in a one-minute period, the frequency correction would be 0.317 Hz. If the direction of rotation is such that the unknown frequency is higher than the measured frequency, the correction is *added* to the base value; otherwise, it is subtracted.

Calibration Against WWV

As discussed in Chapter 3, the National Bureau of Standards broadcasts very accurate time and frequency signals via its radio stations, WWV and WWVH. These broadcasts provide a convenient means of calibrating time-base oscillators, using the arrangement shown in Figure 12–16.

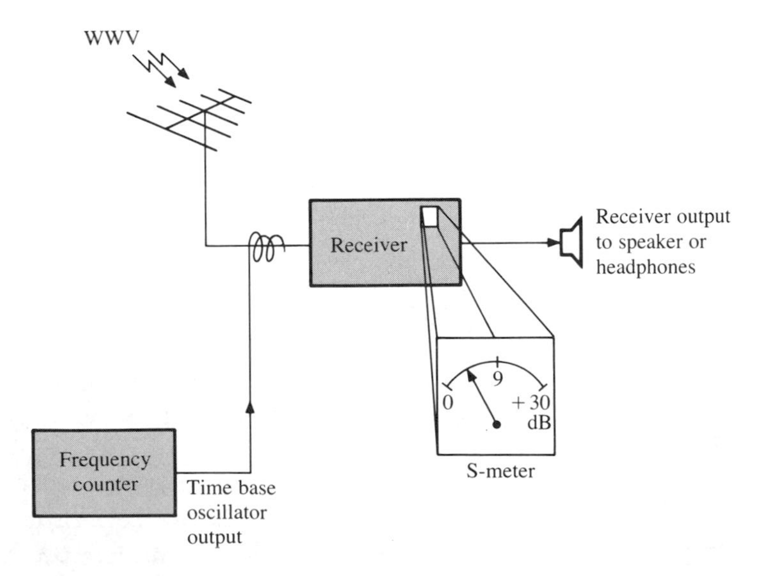

Figure 12–16 Frequency counter calibration against WWV broadcasts.

A shortwave receiver capable of tuning to one or more WWV frequencies (5, 10, or 15 MHz) is required. The WWV signal is listened to using the receiver's speaker or headphones against a harmonic from the time base oscillator. Any frequency difference between the received frequency and a harmonic of the time base oscillator in the audio range produces an audio tone at the speaker—the smaller the difference, the lower the tone, called a **beat note.** For example, if the receiver is set to receive WWV's 10-MHz signal and the counter's time-base oscillator is set at 1 MHz, the 10-MHz signal of WWV is heterodyned against the time base's tenth harmonic.

The time base is adjusted to produce the lowest possible beat note. If the receiver is equipped with a signal strength meter (S-meter), it is possible to refine the adjustment even further. As the frequency difference drops below the audio range ($< 15-20$ Hz), the beat is no longer audible, appearing as a rhythmic swing (frequency "beats" in the S-meter reading) and as audible "clicks" in the speaker or headphones.

The S-meter's sensitivity is adjusted for the meter's most pronounced swing with the potentiometer. The adjustment of the time base oscillator continues until this swing is at a minimum. It is not always possible to stop the swing completely, but a swing of a few cycles per second is usually acceptable. For example, an S-meter needle swing of 12 times per minute is equivalent to a frequency difference of 12/60 seconds, or 0.2 Hz. This difference is N times the *actual* frequency error of the time base, where N is the harmonic being used. For example, were the tenth harmonic used, the actual frequency error of the time base would be 0.02 Hz. The higher the WWV frequency used, the more accurate the frequency adjustment of the time base.

12–12 RESONANT-REED FREQUENCY METERS

The **resonant-reed** (Frahm) **frequency meter** (Figure 12–17) is one of the more simple and rugged frequency-measuring instruments in use. This type of meter consists of a series of thin metal reeds of equal length but with slightly different weights attached to one end. The added weight makes each reed mechanically resonant at a particular frequency. The unweighted ends are attached to a common baseplate. The reed/baseplate assembly is mounted in the field of an electromagnetic coil. The given alternating current to be measured, such as the 60-Hz power line frequency, is passed through the coil, causing the baseplate to vibrate at the frequency of the applied current. As a result, though all the reeds vibrate lightly, the reed tuned to that particular frequency vibrates with a large amplitude.

If the driving frequency is midway between the frequency of two adjacent reeds, both reeds vibrate at an equal but reduced amplitude. If the coil frequency is between two adjacent reeds, but not midway, both reeds vibrate, but the reed closest to the driving frequency vibrates with the greater amplitude.

Figure 12–17 Resonant-reed (Frahm-type) frequency meter. Courtesy of Biddle Instruments.

The weighted ends of the reeds are visible through a narrow window in the face of the meter housing, and are usually painted white, appearing as a series of white dashes against a dark background. Because of the limits imposed by mechanical inertia, these meters are limited to power-line frequencies, usually either 50 or 60 Hz. Consequently, the resonant-reed device normally has a range limited to a few cycles either side of a nominal center frequency, and provides an accuracy of about ±0.3%. Vibrating-reed frequency meters are usually panel-mounted, and find their greatest use with portable, engine-driven generators (alternators), enabling the operator to correctly set the speed.

12–13 SUMMARY

Next to the measurement of voltage, current, and resistance, the measurement of frequency and time has the most varied methods. Although most frequency measurements are performed with a conventional frequency counter, this chapter discussed the universal counter/timer, which is capable of performing frequency, period and time interval, event counting, and frequency ratio measurements. Low-frequency measurements are performed using period measurement, a reciprocal counter, or a frequency multiplier, while high-frequency measurements are best accomplished with either prescaler, transfer oscillator, or heterodyne frequency converter methods. For low-accuracy power-line frequency measurements, the resonant-reed meter is often used.

As with any instrument, frequency counters need to be calibrated. This chapter described methods of calibration against an accurate frequency source, either directly with another frequency counter, by superheterodyning the counter's time base against a standard, or by comparing the counter against harmonics of WWV/WWVH, operated by the National Bureau of Standards.

12–14 GLOSSARY

aging The deterioration of a component's characteristics over a period of time, until stability is reached.

beat frequency The sum or difference frequency that results when two signals of different frequencies are heterodyned.

capture The condition of a phase-locked loop in which the output frequency begins to change from its free-running frequency to that of the loop's input.

error voltage The DC voltage output of the phase detector proportional to the phase difference of its two inputs.

frequency counter An instrument that counts selected input signal transitions for a fixed period of time and displays the resultant frequency on a multidigit numeral display.

frequency multiplier A device whose output is an integer multiple or harmonic of its input frequency.

gate period The time interval that the count gate is enabled.

lock range In a phase-locked loop, the frequency range over which the output follows changes in the input.

locked Also called phase-locked, the condition in which the frequency of an oscillator's output signal exactly follows the frequency of a reference signal.

oven An insulated container used to maintain a constant temperature surrounding the crystal controlling an oscillator.

oven-controlled oscillator A stable, crystal-controlled oscillator mounted within an insulated container of constant temperature.

phase-locked The condition whereby the phase of an oscillator's output signal exactly follows the phase of a reference signal by comparison of the phase difference of the two signals.

phase-locked loop (PLL) A feedback-loop system in which the output equals the input frequency, except for a constant phase difference.

prescaler A frequency divider inserted ahead of a frequency counter for frequency measurements beyond the counter's upper frequency range.

reciprocal counter A frequency counter that uses separate registers to count elapsed time and events, which are then mathematically processed to display either signal frequency or period.

resonant-reed frequency meter A meter using a series of reeds that each vibrate at a specific resonant frequency. It is used mainly for power-line frequencies of 50 or 60 Hz.

short-term stability The frequency stability of an oscillator taken over a very short time period, typically one second.

systemic error An error caused by the slight mismatch between the rise times and propagation delays of the start and stop channels of a universal counter/timer.

temperature-compensated crystal oscillator (TCXO) A crystal oscillator using special components to compensate for the effects of temperature on oscillator frequency.

time base A crystal-controlled oscillator and associated divider chain that control the timing and accuracy of a frequency counter.

transfer oscillator A variable low-frequency oscillator, whose harmonics are mixed with a signal whose frequency is to be measured.

trigger error Error caused by random noise in the input signal.

universal counter/timer (UCT) An instrument capable of performing frequency, period, frequency-ratio, phase-shift, and rise-time measurements.

voltage-controlled oscillator An oscillator circuit, such as a voltage-to-frequency converter, whose output frequency is controlled by an input voltage level.

zero beat Heard when two signals are made equal in frequency by comparing audio sounds; when no sound is heard, then both signals are mixed.

12–15 PROBLEMS

1. Determine the resolution and maximum frequency that can be displayed on a 7-digit frequency counter if the gate time is
   (a) 0.1 s
   (b) 1.0 s
   (c) 10 s
2. If a frequency counter has a gate time of 100 ms and 84,296 pulses of an input signal are counted, determine the frequency of the input signal.
3. A frequency counter has an accuracy of ± 1 count and a time-base error of 5 ppm. Determine the percent error when measuring a signal whose actual frequency is
   (a) 962 Hz
   (b) 7.15 MHz
4. Repeat Problem 3 if the time-base error is decreased to 0.2 ppm.
5. A 440-Hz frequency is measured using a reciprocal counter. If the counter's time base is 1 MHz and has an accuracy of ± 1 count with a time-base error of 1 ppm, determine the percent error.
6. Repeat Problem 5 if the time-base frequency is changed to 100 kHz.
7. A divide-by-100 prescaler is placed ahead of an eight-digit frequency counter to measure an unknown frequency. If the frequency counter, having a gate time of 0.1 s, reads 1,465,207.1 Hz, determine the unknown frequency.
8. If the gate time of the frequency counter in Problem 7 is changed to 1.0 s, what would the counter read?
9. An unknown frequency is measured using the transfer oscillator method. A frequency counter measures the frequency of the first zero beat as 468,333 Hz. Determine the unknown frequency when the VFO is adjusted for an adjacent zero beat with a frequency of 439,063 Hz.
10. A 10X PLL multiplier is used to measure an unknown frequency. If the frequency counter reads 55,286 Hz, determine the unknown frequency.
11. A universal counter/timer is used to measure the phase shift between two sine waves of the same frequency. The period is measured at 4.634 ms, and the time between the zero crossings on the positive slope of both waveforms is 0.357 ms. Determine the phase shift between the two signals.
12. The time base of a frequency counter is to be compared against a 1-MHz standard. After a suitable warm-up period, the time base is adjusted by observing a Lissajous pattern on an oscilloscope. Although an ellipse is obtained after adjustment, it rotates in a "cartwheel" fashion nine times in a 45-second period. Determine the error from the desired 1-MHz frequency as an absolute error, and in parts per million (ppm).
13. The 100-kHz time base of a given frequency counter is compared against WWV, broadcasting on 2.5 MHz. The adjustment was such that the fluctuations on the receiver's *S*-meter were measured at 24 swings in a 60-second period. Determine the absolute error of the comparison.
14. Repeat Problem 13 if the same observations were made using WWV's 15-MHz frequency.

13

WAVEFORM GENERATORS AND ANALYZERS

13–1 INSTRUCTIONAL OBJECTIVES

At the completion of this chapter, you will be able to

- Explain the differences between oscillators, signal generators, function generators, and synthesizers.
- Explain how an amplifier becomes an oscillator.
- Determine the output frequency of a Wien bridge oscillator.
- Explain the characteristics of both an amplitude-modulated and a frequency-modulated carrier.
- Describe how a triangle wave is converted into sine and square waves of the same frequency.
- Describe how a function generator is swept over a range of frequencies.
- Explain how a phase-locked loop is used to synthesize various output frequencies.
- Describe how a square wave is used to test for distortion in amplifiers.
- Briefly describe the operation and use of spectrum, Fourier, wave, distortion, and audio analyzers.
- Determine the percent modulation of an amplitude-modulated carrier using an oscilloscope or a spectrum analyzer.

13–2 INTRODUCTION

Depending on their intended functions, instruments that generate waveforms are referred to either as oscillators, signal generators, function generators, or synthesizers. In most cases, they are used in conjunction with an oscilloscope to design, adjust, or repair other equipment.

Analyzing signals in the frequency domain is another important measurement technique, which is widely used to provide information about the overall frequency response performance of electrical and physical systems. As a group, these instruments are referred to as *analyzers;* this includes the spectrum analyzer, Fourier analyzer, wave analyzer, distortion analyzer, and audio analyzer. These instruments all measure the parameters of a signal in the frequency domain, but each uses a different technique.

13–3 SIGNAL SOURCES

Signal sources can be divided into four basic groups as follows:

- oscillators
- signal generators
- function generators
- synthesizers

When referring to test equipment, the term *oscillator* is usually reserved for a signal source that generates exclusively sinusoidal waveforms, while a **signal generator** is an oscillator that has the ability to modulate a carrier signal. A **function generator** provides not only sine waves but also square waves, triangle waves, sawtooths, or any other arbitrary waveform. *Synthesizers* are unique in that they generate sine waves digitally by a special process accompanied by excellent frequency stability and resolution.

As the cost of digitally-derived sources continues to decrease, synthesizers are increasingly used in place of oscillators. Also, many function generators produce digitally derived signals, and the distinction between function generators and synthesizers has begun to blur.

Several basic categories of generator can be further subdivided into pulse generators and sweep generators. **Pulse generators,** as their name implies, create pulses of well-controlled amplitude, width, and rate of repetition. Some of the more sophisticated pulse generators also give control of pulse rise and fall times, delay time, and DC bias level. **Sweep generators,** however, produce a waveform the frequency of which can be varied in a range selected between a minimum and maximum value; the generator's frequency is "swept," or varied, across this selected range at a given rate. Sweep generators are most often used to study or adjust the frequency response of test circuits.

13–4 LOW-FREQUENCY SINE-WAVE GENERATORS

An **oscillator** is a signal source of known frequency. It produces an output signal, yet requires no input signal. The only external input connections to an oscillator are those for the DC power source. The output signal normally has a frequency, amplitude, and waveform determined by the circuit type, the power supply voltage, and the component values selected by the designer. One common class of sine-wave oscillators consists of an amplifier with controlled *positive feedback.*

Positive feedback is applied to an amplifier by taking a fraction of the output signal and adding it to the incoming signal. The signal actually applied to the input of the amplifier becomes the sum of the incoming signal and the signal being fed back from the output. By increasing the amount of feedback, it is possible to decrease the level of the incoming signal and still get the same output amplitude. If this is continued, a point is eventually reached at which an input signal is no longer needed to produce and sustain an output signal. The amplifier is said to *oscillate,* because it supplies its own input signal.

Figure 13–1 shows the block diagram of an amplifier stage with a positive feedback network path. The overall gain of the amplifier is given by

$$A_V = \frac{G}{1 - BG} \tag{13–1}$$

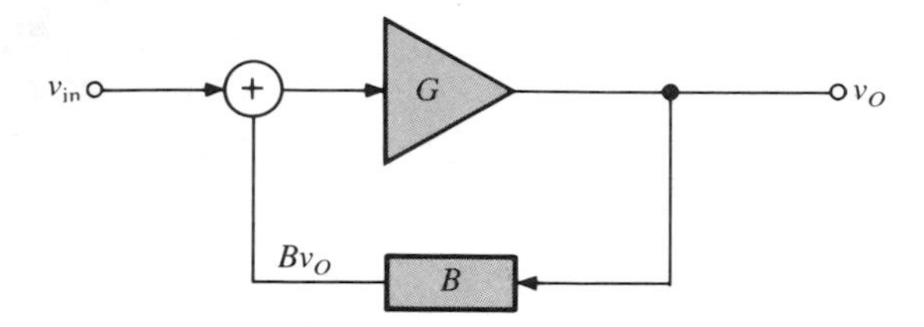

Figure 13–1 Amplifier with positive feedback.

where

G = amplifier voltage gain without feedback
B = fraction of the output signal fed back to the input

If we start with a feedback factor (B) of zero (i.e., no feedback) and slowly increase the amount of feedback (i.e., make B larger), the denominator of Equation 13–1 becomes progressively smaller. As a consequence, the circuit gain (A_V) gets larger. When the product of B and G, known as the *loop gain,* equals one, the denominator of Equation 13–1 equals zero. This is the condition necessary to support and sustain oscillation; no external input signal is required to produce an output.

If the circuit is designed to make the product of the feedback factor and amplifier gain equal to one only at a single frequency, the oscillator produces a *sine wave.* If the loop gain is less than one, the circuit does not oscillate. If the loop gain is much larger than one, the output is distorted. (These conditions are summarized in Table 13–1.) The designer of a sine-wave oscillator usually strives to produce a loop gain value just slightly greater than one at the frequency of oscillation, but substantially less than one at all other frequencies.

Table 13–1 Effects of feedback on amplifier operation

Loop gain (BG)	Output
<1	no oscillation
1	oscillator
>1	distortion

The above conditions require a feedback network that is frequency selective. At audio frequencies less than about 20 kHz, this network is usually composed of a combination of resistors and capacitors; thus, the oscillator is referred to as an *R-C oscillator*. One example of a low-frequency *R-C* oscillator is the *Wien bridge* sine-wave oscillator of Figure 13–2, which uses an operational amplifier as the active element. As shown in Figure 13–3, part of the

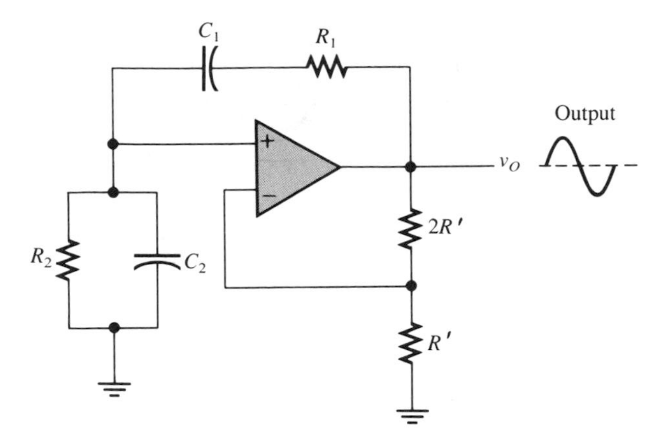

Figure 13–2 Op-amp Wien bridge oscillator.

Wien bridge is a dual R-C network called a *lead-lag* network. Its name comes from the network's output signal, leading the input in phase below a certain frequency and lagging it above that frequency.

The output signal of this circuit is in phase with the input at only one frequency (f_o). There, the magnitude of the output reaches a maximum level of one-third of the input signal:

$$f_o = \frac{1}{2\pi\sqrt{R_1 R_2 C_1 C_2}} \tag{13–2}$$

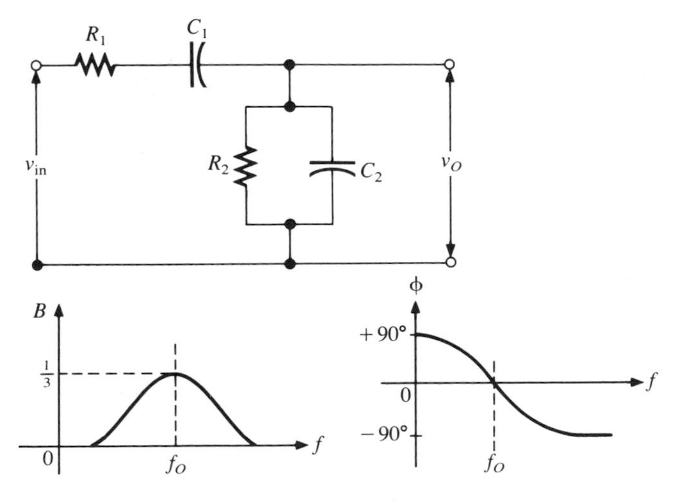

Figure 13–3 Passive lead-lag network.

To minimize the number of different component values possible in Equation 13–2, the resistors R_1 and R_2 are usually made equal, as are both capacitors. When $R_1 = R_2$, and $C_1 = C_2$, Equation 13–2 simplifies to

$$f_o = \frac{1}{2\pi R_1 C_1} \tag{13–3}$$

By using the lead-lag network as the positive feedback path for an amplifier with a gain of 3, a sine-wave oscillator is formed, as shown in the circuit of Figure 13–2. The lead-lag network forms one side of a Wien bridge, and the gain-setting resistors $2R'$ and R' form the other side. If the R-C lead-lag network is omitted briefly, what remains is a noninverting amplifier with a closed-loop gain of 3.

At the oscillation frequency given by Equation 13–2, the Wien bridge is balanced, and the amplifier input voltage is in phase with the output voltage. At all other frequencies, the bridge is unbalanced, and the feedback signal does not have the proper amplitude and phase relationship to sustain oscillation.

Resistors R' and $2R'$ are in the negative feedback path, and the amplifier gain set at 3. Resistor R' is often a device with a *positive* voltage-versus-resistance characteristic, such as an incandescent lamp. This type of characteristic usually regulates output amplitude. If the amplitude of the output voltage (V_o) starts to increase, R' also increases, reducing the gain of the amplifier. If V_o decreases, R' decreases, and the gain of the amplifier is increased. This action maintains the output amplitude at a relatively constant level.

An alternate approach to amplitude stabilization is to make part of $2R'$ a device with a *negative* voltage-versus-resistance characteristic, such as a pair of back-to-back zener diodes.

EXAMPLE 13–1

Find the output frequency of the Wien bridge oscillator of Figure 13–2 with $C_1 = C_2 = 0.047\ \mu\text{F}$; $R_1 = R_2 = 10\ \text{k}\Omega$; $R' = 1.5\ \text{k}\Omega$; and $2R' = 3\ \text{k}\Omega$.

Solution:

From Equation 13–3,

$$f_o = \frac{1}{(6.28)(10\ \text{k}\Omega)(0.047\ \mu\text{F})}$$
$$= 339\ \text{Hz}$$

Since the closed-loop gain is 3 (i.e., $1 + 2R'/R'$), the loop gain equals 1.0, and the circuit oscillates.

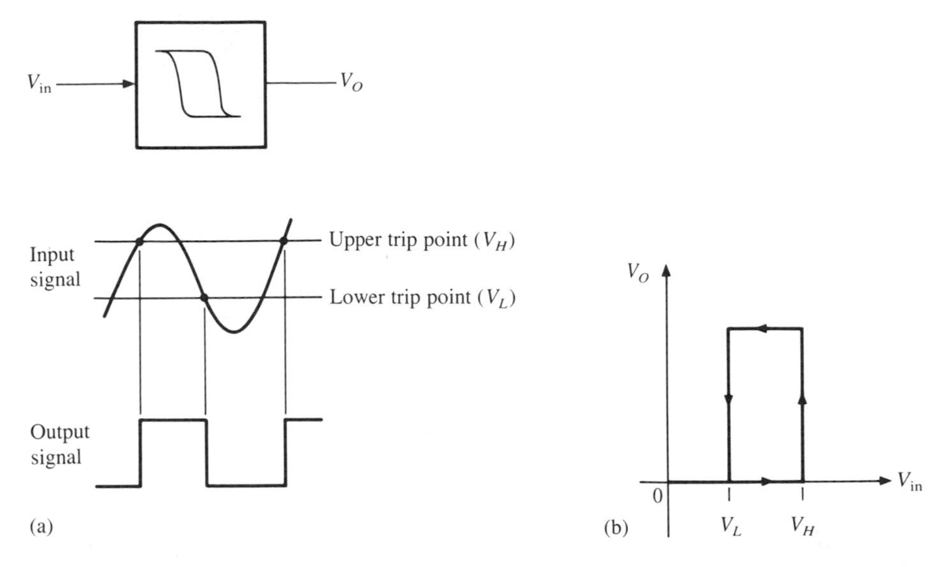

Figure 13–4 Schmitt trigger converts sine wave to square wave. (a) Input and output waveforms. (b) Hysteresis characteristic.

Many audio-frequency sine-wave oscillators include an attenuator and a square-wave converter. The attenuator provides a means of controlling the output amplitude, and the square-wave converter provides an additional waveform. The attenuator can be in the form of either an output amplifier with adjustable gain or a resistive network, such as a potentiometer or variable attenuator pad.

The square-wave converter used with inexpensive oscillators is usually either an amplifier overdriven to clip the tops off a sine wave or the sine wave is passed through a clipper circuit. The latter type of square-wave conversion actually produces only an *approximation* of a square wave, and does not produce a signal with the fast rise and fall times available with laboratory generators. Despite this shortcoming, however, this technique is perfectly adequate for many service-shop applications.

Another type of square-wave converter uses a Schmitt trigger to convert a sine wave to a square wave (Figure 13–4). This method produces a square wave with better rise and fall times.

13–5 HIGH-FREQUENCY SIGNAL GENERATORS

As defined earlier, a signal generator is a sine-wave generator with modulation capabilites. This more or less confines their use to the RF range between 100 kHz and 40 GHz. RF bands over this frequency range are listed in Table 13–2. Such an instrument, as illustrated by the block diagram of Figure 13–5, usually consists of an RF oscillator, and amplifier, a calibrated attenuator, and an output level meter. For setting the frequency, the oscillator section normally has a frequency range switch and a calibrated control to tune across the selected range.

Table 13–2 Radio frequency bands

Band name	Abbreviation	Frequency range
low frequency	LF	30–300 kHz
medium frequency	MF	300 kHz–3 MHz
high frequency	HF	3–30 MHz
very high frequency	VHF	30–300 MHz
ultra high frequency	UHF	300 MHz–3 GHz
super high frequency	SHF	3–30 GHz
extremely high frequency	EHF	30–300 GHz

The amplifier section has an output level control that permits the amplitude of the signal fed to the attenuator to be set at a calibration point, as read by the meter. The attenuator itself reduces the signal, and is usually calibrated in decibels below the calibration level. Though the input to the attenuator may be accurately set, its output is only the level indicated when the output of the signal generator is terminated in its designed load impedance (usually 50 Ω). For loads of less than 50 Ω, a *series resistance* can be added to bring the generator load up to 50 Ω. For loads greater than 50 Ω, a *parallel resistor* can be used to bring the parallel combination down to 50 Ω.

The RF oscillator section usually consists of either a Colpitts or a Hartley oscillator circuit, although other types can be used (Table 13–3). As shown in Figure 13–6, both circuits basically use an inverting amplifier, which can be

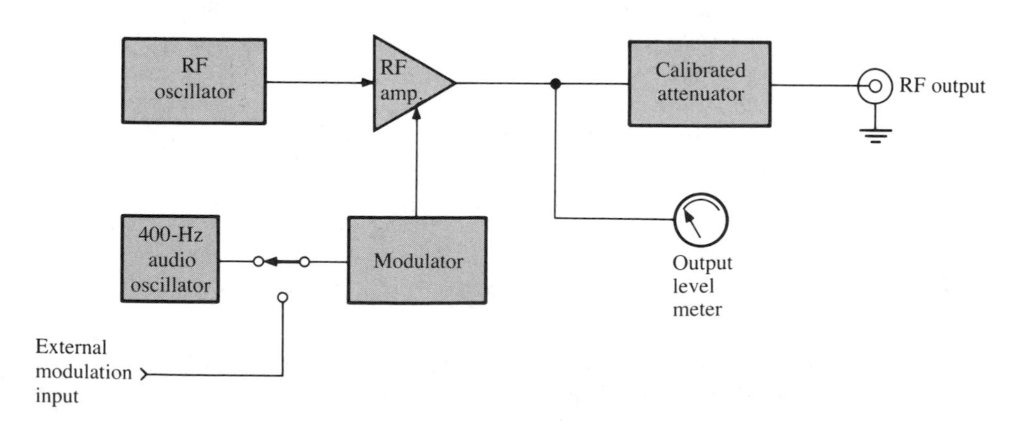

Figure 13–5 RF signal generator block diagram.

either a transistor or an operational amplifier. In addition, a tapped parallel resonant circuit, called a **tank circuit,** provides frequency selectivity and the necessary 180° phase shift. The basic difference between the two oscillators is in the parallel resonant circuits. The Hartley oscillator uses a tap on the inductor to serve as the common connection between the amplifier input and output circuits, while the Colpitts oscillator taps the resonant circuit by splitting the capacitor into two sections.

Table 13–3 Commonly used oscillators

Type	Frequency range
Wien bridge (*RC*)	up to 1 MHz
Phase shift (*RC*)	up to 10 MHz
Colpitts (*LC*)	10 kHz–100 MHz
Hartley (*LC*)	10 kHz–100 MHz
Crystal	100 kHz–50 MHz

Amplitude Modulation

Most sine-wave generators that operate in the RF range provide an **amplitude modulated (AM)** output. Amplitude modulation consists of varying the amplitude of an RF signal at a rate within the audio frequency range. As shown in Figure 13–7, the modulation process is essentially a form of mixing, in that a high-frequency carrier is mixed with a low-frequency tone across a nonlinear device. Consequently, amplitude modulation of the RF (carrier) results, and additional frequency components are generated. The frequency spectrum of the resulting waveform contains the two original frequency components of the carrier and the low-frequency tone, as well as two new frequencies called **sidebands.** The sidebands are spaced at a distance equal to the modulating fre-

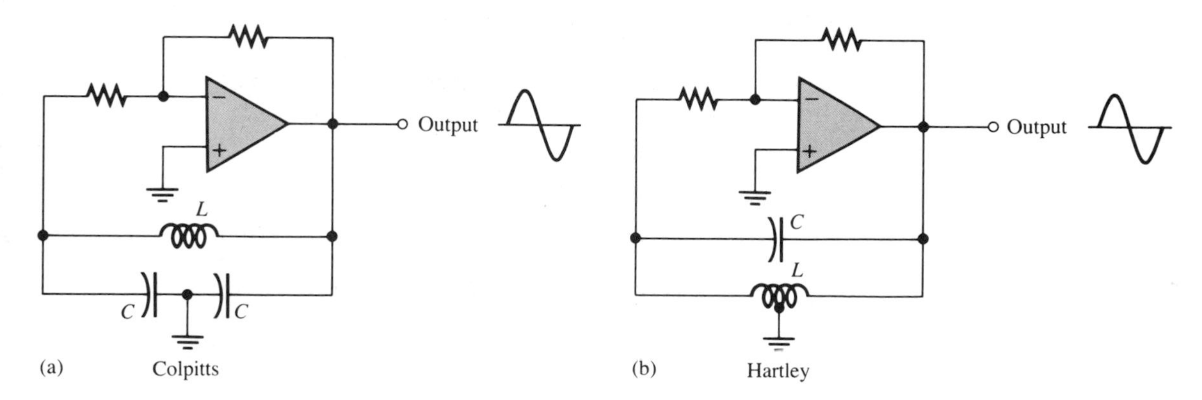

Figure 13–6 *LC* oscillators. (a) Colpitts type. (b) Hartley type.

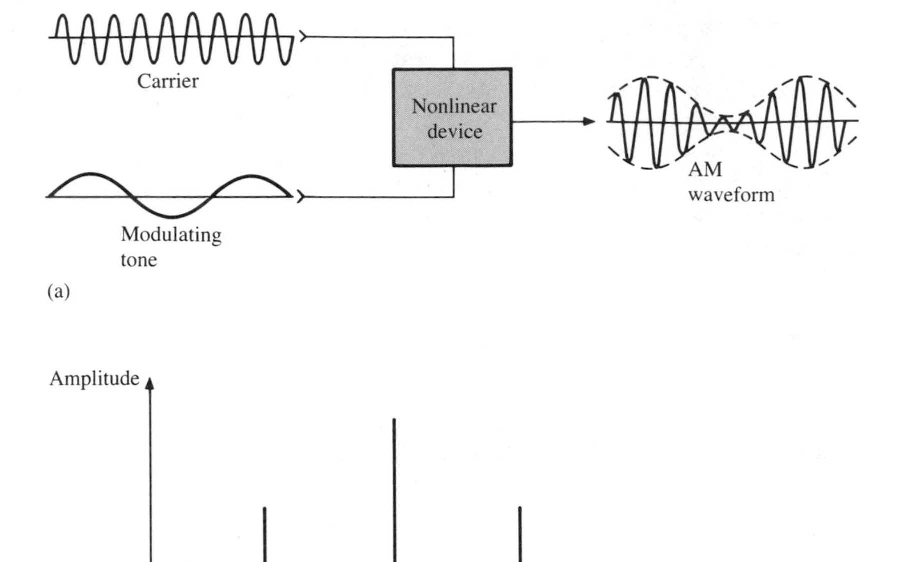

Figure 13–7 Amplitude modulation of a carrier signal. (a) Block diagram. (b) Frequency spectrum.

quency on either side of the carrier frequency. For example, if a 1-MHz carrier is amplitude-modulated by a 1-kHz sine-wave tone, the resulting frequency spectrum, as shown in Figure 13–6, includes sine-wave signals at 1 kHz, 999 kHz, 1 MHz, and 1.001 MHz. The 999-kHz component is called the **difference frequency** or **lower sideband,** while the 1.001-MHz **sum frequency** is the **upper sideband.** Because the low-frequency audio component is easily filtered out, the output consists of only the carrier and the sidebands.

If the modulating tone is not a pure sine wave, there are a series of sidebands at harmonic intervals of the modulating frequencies away from the carrier. This is due to the harmonic content of waveforms other than pure sine waves (Fourier series).

EXAMPLE 13–2

If a 1.5-kHz square wave modulates a 1.2-MHz carrier, determine the frequency of the resulting sidebands.

Solution:

Since the harmonic content of a square wave contains no even harmonics, the sidebands are at $\pm N$ times 1.5 kHz away from the carrier, where N is any *odd* integer. Consequently, the sidebands are located at ±1.5 kHz, ±3 kHz, ±4.5 kHz, and so on, from the carrier. Considering harmonics up to and including the fifth harmonic, the actual sideband frequencies are

lower sidebands: 1.1955 MHz, 1.197 MHz, and 1.1985 MHz
upper sidebands: 1.2015 MHz, 1.203 MHz, and 1.2045 MHz

The amount of change in the amplitude is measured by the **percent modulation.** This is an important characteristic of an AM signal, and is expressed as a percentage between zero and 100% (Figure 13–8).

Based on the minimum and maximum amplitudes of the modulated carrier signal, the equation for computing the percent modulation is

$$\%m = \frac{V_{MAX} - V_{MIN}}{V_{MAX} + V_{MIN}} \times 100\% \qquad (13\text{–}4)$$

EXAMPLE 13–3

Determine the percent modulation of the modulated carrier waveform of Figure 13–8*b* if the maximum and minimum peak-to-peak amplitudes are, respectively, 90 V and 30 V.

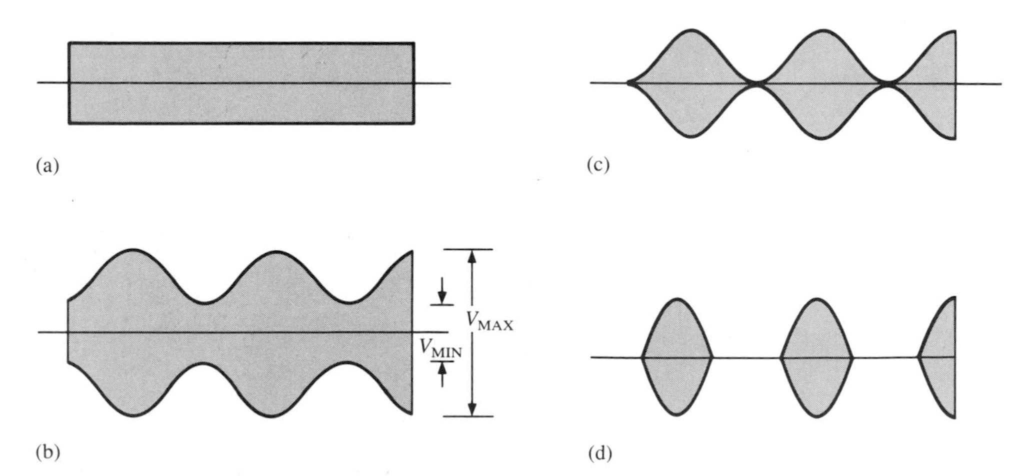

Figure 13–8 Amplitude modulated carrier. (a) Unmodulated. (b) 50% modulation. (c) 100% modulation. (d) Greater than 100% modulation.

Solution:

Using Equation 13–4,

$$\% \text{ modulation} = \frac{90 - 30}{90 + 30} \times 100\%$$

$$= 50\%$$

Distortion of the modulated carrier results when modulation exceeds 100% (Figure 13–8*d*). Fortunately, most generators that provide an AM output are equipped with a control to adjust and maintain the percent modulation.

Frequency Modulation

Signal generators are also available with **frequency modulation (FM)** capability. In frequency modulation, a low-frequency signal varies the frequency of the carrier (1) in proportion to the peak amplitude of the low-frequency signal and (2) at a rate equal to the modulating frequency. Although the amplitude of the resulting signal in the time domain is constant, the resulting frequency spectrum contains many sidebands where the modulated information is contained (see Figure 13–9). These sidebands occur on both sides of the carrier at integer multiples of the modulating frequency. This differs from AM, where there is a single set of side frequencies for each modulating frequency (assuming that the modulating tone is a pure sine wave).

The **frequency deviation** is the difference between the instantaneous frequencies of the modulated and unmodulated carriers. The number of extra sidebands that occur in FM modulation depends on the relationship between

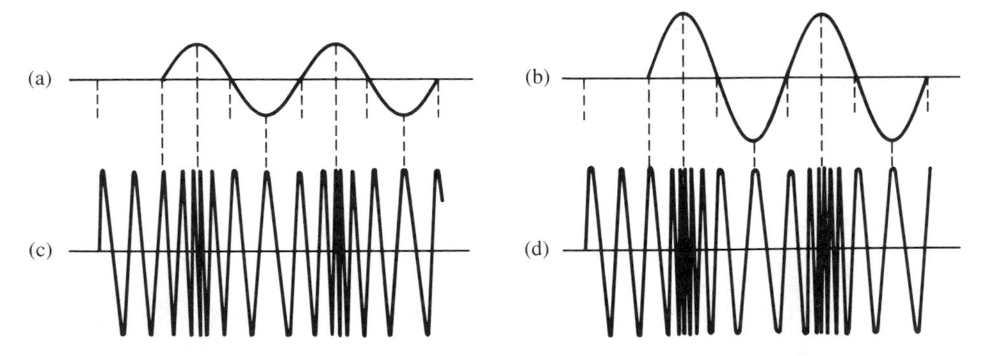

Figure 13–9 Frequency modulation of a carrier signal. The modulating signal of waveforms (a) and (b) are the same frequency, but the amplitude of (a) is less than that of (b). The larger signal causes a greater frequency change in (d) than that shown in (c).

the modulating frequency (f_m) and the resulting frequency deviation (Δf). The ratio of the frequency deviation and the modulating frequency is called the **modulation index.**

$$\text{Modulation index} = \frac{\Delta f}{f_m} \tag{13–5}$$

In an FM system, the modulation index varies with the modulating frequency. For a given modulation index, it is possible to calculate the amplitude of any given pair of sidebands. For certain values of modulation index, a given sideband disappears. This fact is often used in making adjustments to FM equipment. The energy that goes into the sidebands is taken from the carrier; the total power remains constant, regardless of the modulation index (see Figure 13–10).

In communications work, the modulation index often does not exceed 0.6 or 0.7. Furthermore, the most significant extra sideband (the second) is at least 20 dB below the unmodulated carrier level. This provides an effective channel width roughly equivalent to an AM signal. The channel width for FM broadcast service is much wider than that of an AM signal due to its additional significant sidebands. Another number often encountered in FM systems is the **deviation ratio.** This is the ratio of the maximum carrier-frequency deviation to the highest modulating frequency. Unlike AM, FM systems cannot be overmodulated. The level corresponding to a 100%-modulated carrier is set by governmental regulations rather than technical considerations.

In practice, frequency modulation is usually accomplished in the oscillator stage of the signal generator. One commonly used method of frequency-

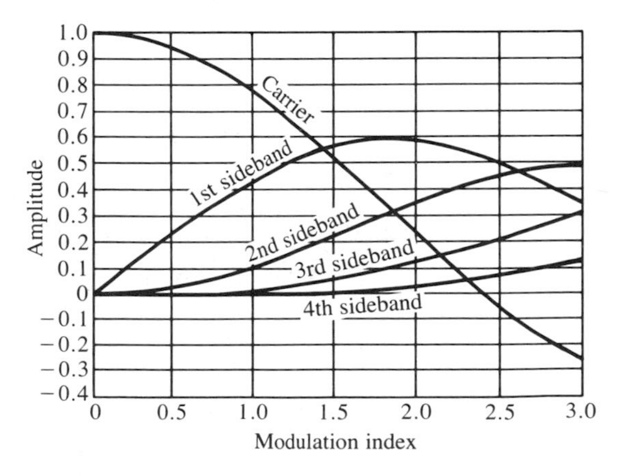

Figure 13–10 Graph of the amplitude of sideband pairs versus modulation index in an FM system.

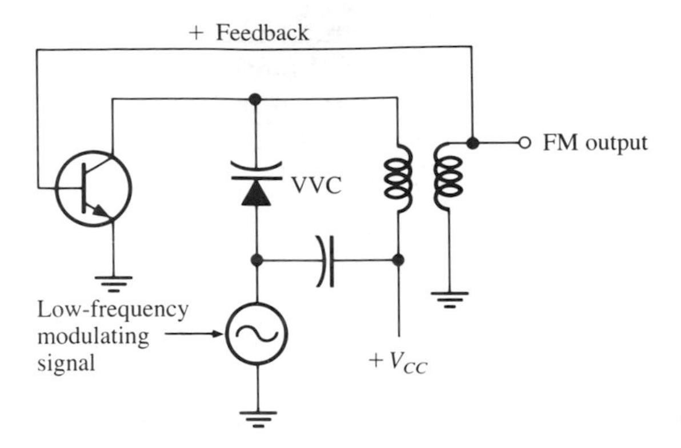

Figure 13–11 FM oscillator using a variable voltage capacitance diode.

modulating in the oscillator stage is that of using a **variable voltage capacitor diode (VVC diode*)** across the tank circuit. The VVC diode is a special semiconductor device designed to make use of the junction capacitance of a diode operated in the reverse-biased mode. In effect, the junction capacitance varies inversely with the applied reverse bias. Using the oscillator circuit of Figure 13–11, the low-frequency modulating signal varies the reverse bias applied to the diode, hence its capacitance. The varying of the capacitance alters in turn the resonant frequency of the tank circuit.

If an FM modulated signal is passed through one or more frequency multiplier stages, the modulation index is multiplied by the same factor as the carrier frequency.

13–6 FUNCTION GENERATORS

The basic function generator usually produces sine, square, triangular, and sawtooth waveforms over a frequency range of about a fraction of a hertz to 2 MHz. Many function generators also include versatile modulation capabilities such as amplitude, frequency, phase, pulse width, and VCO control, and even arbitrary waveform generation. The flexibility of the function generator has made it very popular, and it is rapidly replacing the audio-frequency sine-wave generator in many applications.

The traditional AF and RF generators normally generate a sine wave as the primary signal. The function generator, however, produces a *triangle* wave as its primary signal. All other output waveforms are derived from the triangle wave.

As shown in Figure 13–12, when a capacitor is charged from a constant current source, the voltage across the capacitor is a linear ramp, as opposed to the exponential curve that results when a capacitor is charged through a resis-

*Also called *varactor, epicap,* or *varicap* diodes.

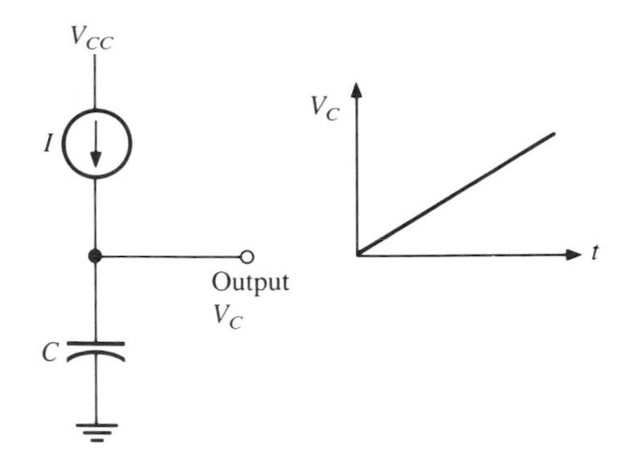

Figure 13–12 Linear ramp voltage generated by charging a capacitor from a constant current source.

tor. To reverse the slope of the ramp, it is necessary only to reverse the direction of the charging current. The rate of charge depends on the size of the capacitor and the amount of charging current. Varying either one changes the slope of the voltage across the capacitor. Either increasing the current or decreasing the size of the capacitor increases the slope. Decreasing the current and increasing the size of the capacitor both decrease the slope.

Figure 13–13 shows the block diagram of a typical function generator. The triangle wave is created by alternately charging and discharging a capacitor through constant current sources. First, the capacitor is charged in the

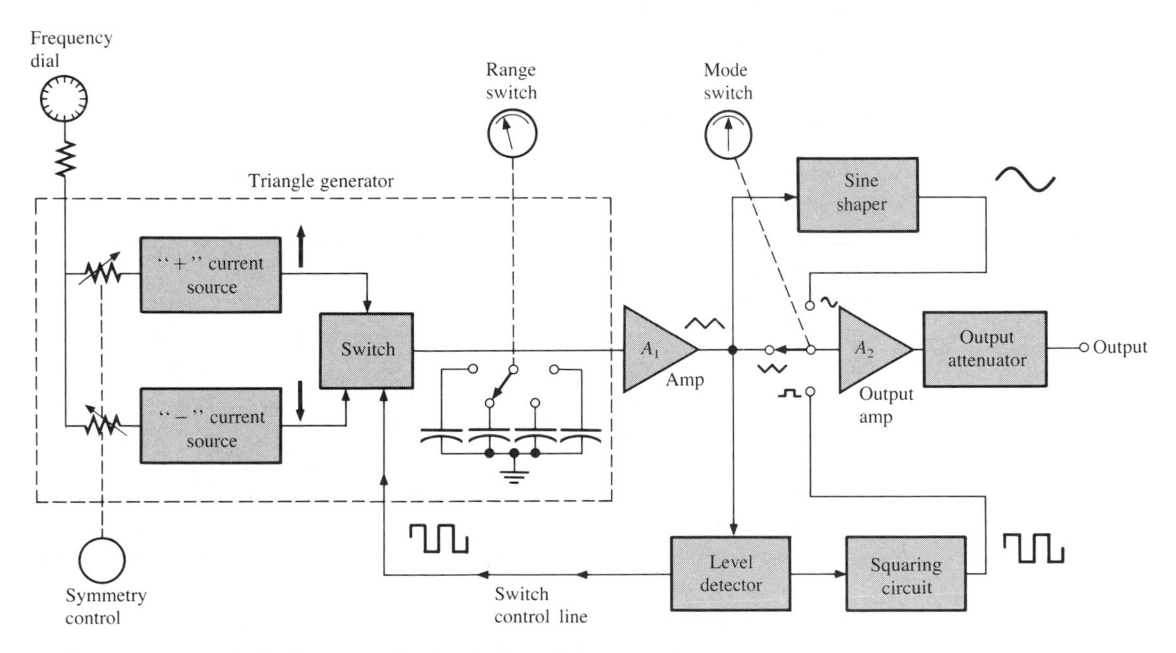

Figure 13–13 Block diagram of a basic function generator.

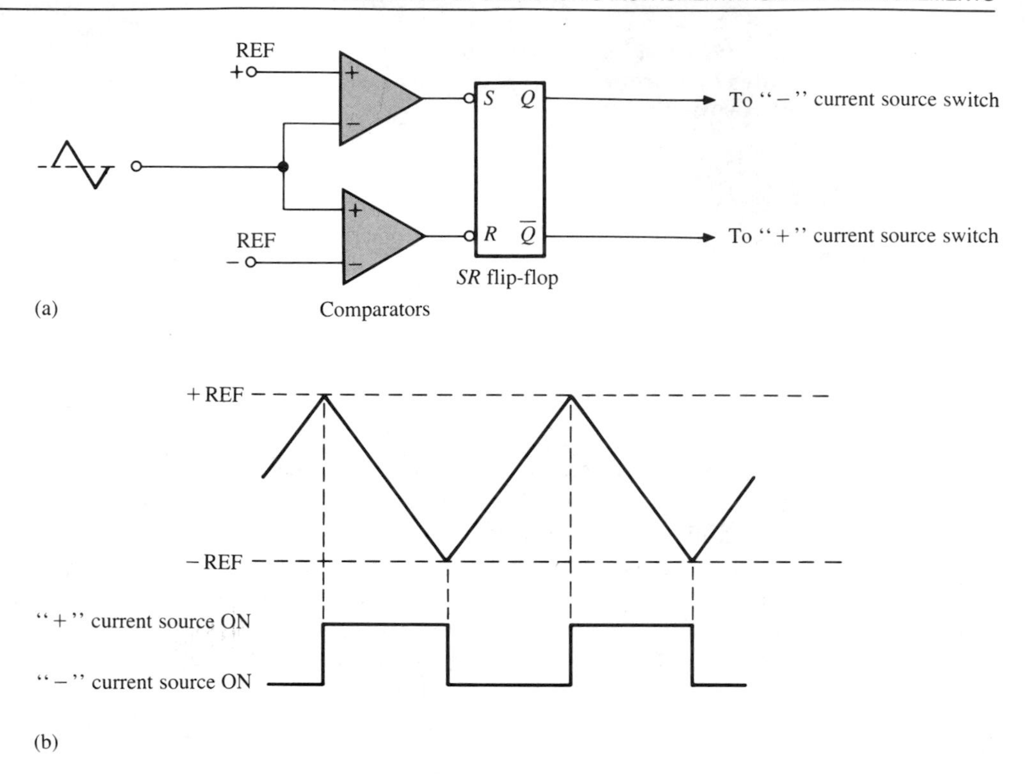

Figure 13–14 Level detector and current source switching. (a) Logic. (b) Waveforms.

positive direction until a given upper level is reached. The level detector switches the capacitor to a constant current source of opposite polarity, and the capacitor is charged in the opposite direction to a given lower level, at which point the process repeats itself (Figure 13–14).

The upper and lower detector limits are usually fixed. Increasing the slope causes the ramp to reach these limits sooner, and thus increases the **pulse repetition rate.** Likewise, decreasing the slope lowers the rate. In practice, to control the frequency of a function generator, both the charging current and the capacitor size are varied.

The triangular wave developed across the capacitor is fed to a buffer amplifier to minimize loading. The output of the buffer drives a level detector, a squaring circuit, and a sine shaper. The sine shaper is a special diode circuit that shapes a triangular waveform into a low-distortion sine wave. Figure 13–15 shows the circuit of a simple sine shaper, essentially a nonlinear voltage divider that distorts the triangle wave to approximate a sine wave. The squaring circuit is bistable, switching between levels each time the level detector switches between current sources. This results in a sharp square wave of the same frequency as the triangle waveform.

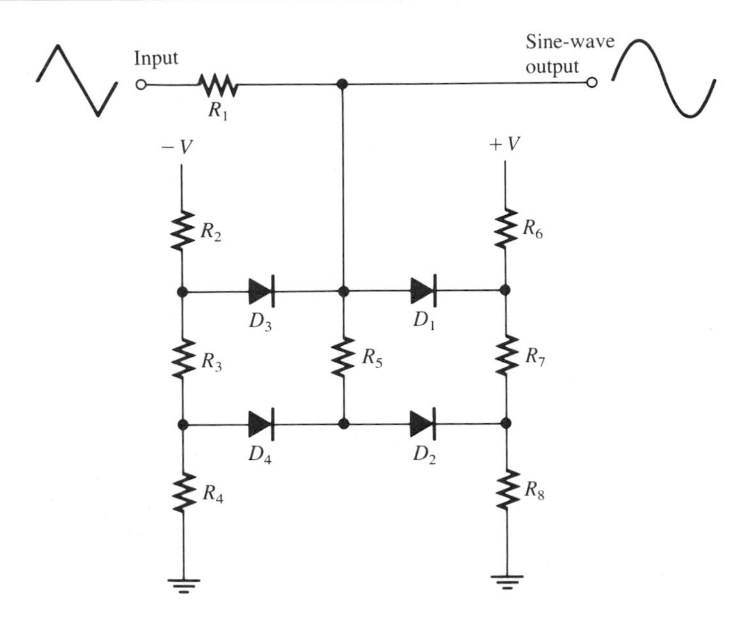

Figure 13–15 Triangle-to-sine wave shaper using diodes.

With no input voltage, all four diodes are reverse biased, and the circuit effectively presents an open circuit to ground. The signal increases in the positive direction; a point is reached at which D_2 becomes forward-biased and R_5 and R_8 are put in series with R_1. This causes the slope of the output signal to decrease. Next, a point is reached at which D_1 is forward biased. This places R_7 in parallel with R_5, and the slope of the output is reduced even further. The reverse action takes place on the downward slope. When the signal passes below zero, the left half of the network distorts the triangle wave in the same manner. By adding additional diodes and resistors, a better approximation of a sine wave is possible.

A waveform selector switch allows the operator to select the desired waveform. The selected waveform is fed to the output amplifier, which in turn drives an output attenuator. The amplifier section contains the variable output amplitude control and the DC offset control. The DC offset control permits the operator to vary the DC reference level of the output waveform. For instance, a 4-V peak-to-peak signal referenced to ground swings between +2 V and −2 V. The same signal with a 1-V DC offset (average value of +1 volt) swings between +3 V and −1 V.

Since some of the waveforms are nonsinusoidal, the amplifier needs a good frequency response lest it attenuate their harmonic components, causing harmonic distortion. Also, since very low-frequency waveforms can be produced, the amplifier is usually DC coupled. For example, given a function generator with a frequency range of less than 1 to approximately 3 MHz, the output amplifier must be DC coupled and have a flat response out to at least 30 MHz. This assumes that harmonic components up to and including the tenth are to be amplified with the same gain. The output attenuator is similar to the output attenuators used in the RF generators described earlier. Some function generators also provide a sweep feature, which is discussed in the next section.

13–7 SWEEP FREQUENCY GENERATORS

All the signal sources discussed so far are available with a sweep option. In addition, a family of sine-wave generators is manufactured especially for this purpose. The ability to provide a signal that is electronically swept over a range of frequencies is particularly useful for measuring the frequency response of amplifiers and filter circuits. By applying a rectified and filtered sample of the output voltage of the circuit under test to the vertical input of an oscilloscope, while the signal derived from the frequency sweep circuit of the generator goes to the horizontal, it is possible to display the frequency response curve (amplitude vs. frequency) on the screen of an oscilloscope (Figure 13–16).

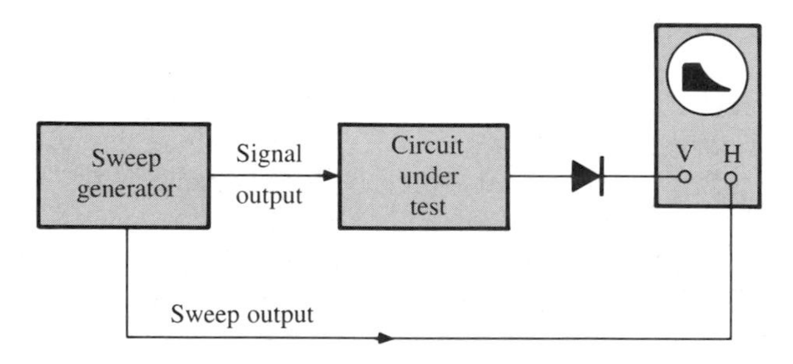

Figure 13–16 Sweep generator method of determining the frequency response of a circuit under test.

Most sweep frequency generators, like the one shown in the block diagram of Figure 13–17, use a VCO whose frequency is varied by a control voltage. Since the output frequency of a VCO is controlled by an input voltage, it can also be referred to as a *voltage-to-frequency (V/F) converter*. For example, if a voltage ramp is applied as the VCO frequency control signal, the oscillator's frequency is swept across its range proportionally to the ramp voltage.

In RF generators, the technique is similar to frequency modulating a carrier. If a varactor diode is used as one of the frequency-determining elements in an oscillator's tank circuit and a voltage ramp is used to vary the

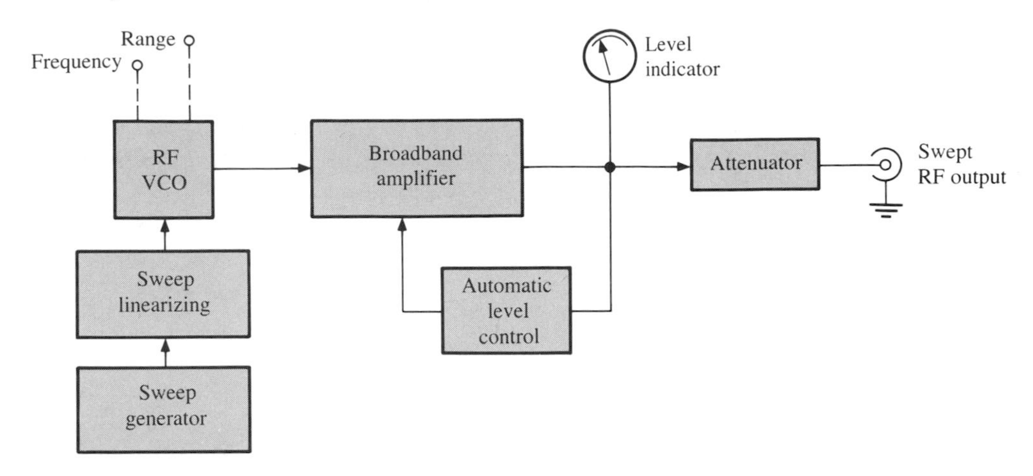

Figure 13–17 Sweep frequency generator block diagram.

reverse bias on the varactor diode, the oscillator frequency can be made to vary with the ramp voltage. As the reverse bias increases, the varactor capacitance decreases, and the oscillator frequency increases. As the reverse bias decreases, the capacitance of the varactor increases, decreasing the frequency.

Because the frequency of the oscillator is not a linear function of the sweep voltage, a linearizing circuit is often included between the sweep voltage source and the oscillator. This is usually a resistor-diode network similar to the sine converter of a function generator. By using diodes to switch resistors into the circuit at various voltage levels, it is possible to approximate the control waveform that provides an oscillator frequency that is a linear function of the sweep voltage source. This technique is called *piecewise-linear approximation*. The required characteristic curve is approximated by a series of straight line segments, and the resistors required to produce the necessary line segments are switched in by the diodes at the proper levels.

Varying the amplitude of the ramp changes the frequency range over which the oscillator is swept. Varying the frequency of the ramp voltage changes the rate at which the oscillator frequency is swept. Most generators provide controls to vary both of these parameters. To keep the output amplitude of the generator constant over the entire frequency range, the generator usually includes an automatic level control (ALC) circuit.

An output voltage directly proportional to the instantaneous frequency (often a ramp) is usually provided for the horizontal input of an oscilloscope. This signal provides a horizontal sweep signal that is a linear function of the instantaneous frequency, which is necessary for a frequency response curve. Laboratory sweep generators provide a means of triggering the sweep from an external source in addition to those controls that initiate the sweep at a particular frequency or make it symmetrical about a selected center frequency.

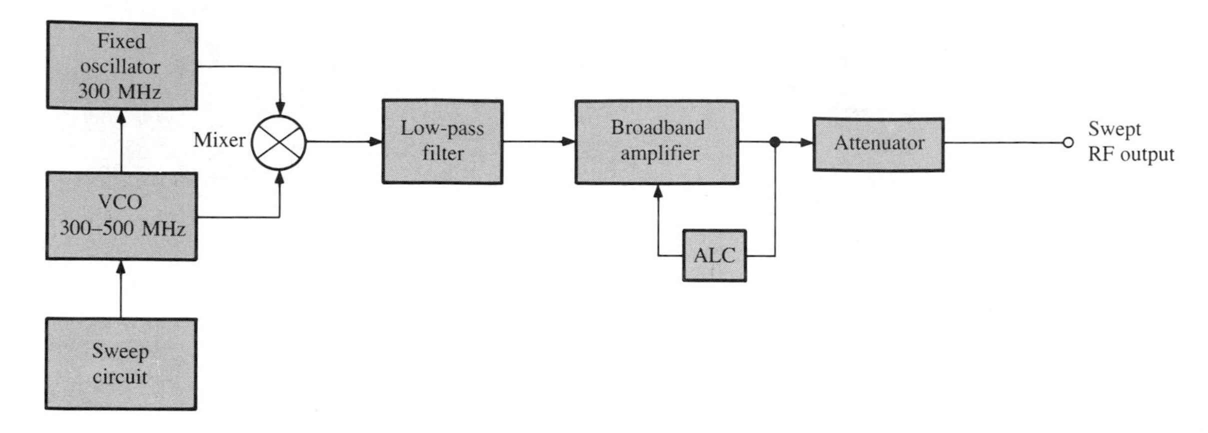

Figure 13–18 Wideband sweep generator block diagram.

Since it is difficult to design an oscillator that can be tuned with a frequency ratio of greater than 2:1, mixing techniques are used when a higher frequency ratio is required (Figure 13–18). Here, the swept oscillator operates at a frequency *above* the required range while it is mixed with a fixed frequency that is also above the required range, producing the actual output frequency. The unwanted frequency components of the mixing process are filtered out. For example, a 400- to 600-MHz sweep oscillator can be mixed with a 350-MHz fixed oscillator to provide an output over the 50- to 250-MHz range. The output range covers a 5:1 ratio, while the swept oscillator's range is less than 2:1.

13–8 FREQUENCY-SYNTHESIZED GENERATORS

Frequency-synthesized generators produce a very accurate and very stable output frequency derived from either multiples or submultiples of one or more crystal-controlled oscillators, a process referred to as *frequency synthesis*. There are two principal methods of frequency synthesis: *direct* and *indirect*. In direct frequency synthesis, the output signal is derived by combining either the outputs of several crystal-controlled oscillators or a single crystal-controlled oscillator with multiple divider-comb generator sections. Various frequency combinations are mixed together to produce the desired frequency, while unwanted mixing components are filtered out. Alternately, the indirect approach uses a spectrally pure VCO locked onto an integer multiple of a crystal-controlled reference oscillator by a phase-locked loop (PLL).

As explained in Chapter 12, the PLL is commonly used to produce a frequency that is an exact multiple of a stable reference frequency. The PLL (Figure 13–19), when used to multiply a frequency, is basically an electronic feedback loop system that constantly compares a submultiple of the output frequency to a *stable* frequency reference, thereby producing an output frequency that is larger than the reference frequency by a given integer multiple.

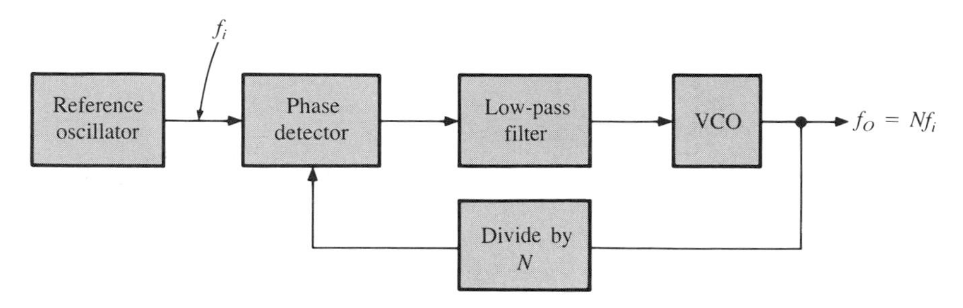

Figure 13–19 The phase-locked loop.

By providing external controls to change the frequency division factor (N), it is possible to vary the output frequency in steps equal to the reference frequency. These controls are in the form of multidecade thumbwheel switches designed to provide a direct readout of the output frequency. However, the PLL has several inherent problems that make it impractical for certain applications. The loop filter is not able to completely filter out all of the unwanted components from the phase detector's output. As a result, there is a small amount of FM noise at the output of the VCO. This makes the PLL unsuited for critical applications that require a spectrally pure signal source. Furthermore, the loop filter limits the rate at which the output frequency can be changed. Consequently, the PLL takes a long time to lock onto a new frequency far from the old. Despite these disadvantages, function generators using indirect frequency synthesis are generally less expensive, require less filtering, and offer greater output power with lower subharmonic components than generators using the direct method.

Figure 13–20 shows the block diagram of a direct frequency synthesis system, which uses a single master reference oscillator, a pair of dividers, a pair of comb generators, and a pair of filters to produce a number of selectable output frequencies. The master oscillator drives a number of **comb generators** either directly or through dividers. A comb generator is a circuit used to

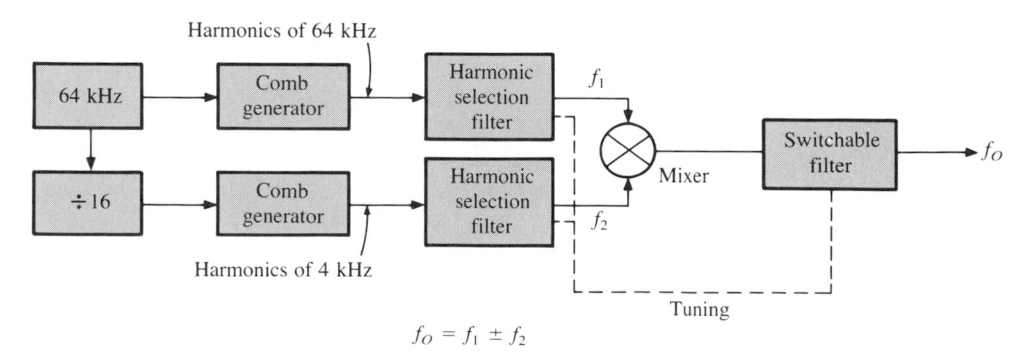

Figure 13–20 Direct frequency synthesizer block diagram.

produce a signal rich in harmonics of the input signal. Each comb generator is followed by a harmonic-selective filter to pass only the desired harmonic.

For example, a comb generator driven by a 64-kHz signal produces an output with frequency components every 64 kHz (64, 128, 192 kHz, and so on). A comb generator driven by a 4-kHz signal (1/16 of 64 kHz) produces an output with frequency components every 4 kHz (4, 8, 16 kHz, and so on). By mixing various harmonics from each comb filter, it is possible to produce a number of output frequencies in increments of 4 kHz. Since the mixing process always produces two additional output frequencies—a sum frequency and a difference frequency—as well as the two input frequencies, it is necessary to follow the mixer with a filter to reject the unwanted components. If 128 kHz were mixed with 8 kHz, the output from the mixer would contain frequency components at 8, 120, 128, and 136 kHz. The output filter would be tuned to pass only the sum (136 kHz) or the difference (120 kHz) frequency component.

More frequency divider-comb generator combinations can be added, the finest resolution is equal to the spacing of the harmonics from the comb generator driven by the lowest frequency. The outputs of fixed frequency crystal-controlled oscillators can be substituted for the outputs of some or all of the harmonic-selective filters.

Since the mixing process produces many unwanted frequencies, great attention must be given to the effective shielding and filtering of the circuit, to eliminate these undesired components. Also, since the frequencies to be mixed can be selected digitally, frequency shifts can be very fast, limited only by the speed of the digital circuits and the pulse response of the output filters.

13–9 PULSE GENERATORS

The performances of certain electronic circuits and systems are easily evaluated by measuring their response to a suitable input pulse. Although used to test analog circuits, the pulse generator probably finds its greatest use in the design and testing of high-speed digital circuits. A pulse is a rectangular waveform similar to a square wave, which is just a pulse with a 50% duty cycle. The **duty cycle** is defined as the ratio of the pulse width to the pulse period:

$$\boxed{\%\ \text{duty cycle} = \frac{\text{pulse width}}{\text{pulse period}} \times 100\%} \tag{13–6}$$

Usually expressed as a percentage, the duty cycle can vary from 1 to 99%. The pulse waveform shown in Figure 13–21 has a duty cycle of 20%.

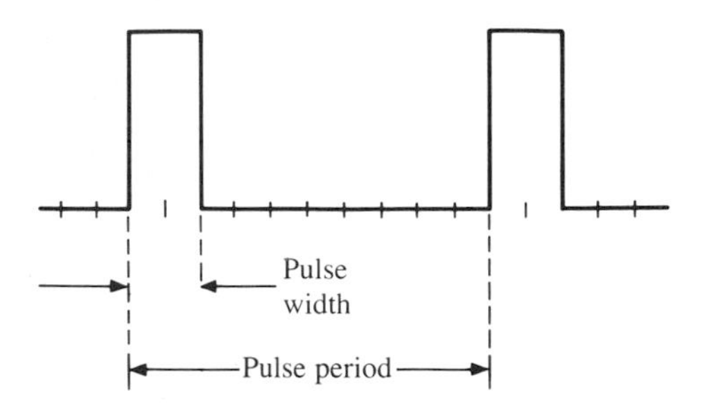

Figure 13–21 Pulse train with 20% duty cycle.

Additional characteristics of a pulse are illustrated in Figure 13–22. These include the following:

- rise and fall times
- pulse width
- preshoot, overshoot, and undershoot
- ringing
- sag, or droop
- pulse repetition rate
- settling time

rise time—the time required for a pulse to rise from 10 to 90% of its normal amplitude.

fall time—the time required for a pulse to drop from 90 to 10% of its normal amplitude.

pulse width—the time the pulse takes from the 50% amplitude point on the leading edge to the 50% amplitude point on the trailing edge.

preshoot—a distortion of the base line preceding an edge.

overshoot—a distortion of the peak value immediately following a rising edge.

undershoot—a distortion of the base value immediately following a falling edge.

ringing—the positive and negative peak distortion, excluding overshoot or undershoot, on the pulse top or base line.

droop, or **sag**—occurs when the peak value gradually decreases during the pulse.

pulse repetition rate (PRR)—the rate at which pulses are produced.

settling time—the time required for a signal to decrease to a given percentage, typically 1 to 5%, of its peak value.

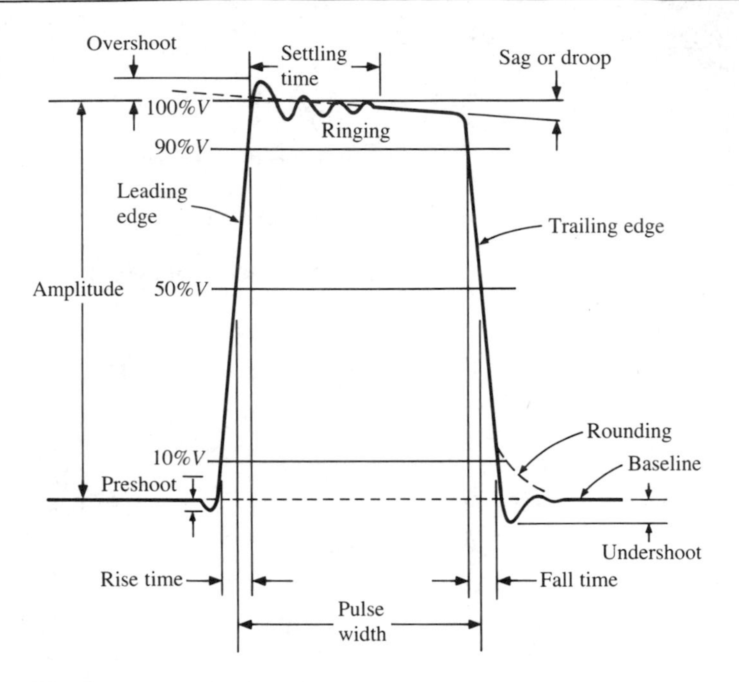

Figure 13–22 Pulse parameters.

A pulse generator, a block diagram of which is shown in Figure 13–23, is an instrument that produces a rectangular train of pulses. It can also generate other waveforms, like the function generator, or it can be designed exclusively for the production of pulses. In a service instrument, some of the pulse parameters are adjustable. These usually include the frequency, PRR, the amplitude, and the duty cycle.

Many pulse generators offer a switch-selected alternative for the rate generator, called the *external trigger*. A signal applied to the external trigger

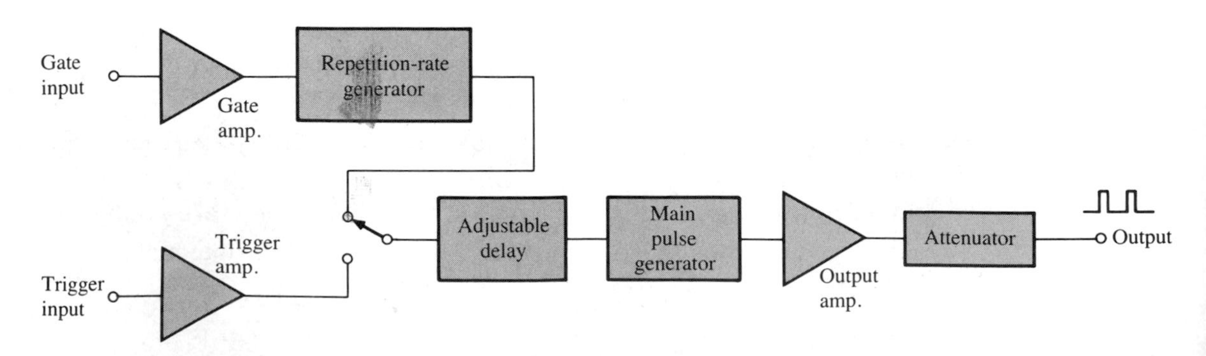

Figure 13–23 Simplified block diagram of a pulse generator.

input goes to a trigger circuit (similar to those found in oscilloscopes and frequency counters), which establishes the triggering amplitude and polarity. Other inputs to the trigger circuit may include a power-line frequency (60 Hz) and/or a manual push-button switch.

Another external amplifier can be used to gate the main repetition-rate generator in a manner similar to the gate mode of a frequency counter. Signals from this gating amplifier permit the repetition-rate generator to generate pulses only when a gating signal is applied. Consequently, a burst of pulses can be generated by keeping the gate open for the time required for the selected number of pulses.

A signal from either the repetition-rate generator or the trigger circuit is applied to an optional delay generator. The delay generator is the equivalent of a monostable multivibrator with an adjustable time period. At the end of the delay generator cycle, the main pulse generator is triggered. The main pulse generator is also a monostable circuit with an adjustable time period. The requirements for this monostable circuit can be very demanding, in that it must cover an extremely wide pulse width range and operate at high duty cycles. In the more expensive instruments, the output stages of the main pulse generator can include independent charge and discharge switches, making rise and fall times adjustable.

The output of the main pulse generator is coupled to one or two DC wideband amplifiers with adjustable gain. In better quality instruments, where the extremely close control of pulse characteristics is required, two amplifiers are used. One amplifier handles positive pulses, and the other handles negative. In this manner, each amplifier can be tailored to its particular function. The wide bandwidth must faithfully reproduce pulses with rise times in the nanosecond range without introducing ringing or other distortions. As a general rule, the output impedance of pulse generators is about 50 Ω.

Pulse amplitude control is usually achieved by changing the gain of the output amplifier, though many units include a step attenuator at the output. Furthermore, a DC offset control is available on some models, so that the pulse baseline can be raised or lowered with respect to zero volts.

A *pulse-burst mode,* available on some pulse generators, allows the user to select the number of pulses generated when a trigger signal is received. With this feature, it is possible to generate anywhere from one to 10,000 pulses in a single burst. Some pulse generators also have a trigger output that produces a trigger pulse either before or after the main pulse, for synchronizing external equipment.

As a diagnostic tool, the pulse or square-wave generator is used to examine whether distortion occurs in a given system, such as an amplifier. By noting the presence or absence of certain characteristics of the output square wave, information is obtained as to the sources of distortion. For example, the leading and trailing edges reveal information about the high-frequency response, while the maximum and minimum amplitude traces represent information about the low-frequency. Figure 13–24 graphically summarizes the basic types of distortion from a square-wave input.

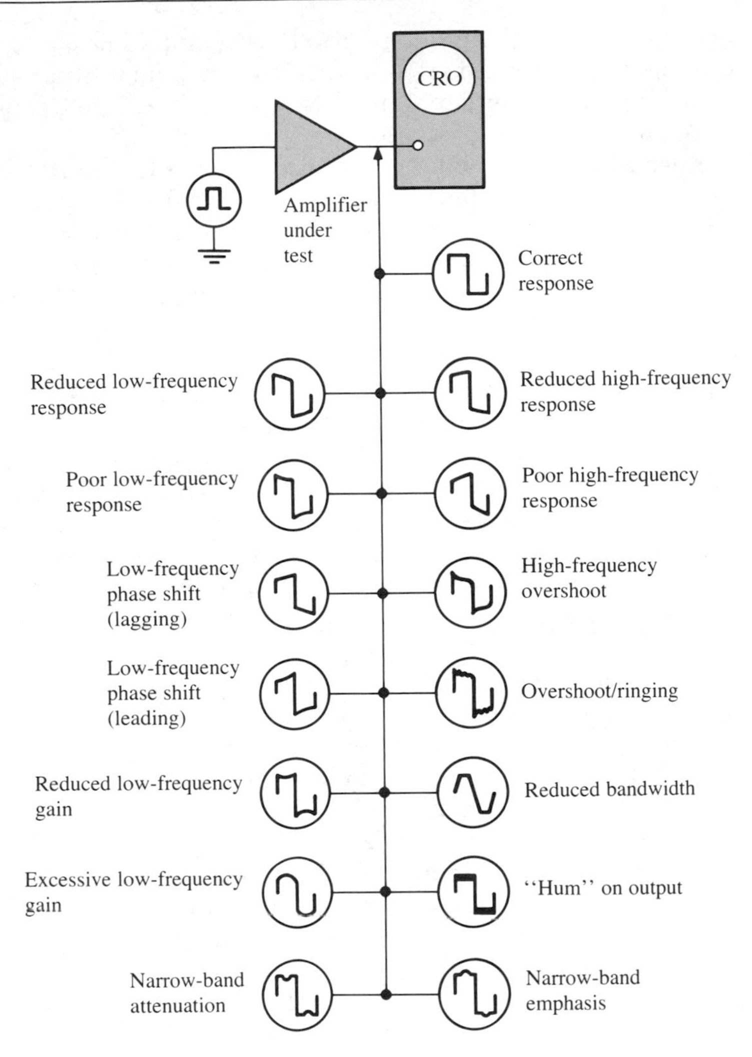

Figure 13–24 Major types of distortion resulting from squarewave testing.

13–10 WAVEFORM ANALYZERS

The oscilloscope is by far the most common instrument used to analyze signals in the **time domain;** that is, signal amplitude versus time. Analysis of signals in the **frequency domain**—signal amplitude versus frequency—is another important measurement technique widely used for providing information about the overall performance of electrical and physical systems.

The following sections discuss several types of instruments used for frequency-response signal analysis. As a group, these instruments are referred to as **analyzers,** which include the following:

- Spectrum analyzer
- Digital Fourier analyzer
- Wave analyzer
- Distortion analyzer
- Audio analyzer

These instruments all measure parameters of a signal in the frequency domain, but each uses a different technique.

Spectrum Analyzer

A **spectrum analyzer** can be thought of as a narrow bandwidth filter that is electrically swept across a fixed range of frequencies while providing a visual display of output amplitude versus frequency. At RF frequencies, it often takes the form of a narrowband swept receiver that provides a visual display on a CRT screen.

Shown in the block diagram of Figure 13–25, the spectrum analyzer in its basic form is a superheterodyne receiver. The tuned circuits of the mixer and the local oscillator (VCO) are swept across a corresponding range of frequencies by the sweep generator. The sweep generator also feeds a signal to the horizontal amplifier, so the horizontal position of the trace always corresponds to the instantaneous frequency at which the receiver is tuned.

A signal appearing at the input of the mixer combines with that of the local oscillator signal, producing a signal the frequency of which is equal to the difference between the input signal and local oscillator frequencies. Tuned circuits at the mixer output and in the intermediate frequency (IF) amplifier pass this difference frequency, while rejecting all other components.

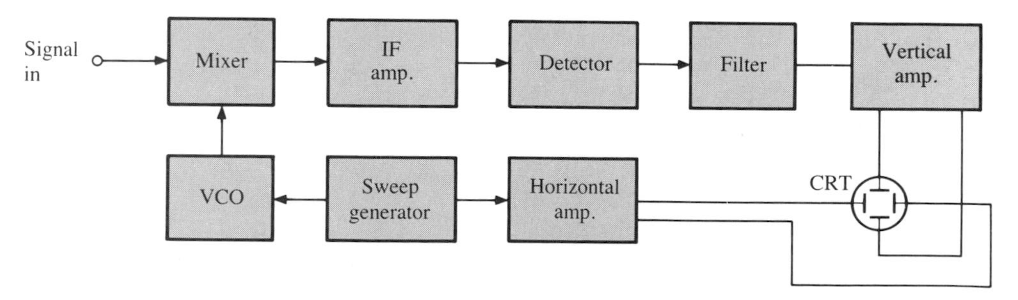

Figure 13–25 Spectrum analyzer block diagram.

For any given VCO frequency, there are two input frequencies that produce the same difference frequency. For example, if the IF amplifier in Figure 13–25 is tuned to 10 kHz, a signal at the mixer input, which is either 10 kHz above or below the local oscillator frequency, is amplified. One of these signals is undesired and is known as an **image frequency.** The image frequency is removed by one or more tuned circuits placed before the mixer, which reject the image, but pass the desired frequency. If the tuned circuit ahead of the mixer input were to track the local oscillator 10 kHz higher, and the local oscillator to sweep a band of frequencies from 500 kHz to 1 MHz, the input of the mixer would be tuned over the range of 510 kHz to 1.01 MHz.

As previously mentioned, the horizontal deflection plates of the CRT are driven by a signal derived from the same sweep generator used to tune the VCO. Therefore, the overall length of the horizontal sweep represents the total sweep frequency spectrum. Using the previous example, if the sweep goes from 510 kHz to 1.01 MHz, the left-hand end of the trace represents 510 kHz, and the right-hand 1.01 MHz. The signal from the output of the IF amplifier is detected (rectified), filtered, and fed to the vertical deflection plates of the CRT. As a result, Y-axis deflection at any intermediate point on the horizontal (frequency) axis represents the signal amplitude at that particular frequency. The spectrum analyzer does not, however, provide information about the phase relationships of the various spectral components.

Figure 13–26 shows how the frequency spectrum of a single-tone amplitude-modulated signal would appear on an ideal spectrum analyzer. The three vertical lines represent the three frequency components present. A slight spreading at the base of the lines is sometimes visible with a very narrow

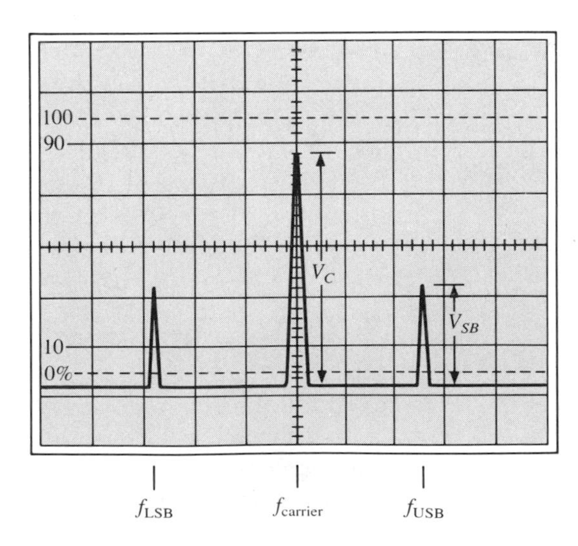

Figure 13–26 Representation of a display of a spectrum analyzer showing a carrier amplitude modulated by a single-frequency tone.

spectrum; this is due to the analyzer's finite bandpass characteristic. The resolution of a spectrum analyzer is determined by the bandwidth of the IF filters, and is usually considered to be the 3-dB bandwidth of the filters.

In the case of a single-tone, amplitude-modulated carrier, more than three vertical components, representing additional frequencies, would indicate distortion somewhere in the system. It is a simple matter to calculate the percentage modulation of an AM signal from the relative amplitudes of the side bands and the carrier:

$$\% \text{ modulation} = \frac{2V_{SB}}{V_C} \times 100\% \qquad (13\text{–}7)$$

where

V_{SB} = amplitude of either sideband component
V_C = amplitude of the carrier

Fourier Analyzer

Any periodic nonsinusoidal waveform can be represented by a unique series of harmonically related sine waves having definite amplitude and phase relationships. The mathematical representation of the components of a nonsinusoidal waveform is called a *Fourier series expansion.* When the magnitudes of these harmonic components are plotted as a function of frequency, we have a **Fourier spectrum.**

The Fourier spectrum of any periodic signal is displayed on a **Fourier analyzer.** It uses digital sampling and transformation techniques to form a Fourier spectrum display that has both phase and amplitude information. The actual calculations are based on a computer algorithm called the *Fast Fourier Transform* (FFT). This algorithm calculates the magnitude and phase of each frequency component from a block of time domain samples of the input signal.

The block diagram of a typical Fourier analyzer is shown in Figure 13–27. First, the signal is sent through a low-pass filter to remove any out-of-band frequency components. Next, the filtered signal is sampled and digitized at regular intervals, until enough samples have been accumulated. The computer produces frequency domain results from these time domain samples. The results are stored in memory, from which they can be displayed on a CRT, plotted, or processed further. The Fourier analyzer has several advantages over the swept spectrum analyzer, such as low-frequency coverage, high-frequency resolution, and direct transfer function measurements.

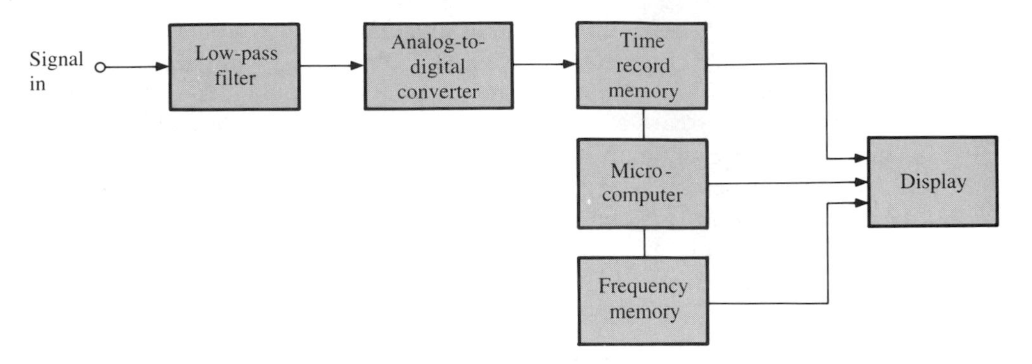

Figure 13–27 Fourier analyzer block diagram.

Wave Analyzer

The **wave analyzer** is sometimes referred to as a frequency selective voltmeter. It is basically a tunable, finite-bandwidth bandpass filter driving a meter circuit. The filter is tuned over a given range of frequencies, and the amplitudes of signal components are displayed on the meter as they come within the filter's passband. The frequency is read from either a mechanical dial or digital display.

In practice, wave analyzers have three basic uses:

1. measurement of the amplitudes of individual components of a complex frequency system
2. measurement of signal amplitudes in the presence of noise and interfering signals
3. measurement of energy in a specific, well-defined bandwidth

Distortion Analyzer

Nonlinear elements in a system create harmonic frequencies from a pure sine wave; this is known as **harmonic distortion.** The amount of distortion due to a particular harmonic, as a percentage of the fundamental component, is given by

$$\text{percent } N\text{th harmonic distortion} = \frac{V_N}{V_1} \times 100\% \tag{13–8}$$

where

V_N = RMS voltage of the Nth harmonic
V_1 = RMS voltage of the fundamental component

Rather than specifying the amount of distortion at a particular harmonic, a more generalized (but effective) measure, called **percent total harmonic distortion (THD)**, takes into account all harmonic components above the fundamental. In terms of the percent distortion at each harmonic, percent THD is determined from

$$\%\text{THD} = \sqrt{(\%\ 2\text{nd})^2 + (\%\ 3\text{rd})^2 + \ldots\ldots} \qquad (13\text{–}9)$$

EXAMPLE 13–4

The output spectrum of a given waveform shows the following components up to the fifth harmonic:

$V_1 = 1.0$ V
$V_2 = 0.2$ V
$V_3 = 0.1$ V
$V_4 = 0.02$ V
$V_5 = 0.005$ V

Determine (a) the percent harmonic distortion of the third harmonic and (b) the percent total harmonic distortion.

Solution:

(a) Using Equation 13–8, the percent third harmonic distortion is

$$\text{third harmonic distortion} = \frac{0.1\ \text{V}}{1.0\ \text{V}} \times 100\% = 10\%$$

(b) Using Equation 13–9, the percent total harmonic distortion is

$$\%\text{THD} = \sqrt{(20\%)^2 + (10\%)^2 + (2\%)^2 + (0.5\%)^2} = 22.5\%$$

Rather than the harmonic components of a given waveform being individually determined, a **distortion analyzer** is often used to determine the THD. The distortion analyzer is essentially a narrowband notch filter followed by a broadband detector and a meter circuit. First, the analyzer measures the amplitudes of the fundamental and harmonic components, as well as of any noise that might be present. The notch, or band reject filter, is switched in to remove the fundamental. The ratio of the second measurement to the first measurement is generally accepted as a good approximation of total harmonic distortion. Mathematically,

$$\%\text{THD} \simeq \frac{\sqrt{(\text{harmonics})^2 + (\text{noise})^2}}{\sqrt{(\text{fundamental})^2 + (\text{harmonics})^2 + (\text{noise})^2}} \tag{13–10}$$

In order that the resulting measurement be a close approximation to the true value, the harmonic content of the signal driving the system being studied must not be more than one third of the distortion expected to be introduced by the system. An approximation error of 0.5% can be expected for THD levels of 10%.

Currently, desktop computers combined with automatic spectrum analyzers provide a rapid means of measuring true harmonic distortion levels. The fundamental and its harmonic components are rapidly measured one at a time, and the percent THD is internally computed using Equation 13–9.

Audio Analyzer

The **audio analyzer** is a general-purpose audio test instrument that performs several low-frequency measurements, as well as measuring distortion. It contains both a true RMS voltmeter and a DC voltmeter to measure complex waveform levels. In addition, it includes a low-distortion audio oscillator as a signal source. It automatically performs signal-to-noise (S/N) ratio measurement, and includes a built-in frequency counter to measure the frequency of the input signal.

13–11 SUMMARY

This chapter discussed signal sources in terms of audio oscillators, function generators, signal generators, and synthesizers. Audio oscillators are sine-wave generators with frequencies less than 20 kHz. RF signal generators usually have the ability to modulate either amplitude or frequency of a carrier signal. Function generators, based on a triangle wave, generate other waveforms as well. Frequency synthesizers digitally generate their waveforms. Although signal sources can be of a fixed frequency, which can be adjustable, some can also be electronically swept over a range of frequencies.

This chapter also discussed those instruments that measure signal components as a function of frequency or frequency domain. As examples, this chapter discussed the purpose and characteristics of spectrum, Fourier, wave, distortion, and audio analyzers.

13–12 GLOSSARY

amplitude modulation (AM) A type of modulation in which the amplitude of the carrier signal varies in accordance with the modulating signal.

audio analyzer An instrument that measures distortion in the audio frequency range.

comb generator An oscillator current which produces a signal rich in harmonics of the input signal.

deviation The difference between the instantaneous frequency of the modulated signal and its carrier.

deviation ratio The ratio of the maximum carrier frequency deviation to the highest modulating frequency.

difference frequency Also called the **lower sideband,** the output frequency of a mixer circuit equal to the difference of its two input frequencies.

distortion Any undesired change in the waveform of the signal, especially in shape or phase.

distortion analyzer An instrument that measures total harmonic distortion by determining the harmonic components of a given waveform.

droop Also called **sag,** occurs when the peak value gradually decreases during the pulse width.

duty cycle Expressed as a percentage, the ratio of the pulse width time to the signal period.

fall time The time required for a pulse to decrease from 90 to 10% of its normal amplitude.

Fourier analyzer A computer-driven instrument that determines the Fourier-series components of any periodic waveform.

Fourier spectrum A graph of the Fourier series components of a waveform or signal.

frequency deviation See **deviation.**

frequency domain Signal amplitude versus frequency.

frequency modulation (FM) Modulation of the carrier such that the carrier frequency is proportional to the amplitude of the modulating signal.

frequency synthesis The generation of a stable output frequency derived from multiples or submultiples of one or more crystal-controlled oscillators.

function generator An oscillator producing waveforms of varying waveshapes and frequencies.

harmonic distortion A form of distortion caused by a signal passing through a nonlinear system, in which harmonics are added to the fundamental signal.

image frequency The unwanted sideband frequency in a mixer circuit, which is usually removed by a tuned circuit.

lower sideband See **difference frequency.**

modulation index The ratio of the frequency deviation of an FM system to the frequency of the modulated wave.

oscillator A signal source of known frequency.

overshoot A distortion of the peak value immediately following a rising edge.

percent modulation The ratio of the peak to the minimum variation of the amplitude-modulated carrier.

preshoot A distortion of the pulse base line before the leading edge.

pulse A brief, sharp change in a normally constant current, voltage or other quantity, characterized by a rise and fall time of finite duration.

pulse generator A device which generates pulses of controlled amplitude, width, and repetition rate.

pulse repetition rate (PRR) The rate at which pulses are produced.

pulse width The time needed for the pulse to travel from the point of 50% amplitude on the leading edge to the 50% point on the trailing edge.

ringing The transient oscillation of either the pulse's top or base line.

rise time The time required for a pulse to rise from 10 to 90% of its normal amplitude.

sag See **droop.**

sidebands Frequency components of a modulated carrier above and below the carrier frequency.

signal generator A sine-wave generator with the ability to modulate the carrier signal.

spectrum analyzer An instrument with a CRT that displays the frequency components of a complex signal.

sum frequency Also called the **upper sideband,** the output frequency of a mixer circuit equal to the sum of its two input frequencies.

sweep To change steadily with time.

sweep generator A device which generates a waveform whose frequency can be varied at a selected rate between a minimum and maximum value.

tank circuit A parallel resonant *L-C* circuit.

time domain Signal amplitude versus time.

total harmonic distortion (THD) A measure of nonlinearity that is the ratio of the power of the higher order harmonics of the output signal of a system to the fundamental component of that signal.

undershoot A distortion of the pulse base value immediately following the falling edge.

upper sideband See **sum frequency.**

variable voltage capacitor diode A reversed biased diode whose junction capacitance varies in proportion to the magnitude of the reverse bias. Also called a *varactor, epicap,* or *varicap diode.*

wave analyzer An instrument that measures amplitudes of the harmonic components of complex signals.

13–13 PROBLEMS

1. For the Wein bridge oscillator of Figure 13–28, determine the value of the resistors required to cause oscillation at 1.75 kHz if the resistors are equal in value.

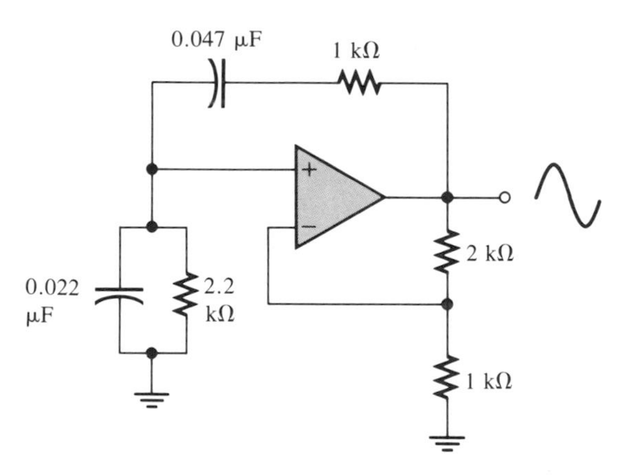

Figure 13–28 Circuit for Problem 1.

2. A 610-kHz carrier is amplitude-modulated by a 500-Hz sine wave. Determine
   (a) the frequency of the upper sideband
   (b) the frequency of the lower sideband

3. A 900-kHz carrier is amplitude modulated by a 1-kHz signal. Determine the frequencies of the sidebands if the waveform contains only even harmonics.

4. For the amplitude-modulated waveform of Figure 13–29, determine the percent modulation.

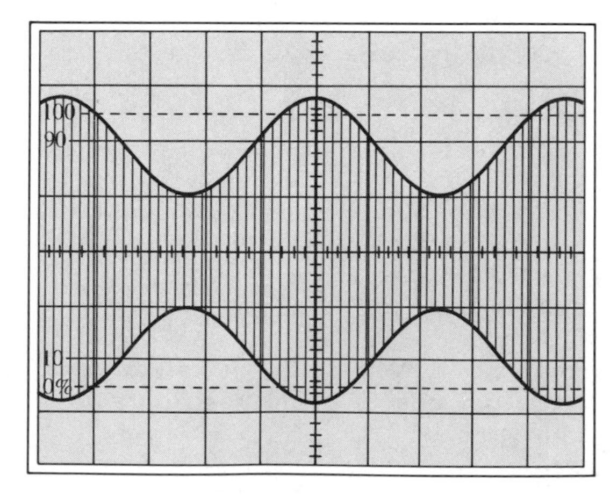

Figure 13–29 Waveform for Problem 4.

5. A 96.5-MHz carrier is frequency modulated by a 900-Hz sine-wave tone. If the carrier frequency varies by ± 7 kHz and the maximum deviation on either side of the carrier frequency by law is 37.5 kHz, determine
(a) the frequency deviation
(b) the modulation index
(c) the deviation ratio

6. If one cycle of a 7.5-kHz pulse train has a pulse width of 60 μs, determine the percent duty cycle.

7. Using the spectrum analyzer display of Figure 13–26, determine the percent modulation.

8. The output spectrum of an amplifier with a 1-kHz sine-wave input signal shows the following harmonic components:

at 1 kHz, V = 0.852 V RMS
at 2 kHz, V = 0.154 V RMS
at 3 kHz, V = 0.096 V RMS
at 4 kHz, V = 0.071 V RMS
at 5 kHz, V = 0.055 V RMS
at 6 kHz, V = 0.048 V RMS

Determine
(a) the percent harmonic distortion at each harmonic
(b) the total harmonic distortion up through the sixth harmonic

14

RF AND FIBER-OPTIC MEASUREMENTS

14–1 INSTRUCTIONAL OBJECTIVES

At the completion of this chapter, you will be able to

- Explain the operation and function of the RF probe.
- Describe how RF power is measured using an RF ammeter and a noninductive resistor.
- Explain how to measure RF power using calorimetry and photometric comparison methods, as well as with a bolometer.
- Describe the construction and operation of a directional wattmeter to measure both power levels and the voltage standing wave ratio on a transmission line.
- Explain how the absorption wavemeter and dip meter are used to measure frequency.
- Describe the purpose of a noise figure meter.
- Discuss those measurements made on fiber-optic equipment.

14–2 INTRODUCTION

The measurement of radio frequency (RF) signals often greater than 60 kHz and even in the microwave range above 1 GHz requires techniques and instruments that must not affect the performance of the system under test. This chapter discusses the measurement of RF voltages using an RF probe connected to an electronic voltmeter, using voltage and current methods, and photometric comparison, bolometer, calorimetric, and directional wattmeter methods. The passive measurement of frequency is accomplished with either absorption wavemeters or dip meters. The measurement of RF noise using a noise figure meter is discussed. Finally, this chapter discusses the growing use of fiber optics and the equipment used in the testing of fiber-optic systems.

14–3 THE RADIO FREQUENCY (RF) VOLTMETER

The long leads and relatively high input impedances of conventional voltmeters make them unsatisfactory for the measurement of radio frequency signals. As shown in Figure 14–1, an **RF probe** is used in conjunction with a conventional electronic voltmeter (on a DC voltage range) for general RF measurements. The probe circuit is a shunt rectifier network that charges a capacitor to a voltage equal to the peak value of the applied signal. The series resistor in conjunction with both the cable capacitance and the input capacitance of the voltmeter forms a lowpass filter, blocking RF from reaching and damaging

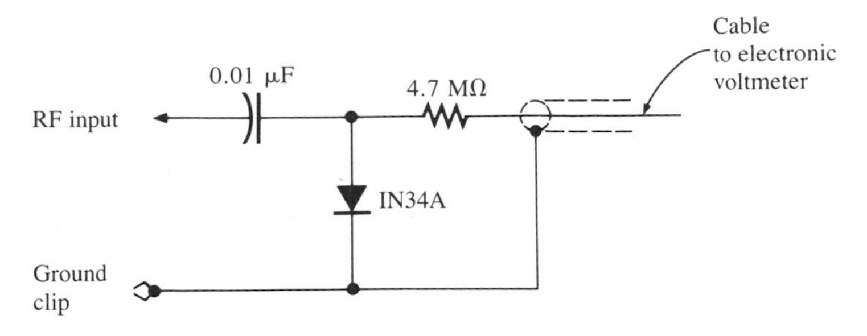

Figure 14–1 RF probe circuit.

the meter. This same resistor, in conjunction with the input resistance of the voltmeter, forms a voltage divider, reducing the peak rectified voltage down to a DC level equivalent to the RMS value of the applied signal. The diode used must have a very low junction capacitance and a very high reverse resistance.

To minimize the effects of long leads and of the input capacitance of the voltmeter, the rectifier circuit is installed in the probe itself. This provides the shortest possible RF path and still allows the meter to be located some distance from the probe. Some RF probes are active, with a transistor amplifier as well as the rectifier circuit. Unlike the passive probe of Figure 14–1, active RF probes require that supply voltages be brought to the probe.

14–4 RF POWER MEASUREMENT

The voltage and current levels at RF and microwave frequencies may vary along a transmission line (even when there is no loss associated with the line), but the power delivered remains constant. For this reason, power is often the best measure of signal levels. Power meters are used to measure the output of transmitters, calibrate signal generators, measure the transmission characteristics of various devices, and level signals. The following sections discuss several methods that are used to measure RF power.

Voltage and Current Methods

For accurate measurements of both audio frequency (AF) and RF power at frequencies below 100 to 150 MHz, a thermogalvanometer or *thermocouple ammeter* (see Section 6–5) is placed in series with a known *noninductive* resistor (usually 50 Ω), and the dissipated power is calculated by the basic power equations

$$P = VI \tag{14–1a}$$

or

$$P = I^2R \tag{14–1b}$$

or

$$P = \frac{V^2}{R} \tag{14–1c}$$

Photometric Comparison

Another simple technique, used for measuring RF power at frequencies of 30 MHz or less, involves comparing the brightness of an incandescent lamp driven first by the RF source and then by a 60-Hz source adjusted to produce the same light intensity. The 60-Hz power, which is easy to measure with an ammeter and a voltmeter, is equal to the unknown RF power. This simple method requires only a voltmeter, an ammeter, an incandescent lamp, and a photographic exposure meter.

EXAMPLE 14–1

The output power of a 7-MHz transmitter is to be measured using photometric comparison. With an incandescent light bulb coupled to the output of the transmitter, its intensity was measured using a photographic light meter at a distance of three feet. The same light bulb was connected to the 60-Hz power line through an autotransformer, which was adjusted so that the intensity of the light bulb at a distance of three feet was the same as in the first measurement. At this point, the voltage across the light bulb was measured to be 57 V, while the current through the bulb was measured as 1.2 A. Determine the output power of the transmitter.

Solution:

Using the power formula $P = VI$,

$$\begin{aligned} P &= (57\text{ V})(1.2\text{ A}) \\ &= 68.4\text{ W} \end{aligned}$$

The Bolometer

When RF energy is absorbed into a given material, the temperature of the material increases. If the material absorbing the RF energy has a nonzero temperature coefficient, its resistance also changes. Such a temperature-sensitive, resistive-element material, when used to measure RF power, is called a **bolometer.** At power levels below 10 mW, both RF and microwave power levels are often measured with a bolometer as part of a Wheatstone-bridge circuit. The bolometer device is placed within the RF energy field. Changes in the RF field strength produce temperature changes in the device, and cause a corresponding change in the resistance of the device.

Figure 14–2 illustrates two types of bolometer sensors based on this simple principle: the **thermistor** and the **barretter.** The thermistor is made of metallic oxide materials that exhibit semiconductor characteristics. Consequently, the resistance of the thermistor sensor *decreases* with an increase in temperature, resulting in a *negative temperature coefficient.* The barretter, on the other hand, is made from a short piece of fine platinum wire of a *positive* temperature coefficient. One advantage of a barretter over a thermistor is that the relationship between power absorbed and resistance is linear. However, thermistors are more rugged than the barretter, both physically and electrically, and are more commonly used.

Bolometers are mounted in a special section of transmission line, either a coaxial cable or a waveguide, in such a way as to present a near perfect impedance match, as shown in Figure 14–3. This arrangement is called a *bolometer mount,* and provides all necessary connection for the RF and the bridge circuit.

The bolometer is normally connected in one leg of a Wheatstone bridge, as shown in Figure 14–4, and either a DC or a low-frequency AC excitation voltage is applied to the bridge. The excitation voltage causes a bias current to flow through the bolometer and permits the bridge to be balanced with no RF power applied. The application of RF power heats the bolometer and unbalances the bridge. Withdrawing a like amount of bias power from the bolometer element returns the bridge to balance. The amount of bias power that must be removed to re-establish bridge balance can be converted to a power reading.

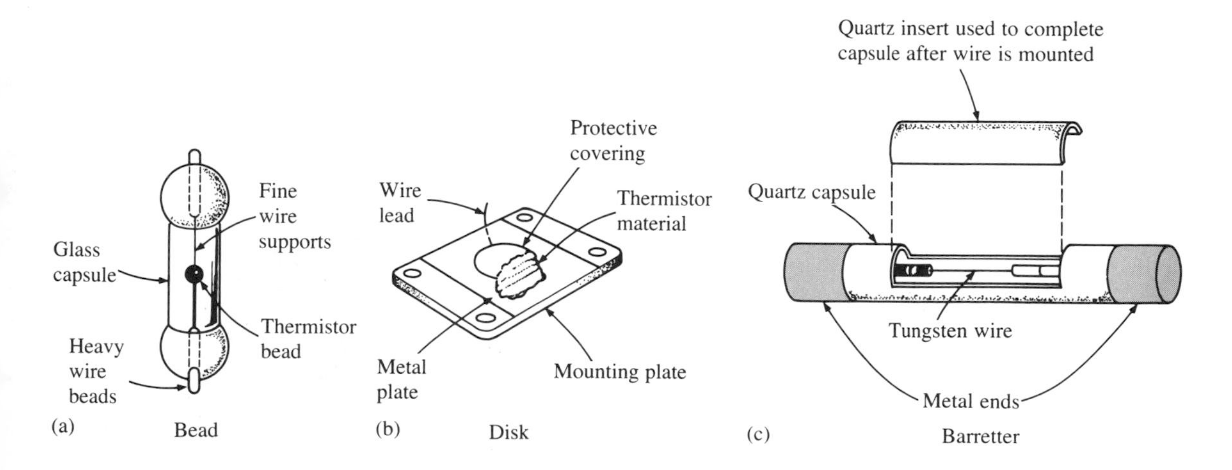

Figure 14–2 Thermistors. (a) Bead type. (b) Disk type. (c) Barretter.

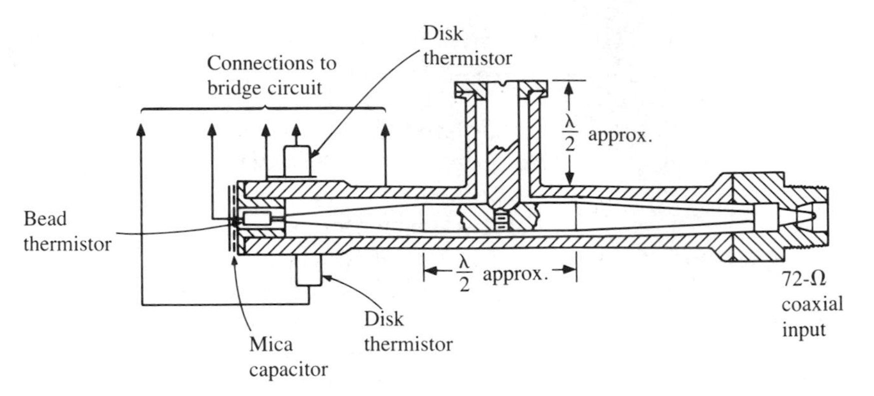

Figure 14–3 Coaxial bolometer mounting.

Calorimetric Measurement

At power levels above about 10 mW, bolometer elements are not used directly, since they can be damaged by excessive heating. Instead, the dissipated RF power is used to heat a resistor, and part of the resulting heat is transferred to the bolometer element. This system is referred to as *calorimetry,* or *calorimetric measurement.*

Calorimetric power meters based on this calorimetric technique can use either a dry or a fluid medium for heat transfer. Dry calorimeters use a static thermal path between the resistor and the temperature sensor, for example, mounting the heat-dissipating resistor and the temperature sensor on the same

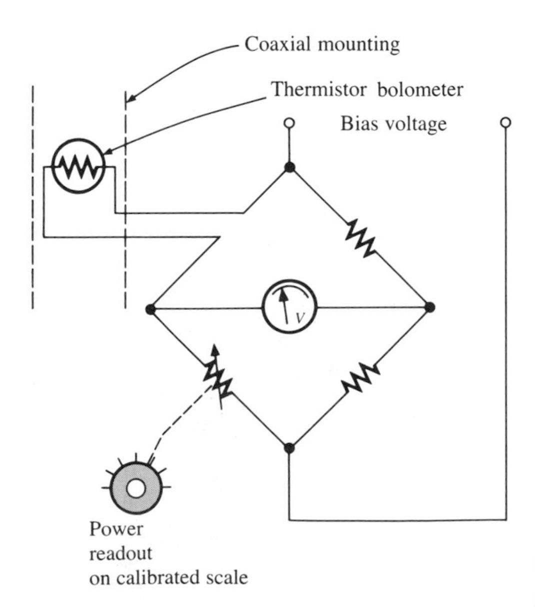

Figure 14–4 Bolometer power measurement bridge.

heat sink. This arrangement requires time for the sensor to come to the same temperature as the resistor, and thus tends to be slow. Fluid calorimeters, however, use a moving stream of oil to transfer heat more quickly to the sensor.

14–5 DIRECTIONAL WATTMETERS

The RF power in a transmission line can be broken down into two components. One component travels towards the load and is called the **incident** or **forward power.** The second component travels towards the source from the load and is called the **reflected power.** The square root of the ratio of the reflected to forward power is called the **reflection coefficient,** and varies from zero to one.

If we were able to measure the voltage between the two leads of a transmission line at selected points along the line, we would find that the voltage periodically reaches alternate maximum and minimum values in a pattern similar to a rectified sine wave. The **voltage standing wave ratio (VSWR)** is a measure of the ratio of the maximum to minimum RMS voltage levels on the line, so that

$$\text{VSWR} = \frac{V_{\text{max}}}{V_{\text{min}}} \tag{14–2}$$

As a practical means, however, the VSWR, or SWR for short, is measured with an in-line wattmeter.

Shown in Figure 14–5 is a circuit of an in-line directional wattmeter that measures the voltage standing wave ratio as well as the forward and reflected power levels at that point on the transmission line. If the VSWR meter is used with a transmission line with the proper characteristic impedance (e.g., 50 or 75 Ω), it can be calibrated to read power. Known as an *SWR meter,* this circuit makes use of the mutual inductance of the windings of a toroidal-wound transformer to establish a bridge circuit. Capacitors C_1 and C_2 form a capacitive voltage divider, with the voltage across C_2 in phase with the voltage on the transmission line. The phase of the voltage across resistor R is a function of the current flowing along the transmission line. From the reflection coefficient K, the VSWR can be determined from

$$\text{VSWR} = \frac{1 + K}{1 - K} \tag{14–3}$$

where $K^2 = P_{\text{reflected}}/P_{\text{forward}}$. As the reflection coefficient varies from zero to one, the VSWR varies from 1 to infinity.

Figure 14–5 A unidirectional power meter.

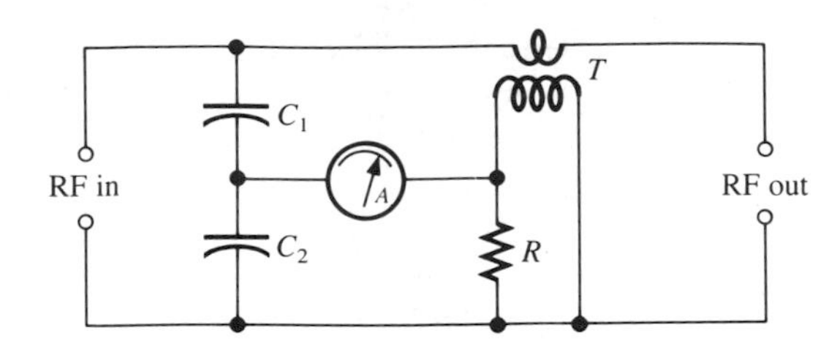

The VSWR bridge is adjusted so that if a pure resistance, equal to the impedance of the bridge (usually 50 Ω), is connected to the "RF OUT" terminals, the voltages across R and C_2 are out of phase and the voltmeter reading is a minimum. Since the reflection coefficient is zero, the VSWR is 1:1, indicative of a perfectly matched line. Consequently, all the power from the RF source is delivered to the load, and the reflected power is zero. Any other value of resistance or impedance connected to the output terminals causes an upscale reading proportional to the reflected (reverse) voltage. By reversing the secondary terminals of the transformer, the incident or forward power component of the voltage can be measured at that point on the transmission line. If the VSWR meter is used with a transmission line with the proper characteristic impedance (e.g., 50 or 75 Ω), it can be calibrated to read power. The **net power** is equal to the incident power minus the reflected power. Instruments of this type usually have a limited frequency range over which they meet their designed accuracy.

14–6 PASSIVE FREQUENCY MEASUREMENT

The measurement of radio frequencies and above generally requires some form of indirect measurement, because capacitive coupling and the resultant loading usually alter the measurement. In effect, oscillators and other tuned circuits are "de-tuned," and operate at frequencies different from their intended ones. Therefore, *passive* measurements are generally made, in which the instrument is not electrically connected to the circuit under test. Frequency counters, absorption wavemeters, and dip oscillators are often used, and are discussed in the following sections.

The Absorption Wavemeter

The **absorption wavemeter,** shown in Figure 14–6, is an extremely simple, but low-accuracy, instrument used for the measurement of frequency. It is entirely passive, deriving its operating power from the circuit under test. It consists of a calibrated tuned circuit that is inductively coupled to an energized

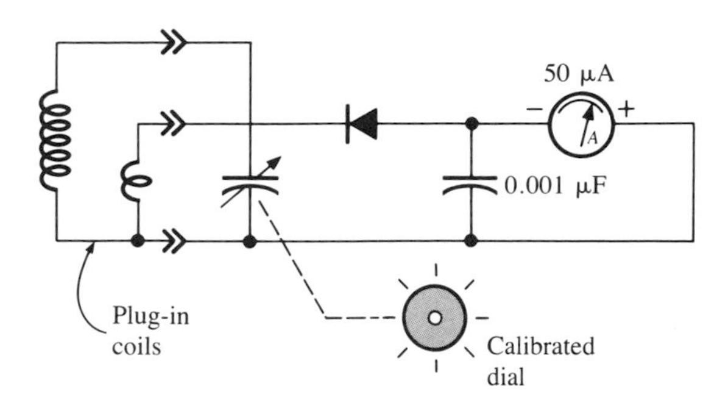

Figure 14–6 Circuit for an absorption wavemeter.

circuit under test. When both the wavemeter and circuit under test are tuned to the same frequency, the absorption wavemeter absorbs power from the energized circuit and drives an indicator (such as a small lamp or a sensitive meter). The frequency is read off the calibrated dial connected to the variable capacitor.

The absorption wavemeter measures the frequency of the RF energy present, and not necessarily the resonant frequency of the circuit from which it is absorbed. Since the wavemeter responds only to the frequency to which it is tuned, there is no chance of the harmonic ambiguity that plagues some other frequency-measuring methods. The accuracy of an absorption wavemeter is low compared to that of most other methods of frequency measurement, and its coupling tends to slightly detune the circuit under test. However, it is extremely useful in many routine service checks and adjustments involving RF transmitters and oscillators.

The Dip Meter

The **dip meter** is a calibrated, tunable oscillator. When the dip meter and the circuit under test are inductively coupled and tuned to the same frequency, the circuit under test absorbs some energy from the dip meter, causing the reading of a meter in the oscillator circuit to "dip."

In practice, dip meter operation is the reverse of that of the absorption wavemeter. The dip meter *supplies* the RF power, and the circuit being checked absorbs energy. Consequently, when a dip meter oscillator is turned on, the circuit being checked must be de-energized.

Original dip meters used a vacuum tube oscillator circuit and were called **grid dip oscillators (GDO)** because the meter indicated a drop in grid signal. Modern dip-meter circuits use solid state devices, like the FET circuit of Figure 14–7, but are still often referred to as "grid dip meters."

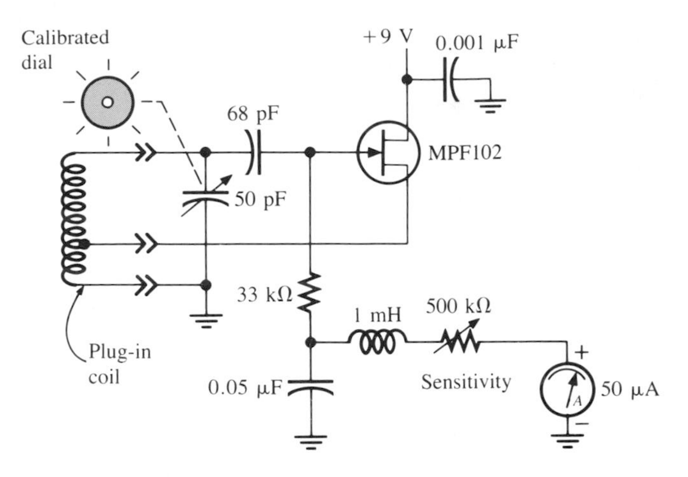

Figure 14–7 FET dip meter circuit.

As shown in Figure 14–8, most dip meters are supplied with a series of plug-in coils, permitting coverage of a very wide frequency range. Dip meters are particularly useful for adjusting tuned circuits and antennas before they are energized for the first time. Most dip meters have a switch permitting the oscillator to be switched off so that the same instrument can also be used as an absorption wavemeter.

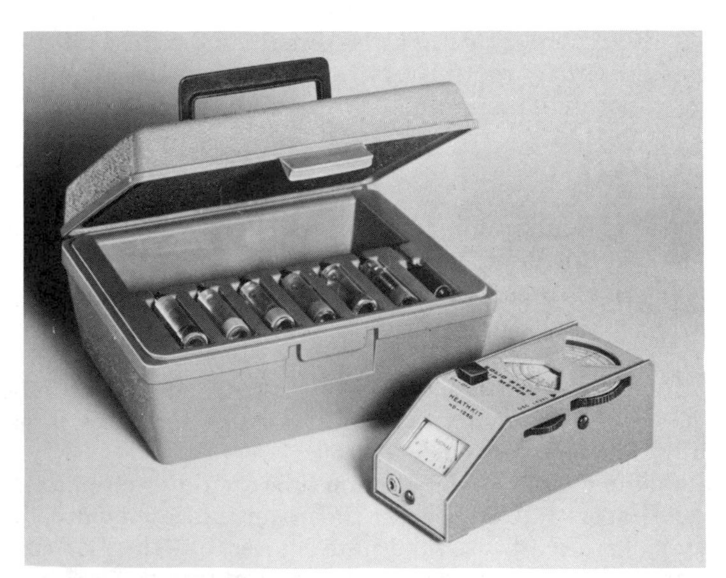

Figure 14–8 Commercial dip meter. Courtesy of HEATH Company, St. Joseph, MI.

The Frequency Counter

Chapter 12 discusses frequency measurements using a frequency counter in the traditional sense, by direct connection to the circuit under test. However, at radio frequencies where high power is involved, the RF energy, if connected directly to the counter's input, would destroy the counter. The most common method is to use a short piece of stiff wire connected to the counter's input. In effect, the wire acts as an antenna and transfers a modicum of RF energy to the counter. The counter is placed near the source of RF radiation and the frequency read from the counter's display.

14–7 THE FIELD STRENGTH METER

A very basic measurement associated with radio communications of any type is the measurement of radiated field or signal strength. This measurement is normally made by measuring the voltage induced in a small antenna. The relationship between this induced voltage and the field strength is either determined by calculation or read directly from a special meter scale calibrated in terms of microvolts per meter.

A very simple field-strength meter used for making *relative* measurements in the immediate vicinity of the transmitting antenna can consist of little more than a small antenna, a tuned circuit, a diode detector, and a meter movement, as shown in Figure 14–9. It is essentially a basic radio receiver equipped with an indicator that provides a visual representation of relative signal strength. The meter scale is marked in arbitrary divisions, for example, from 0 to 100.

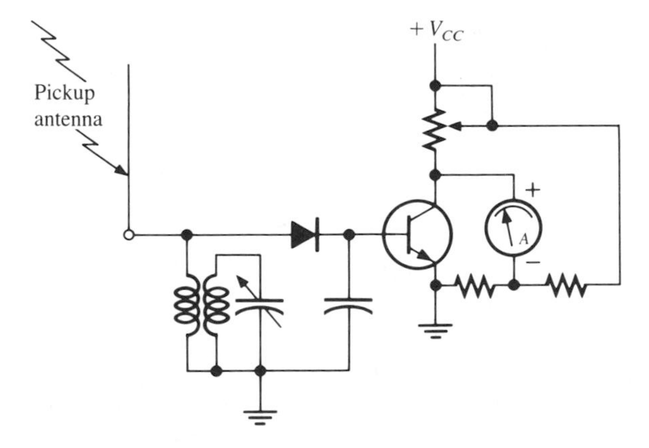

Figure 14–9 Field strength meter circuit.

A sample of the radiated signal is picked up by the antenna and rectified by the diode, and the resulting DC current is applied to the meter movement. The pick-up antenna must always have the same polarization as the antenna being checked. If the source antenna is vertical, then the pick-up antenna of the field strength meter must also be placed in a vertical position. Such an instrument is very useful for making transmitter and antenna system adjustments, either at a fixed direction or distance from the source or in a 360° pattern at a fixed distance. However, this simple instrument is useful only in the immediate vicinity of the transmitting antenna, where the radiated field is strong.

More elaborate instruments, which can measure absolute field strengths less than 1 μV/m, consist of an especially designed superheterodyne receiver that has a linear response to signal amplitude, and includes a built-in precision attenuator and an output indicator.

14–8 NOISE FIGURE METERS

As a signal passes through a receiving system, additional noise is introduced by the system itself. The **noise figure** of a system or device is a measure of the degradation in the **signal-to-noise (S/N) ratio** that occurs as the signal passes through the system or device. In effect, it is a figure of merit used to express how well a receiving system can process weak signals.

Noise figure, usually in units of decibels, is often expressed as the ratio of the total output noise power (at a source temperature of 290 K, or 17 °C) to the output noise power that would be present if the system or device were free of noise and did not add any noise to the signal, or

$$F\ (\text{dB}) = 10 \log_{10} \left(\frac{S_{IN}/N_{IN}}{S_{OUT}/N_{OUT}} \right) \tag{14–4}$$

where

S_{IN} = input signal power level
N_{IN} = input noise power level
S_{OUT} = output signal power level
N_{OUT} = output noise power level

A special calibrated broadband noise source is normally used in conjunction with the noise figure meter. The effective temperature of the source is specified because thermal agitation of electrons in the source impedance causes noise energy to be radiated as well.

Modern noise figure meters, such as the HP8970A shown in Figure 14–10, are controlled by a microprocessor. A typical meter measures noise

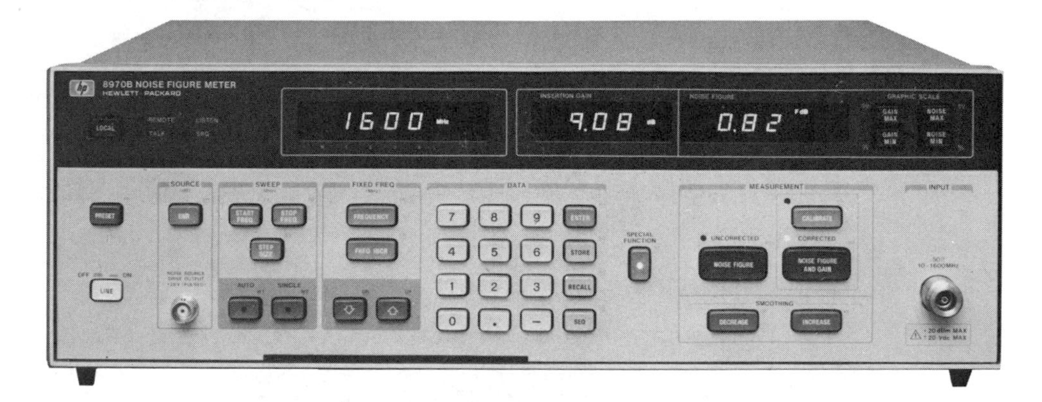

Figure 14–10 Model 8970B Noise Figure Meter. Courtesy Hewlett Packard.

power input to and output from the device under test at two different source temperatures, and then calculates the noise figure and gain of the device. It even automatically corrects for any noise introduced by the measurement system. The resulting noise figure can be read from an LED display, or, for swept frequency measurements, from an oscilloscope as a curve of frequency versus noise figure.

14–9 FIBER-OPTIC TEST EQUIPMENT

Use of fiber-optic transmission lines is the state-of-the-art method for transmitting signals from one location to another. Fiber-optic lines are small, cylindrical waveguides made from glass fibers that operate at optical (4.3 to 7.5×10^5 GHz) rather than microwave frequencies (1 to 100 GHz). At the transmitting end, a modulated light source converts electrical signals to equivalent light intensities. At the receiving end, a photosensitive device converts the modulated light beam back to an electrical signal.

By comparison, fiber-optic cables are smaller and lighter than those using copper conductors, and offer complete electrical isolation. In addition, they have an extremely wide bandwith and are immune to both electrical noise pickup and crosstalk from adjacent and nearby circuits. The growth in fiber-optic communications has been remarkable, and, as a result, specialized test equipment that includes light sources, power meters, and receivers has been developed to support the needs of the industry.

Optical Signal Sources

The *light emitting diode,* or *LED,* is the most often used light source for short-range systems with bit rates below 100 MHz. For long-range, high-speed communication, semiconductor lasers are normally used.

The *optical signal source,* similar to the simple circuit of Figure 14–10, is a specialized piece of test equipment that delivers accurate and repeatable calibrated signals for evaluating the performance of fiber-optic components and receivers. The Hewlett-Packard model HP8150A is a good example; it is based on an electrical-to-optical transducer that can be modulated by signals from DC to approximately 250 MHz, and has a typical output power range between 1 nW and 2 mW.

Optical Pulse Power Meter

The optical pulse power meter allows the user to make both peak and average pulse power measurements in a fiber-optic system. Such an instrument permits the characterization of fiber optic components and transmitters and the determination of upper and lower power levels.

An important component of this power-measuring system is the transducer that converts light energy to an electrical signal. It must produce an electrical signal that corresponds directly to the optical input waveform. In addition to its primary task as part of the power measurement system, the output of this transducer can also be applied to other instruments such as an oscilloscope.

Optical Receiver

The **optical receiver** is a linear transducer that converts optical signals into their electrical equivalents. It is intended to be a "front-end" interface to conventional electronic test equipment, and also a general-purpose receiving device for testing fiber-optic modules and systems.

A typical optical receiver (Figure 14–11) employs a PIN (positive-intrinsic-negative) diode or avalanche photodiode as the receiving element. It is useful over an operating wavelength range between 550 and 950 nm. In a PIN diode, the normal *p*- and *n*-regions are separated by a very lightly doped *n*-region. This provides a very low reverse leakage current and a wide junction width,

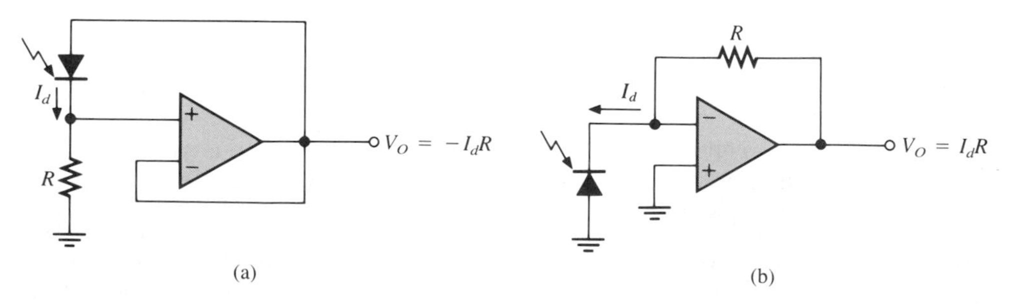

Figure 14–11 Fiber optic receivers using a PIN diode. (a) Unity-gain amplifier. (b) Transimpedance amplifier.

reducing junction capacitance. As a result, the diode's response speed is improved. If high-detection sensitivity is required, avalanche photodiodes are sometimes used; being temperature sensitive, however, they are somewhat noisier due to the avalanche process.

14–10 SUMMARY

This chapter discussed the field of RF measurements. These are measurements of voltage using an RF probe connected to an electronic voltmeter, the measurement of RF power using voltage and current methods, and photometric comparison, bolometer, calorimetric, and directional wattmeter methods. The passive measurement of frequency is accomplished with either absorption wavemeters or dip meters. The measurement of RF noise using a noise figure meter is also discussed. Finally, the chapter discussed the growing use of fiber optics and the equipment used in the testing of fiber-optic systems.

14–11 GLOSSARY

absorption wavemeter An instrument for the passive measurement of frequency that works by absorbing energy from the circuit it is coupled to.

barretter A short piece of platinum wire with a positive temperature coefficient used in the construction of bolometers.

bolometer An instrument that measures RF energy. Its intrinsic resistance changes when heated by the RF energy.

dip meter See **grid dip oscillator.**

forward power Also called **incident power,** the power traveling in a transmission line from the source to the load or antenna.

grid dip oscillator (GDO) Also called a dip meter, an instrument that passively measures resonant frequency by supplying energy to a de-energized circuit under test.

incident power See **forward power.**

net power Equal to the difference between the forward and reflected power levels in a transmission line.

noise figure A figure of merit proportional to the ratio of the signal-to-noise ratio of the input of a system to the signal-to-noise ratio at the system's output.

optical receiver A circuit using either a PIN or an avalanche diode to convert changes in light intensity through a fiber optics cable into corresponding electrical signals.

reflected power The power traveling in a transmission line from the load or antenna to the source.

reflection coefficient The square root of the ratio of the reflected to forward power in a transmission line.

RF probe A rectifier circuit used with a high-impedance voltmeter to measure RF voltages.

signal-to-noise (S/N) ratio The ratio of the signal level at a given point in a circuit to the noise level at that point.

standing wave ratio (SWR) See **voltage standing wave ratio (VSWR).**

thermistor Contraction for *thermal resistor,* a semiconductor device whose resistance is inversely proportional to temperature.

voltage standing wave ratio (VSWR) A figure of merit of a transmission line indicating the degree of mismatch between the transmission line and its load. It is the ratio of the maximum to minimum voltage at a given point on the transmission line.

14–12 PROBLEMS

1. Determine the VSWR on a transmission line that has a maximum voltage of 125 V RMS and a minimum of 60 V RMS.
2. Voltage readings are taken at several points along a transmission line. The maximum voltage reading is 72 V RMS. If the VSWR is 1.7:1, determine the minimum voltage on the transmission line.
3. A transmission line has maximum and minimum voltage readings of 105 and 76 V RMS, respectively. Determine the reflection coefficient.
4. Determine the VSWR of a transmission line if the reflection coefficient is 0.27.
5. If, when connected to a 50-Ω load, the VSWR of a transmission line is 4.29:1, determine
   (a) the percentage of the forward power reflected by the load
   (b) the percentage of the forward power absorbed by the load
6. A 50-Ω transmission line connects the output of a 75-W transmitter to a 50-Ω load. If the VSWR is measured at 1.8:1, determine
   (a) the power reflected from the load
   (b) the net power
7. The following measurements were made on a receiver's detector stage:
   input signal level

   without noise = 100 μW
   with noise = 109 μW

   output signal level

   without noise = 92 μW
   with noise = 115 μW.

   Determine the noise factor, in dB.
8. What is the signal-to-noise ratio at the input of a receiver's detector stage if the noise figure is measured as 8.2 dB and the signal-to-noise ratio at the output is +12.5 dB.
9. If the PIN diode of the fiber-optic receiver shown in Figure 14–12 has a response of 0.7 μA/μW, determine the output voltage of the receiver if
   (a) the diode dark current is 0.5 μA
   (b) the power level of the light source striking the PIN diode is 9 μW.

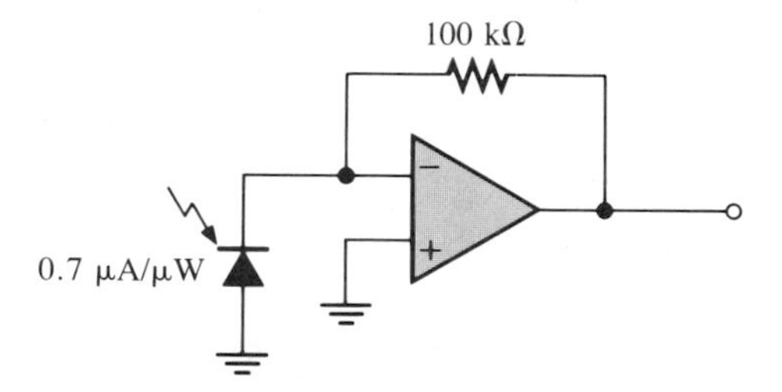

Figure 14–12 Circuit for Problem 9.

10. If a PIN diode has a response of 0.48 μA/μW for the circuit of Figure 14–13, and the output voltage equals 395 mV, what is the power level of the light source received by the PIN diode?

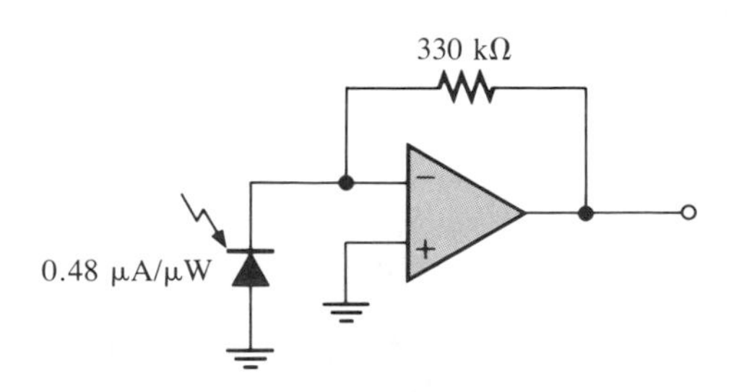

Figure 14–13 Circuit for Problem 10.

15

LOGIC ANALYZERS

15–1 INSTRUCTIONAL OBJECTIVES

After completing this chapter, you will be able to

- Explain the purpose and operation of a logic analyzer.
- Discuss the trigger modes of a basic logic analyzer.
- Define qualifier lines, trigger delay, word event delay, and clock delay.
- Describe the purpose and operation of a logic probe, pulser, clip, and comparator.
- Describe the purpose and operation of a digital signature analyzer.

15–2 INTRODUCTION

The analysis of two synchronous digital signals can usually be accomplished with a dual-trace oscilloscope, provided its bandwidth is high enough. However, the analysis of two or more digital signals, whether synchronous or asynchronous, is more conveniently carried out using those instruments specifically designed for that purpose.

The logic analyzer and digital signature analyzer providing CRT-driven displays of the results are invaluable for the testing and monitoring of complex systems, whether computer-based or not. For troubleshooting the response of individual integrated circuit devices, such as gates and flip-flops, without removing them from the circuit under test, small hand-held instruments like the logic probe, logic clip, and logic pulser are helpful. Also, if the integrated circuit device can be removed from the circuit, then the logic comparator can be used to determine whether the device is good. This chapter discusses the operation and use of each of these instruments.

15–3 LOGIC ANALYZERS

The **logic analyzer** shown in Figure 15–1 is to digital circuits what the oscilloscope is to analog circuits. The basic logic analyzer is a multiple input device that, upon being triggered, stores the sequential states of a number of digital signals as logic 1s and 0s. This provides the user with a record of the events on the monitored lines over a period of time before and/or after the trigger point.

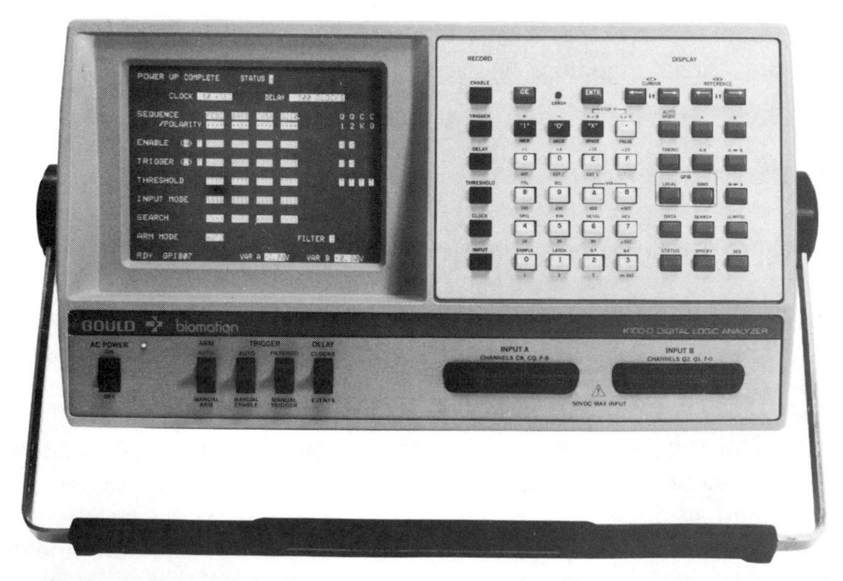

Figure 15–1 Model K1000-D logic analyzer. Courtesy Gould, Inc., Design and Test Systems Division.

Although logic analyzers can be used to monitor the operation of any logic circuit, they find their greatest use in microprocessor-based systems, where it is often necessary to trace the execution of many sequential operations. For example, when a computer's address bus is monitored by a logic analyzer during the instruction-fetch cycle, a fixed number of program addresses are stored in the sequence in which they appeared on the bus. This enables the user to trace through this section of the program's operation to determine whether the proper paths have been followed. By proper selection of the trigger signal, any section of a logic sequence can be stored; the number of steps being limited by the size of the analyzer's memory.

The Basic Trigger Mode

The trigger section of the logic analyzer lets the operator select the starting point for the storage of data samples. It continuously compares the incoming data bits to a user-selected data word, called a **trigger data word** (e.g., binary bit pattern of eight bits). The trigger data word in low-priced analyzers is selected by front panel-mounted, 3-position switches, one switch for each bit. The positions are marked 1, 0, and *X*, where *X* stands for "don't care." In the "don't care" position, the incoming data channel satisfies the trigger requirements regardless of whether it is a logic 1 or 0. Higher-priced analyzers, however, are programmed from a keyboard.

The trigger circuit of most logic analyzers also monitors one or more **qualifier lines.** The data on these lines is not stored in memory, but is combined with the trigger data word to establish the trigger requirements. In effect, the length of the trigger data word is increased by the number of qualifier bits. For example, if a qualifier line were connected to the microprocessor's READ line, data storage could be triggered only when the READ line was at the selected level and the proper trigger word present on the data inputs. Once triggering is initiated, the trigger circuit has no effect.

Additional Trigger Modes

In addition to the basic triggering mode just discussed, most logic analyzers have modes that further increase versatility.

Using the B&K Model LA1025, Figure 15–2 illustrates a trigger scheme that monitors 16 data sampling lines, or *channels,* as well as the qualifier bits. When the incoming data samples match the user-selected, 16-bit trigger word and the qualifiers are satisfied, a trigger signal occurs, and the storage of data samples begins. In this mode, the first word stored in the logic analyzer's memory is the trigger word itself.

This basic trigger scheme can often be modified to select any combination of event-delay, clock-delay, or trigger-position modification. The **event delay** is a circuit that delays the output of the trigger circuit by a fixed number of **trigger events.** For example, if the event delay were set to 9, the trigger word/qualifier combination would be ignored for the first nine times that it occurred. Data storage would be initiated on the 10th occurrence. As with the basic

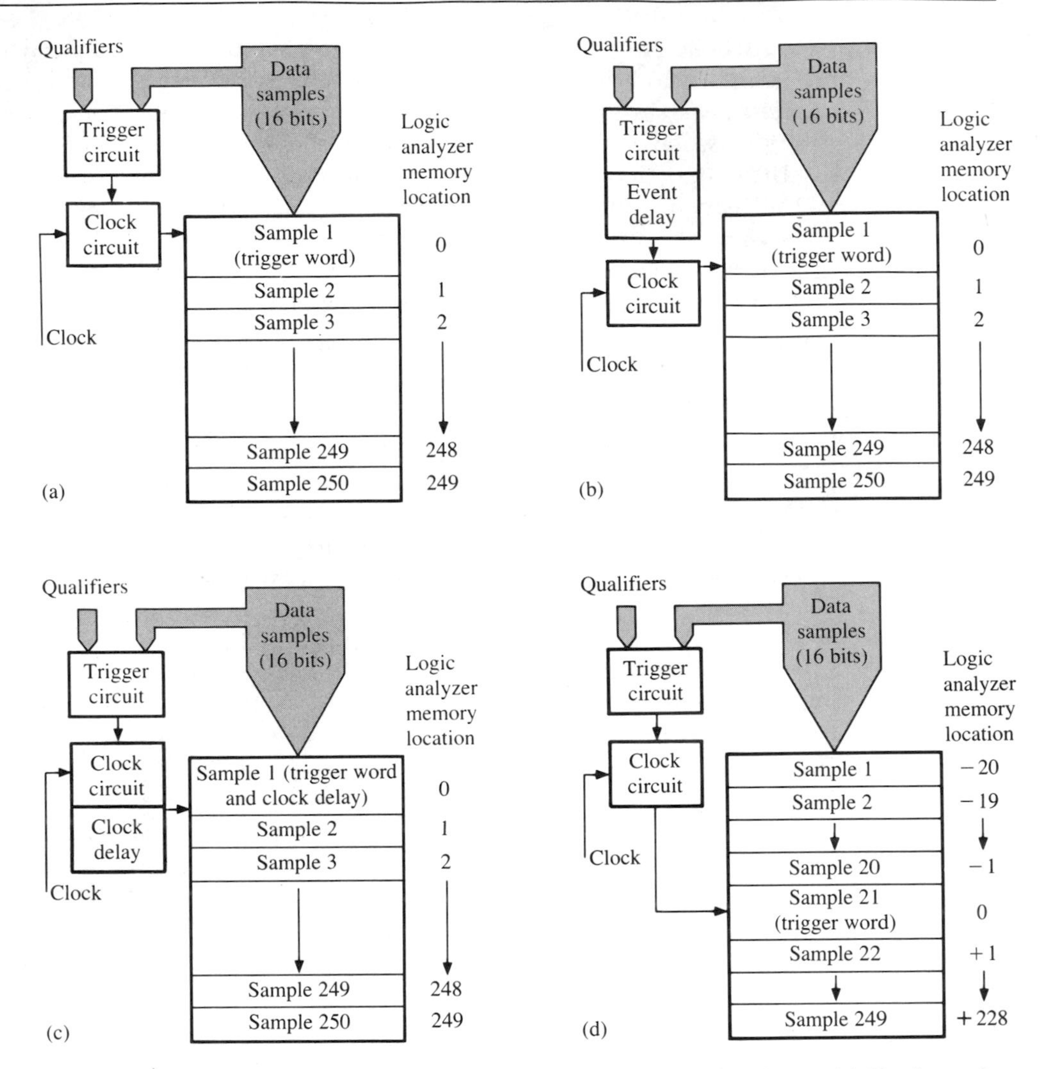

Figure 15–2 Trigger modes of B+K Model LA1025 logic analyzer. (a) Basic mode. (b) Event-delay mode. (c) Clock-delay mode. (d) Triggered word position mode.

triggering mode, the first data word stored in memory would be the data present at the inputs when triggering occurred. This is illustrated in Figure 15–2*b*. This feature is particularly useful for troubleshooting problems in a program loop that do not occur on the first pass.

Clock delay is a feature that permits data storage to begin a set number of data words following the trigger signal. This is illustrated in Figure 15–2*c*. For example, if the clock delay were set to 20, data storage would not commence until the 21st clock pulse following the trigger signal. The event delay takes priority over the clock delay and must be satisfied before the clock delay begins. Since the storage area of the analyzer is limited, by combining the event delay and the clock delay, the user can "zero in" on the portion of the program of interest. A "±" trigger selector allows the operator to place the trigger word anywhere within the memory. This means that data words that occurred before the trigger pulse was produced can be stored. This mode is shown in Figure 15–2*d*.

Clocking

Once the trigger circuit initiates the storing of data samples, it no longer has any control. The actual storing of data samples is controlled by and synchronized with the clock circuit. Data storage continues under clock control until the memory is full. The clock circuit tells the memory when the incoming data samples are to be stored. Most logic analyzers permit the operator to select either an internal or an external clock source. When the external clock source is used, data storage can be selected to occur on either the rising or falling edge of the clock pulse. The "external clock pulse" referred to here can be any control signal. It is not necessarily a computer system clock signal in the usual sense, although it can be.

When the internal clock is used, data is stored at one of a number of user-selected sampling rates. This is equivalent to snapping a picture of the incoming data at fixed intervals. This feature permits the user to observe a circuit that has no suitable clock. In addition, there is usually also a manual trigger push-button to begin storage. When a program halts before the memory is full, operation of a manual push-button completes the sampling process.

Data Display

Simple logic analyzers display the stored data one word at a time on LED or LCD displays, along with the memory address of the stored data. The operator can step through the data in either direction to examine each of the stored data samples. More expensive logic analyzers display multiple data words on a CRT. The data displayed on the CRT scrolls as the operator advances through the analyzer's memory. In this mode, the instrument serves as a "state analyzer." The CRT also has the option of a waveform display similar to that of a multi-channel oscilloscope. This provides the user with a timing diagram with one trace for each data input line, and enables the instrument to function as a "timing analyzer." The more sophisticated logic analyzers are microprocessor-based, permitting an even wider variety of information to be displayed. Some of the possibilities include a trigger menu, a file menu, instruction disassembly, signature analysis, a search function, activity histograms, and data correlation.

A *trigger menu* permits the selected sampling conditions to be displayed. These include the active probes, the trigger word, the qualifiers, and the event and clock delays. The *file menu* allows access to disk files for storing and retrieving collected data. *Instruction disassembly* is used to trace program execution in microprocessor-based systems by collecting the stream of binary data words on the data bus and converting them back into mnemonics, in a format similar to an assembler listing. The **signature analyzer** matches the waveform at a particular point in a defective system against the waveform at the same point in a properly operating system. The *search function* allows a search through the sample memory for specific data patterns. The *histogram function* is used to measure activity in a certain area of a system. *Data correlation* is a statistical tool for comparing recently collected data to data previously collected and stored as a disk file.

15–4 LOGIC PROBES

A **logic probe,** such as that shown in Figure 15–3, is an inexpensive hand-held device that greatly simplifies the tracing of logic levels and pulses in digital circuits. Indicator lights in the body of the probe indicate HIGH (logic 1) or LOW (logic 0) logic levels, intermittent pulse activity, and normal pulse activity. The better logic probes also detect intermediate logic levels, such as an open input on a TTL gate or an open-collector output without a pull-up resistor. This feature is useful for troubleshooting tri-state logic outputs and for detecting unconnected inputs.

The circuitry for the logic probe is contained within the body of the probe itself, while the positive supply voltage ($+V_{CC}$) and ground return are normally derived from the circuit under test through a cord exiting from the handle of the probe.

Logic probes are also ideal for detecting short-duration pulses and low-repetition rate pulses that are difficult to observe on an oscilloscope. For example, pulse widths greater than 10 ns light the "pulse" LED for 50 ms or more. Negative pulses cause the LED to go off momentarily. Less expensive logic probes, such as the circuit of Figure 15–4, often have three LEDs as indicators. One LED is on for a HIGH logic level, and off for LOW. The second LED is on for LOW logic levels and off for HIGH, while the third LED is driven by a "pulse stretcher" (monostable multivibrator) circuit, and flashes with pulses of a duration of at least 30 to 50 ns.

Most logic probes currently manufactured have a switch that selects the logic family to be tested. A typical probe is able to handle TTL, DTL, RTL, ECL, CMOS, MOS, and NMOS logic. In addition, many logic probes have a memory feature to freeze and store single-occurrence pulses.

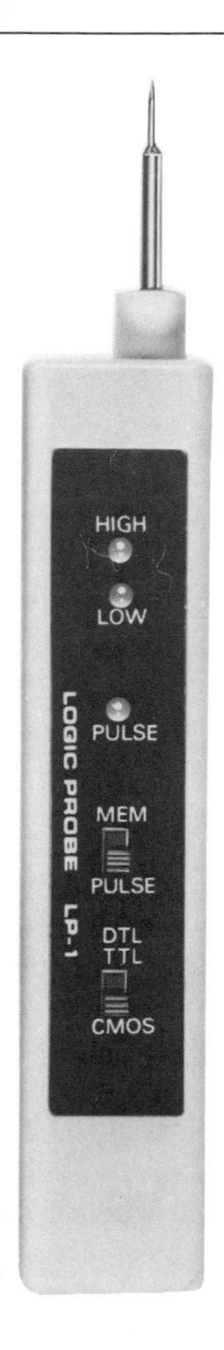

Figure 15–3 Hand-held logic probe. Photograph courtesy of Global Specialties, an Interplex Electronics Company.

15–5 LOGIC CLIPS

A **logic clip** of Figure 15–5, frequently called by the slang term **glomper clip,** is a device that clips to the top of a dual in-line (DIP) integrated circuit (IC) device. It makes contact with each of the IC's pins and continuously shows the logic level at all pins. A series of LEDs, one for each pin, serves as the HIGH and LOW logic level indicator. A lighted LED indicates a logic 1, and an unlit LED is a logic 0.

Test pins are normally buffered to minimize loading of the circuit under test. The more expensive logic clips require no external power supply connections; they power themselves from the circuit under test by automatically

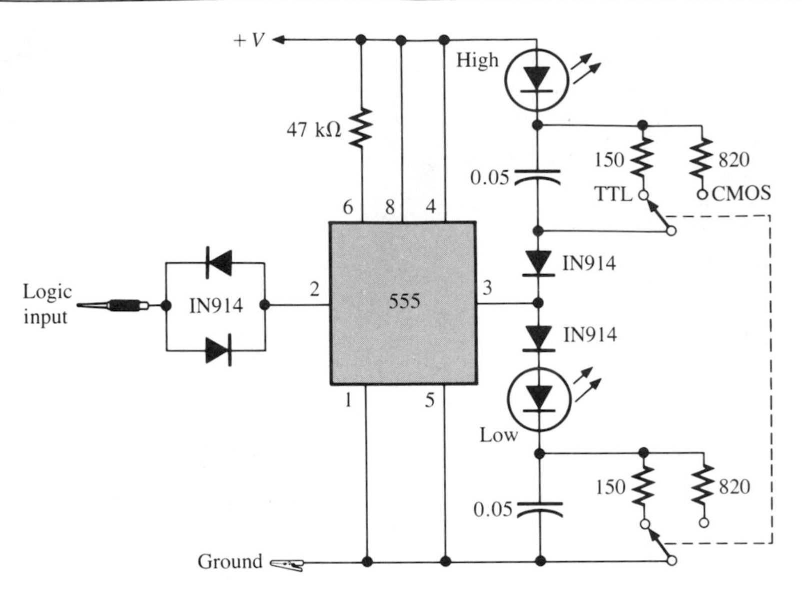

Figure 15–4 Simple logic probe circuit.

locating the positive supply and ground pins of the integrated circuit. A typical logic clip has each pin internally connected to a decision gate network, threshold detector, and a driver amplifier connected to an LED.

The decision gate network does the following:

1. Finds the positive supply voltage pin and connects it to the clip voltage bus
2. Finds all logic 1 pins and activates the corresponding LEDs
3. Finds all open circuits and activates the corresponding LEDs
4. Finds the ground pin, connects it to the clip ground bus, and blanks the corresponding LED

The threshold detector compares the input voltage to an internal reference. If the voltage is above the threshold voltage, the LED is activated. If the voltage is below the threshold, the LED is not activated. An amplifier at the output of the threshold detector drives the LED.

15–6 LOGIC PULSER

The **logic pulser,** like that shown in Figure 15–6, is a pulse generator that acts as a signal source for the testing of logic elements. It can be a single-shot device or, in some cases, it can generate a continuous stream of pulses.

Figure 15–5 Logic clip. Photograph courtesy of Global Specialties, an Interplex Electronics Company.

A typical logic pulser generates one pulse each time a fingertip push-button switch is depressed. The pulser probe tip is touched to a circuit node to act as a stimulus (input signal) for a logic element, and a logic probe is used to observe the resulting output. The pulser has a high source and sink current driving capability, and can override IC outputs originally in either the HIGH or the LOW state.

With higher-quality probes, the user need not be concerned with the state of the test node. Pressing the push-button switch automatically drives a logic output or input from LOW to HIGH or HIGH to LOW. The output pulse width is intentionally kept narrow to limit the amount of energy delivered to the device under test, thereby eliminating the possibility of damage. As a general rule, the pulser has a three-state output. In the "off" state, the probe exhibits a high output impedance, ensuring that circuit operation is unaffected by the probe until the "pulse" switch is pressed. Pulses can be injected while the circuit is in operation, and no disconnections are needed. Like the logic probe, the pulser's power is usually derived from the circuit under test. Used together, a pulser and logic probe make a simple yet very effective troubleshooting combination for just about any kind of logic circuit.

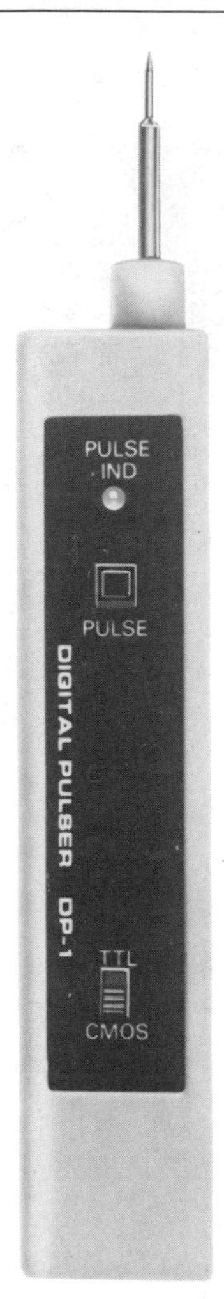

Figure 15–6 Logic pulser. Photograph courtesy of Global Specialties, an Interplex Electronics Company.

15–7 LOGIC COMPARATOR

The **logic comparator** is an instrument that tests digital ICs by comparing them to a good IC of the same type. The comparator clips onto powered ICs, and the reference IC shares the power and input signals with the test IC. When comparing the outputs of the two ICs, any level difference that exists for more than about 0.3 μs lights an LED for a minimum of 0.1 s, indicating a fault.

Before use, the outputs of the particular IC to be tested are selected by a group of miniature switches, which tell the comparator which pins of the reference IC are inputs and which outputs. Logic comparators are usually available for 14- and 16-pin ICs.

15–8 DIGITAL SIGNATURE ANALYZER

A **signature analyzer** detects problems in a faulty piece of equipment by comparing the data stream at a node against the data stream at the same node in a good piece of equipment. One type accomplishes this by generating a compressed, four-digit "fingerprint," or signature, of the data from a working system and comparing this against the "signature" of the system under test. Fault-finding is reduced to tracing signal flow and comparing measured signatures to known good signatures either recorded on paper or stored in a computer.

15–9 SUMMARY

This chapter discussed methods used in the analysis and troubleshooting of complex digital circuits. For the testing and monitoring of complex systems, whether computer-based or not, the logic analyzer and digital signature analyzer are important tools. For troubleshooting the response of individual integrated circuit devices (such as gates and flip-flops) without their removal from the circuit under test, small hand-held instruments like the logic probe, logic clip, and logic pulser are helpful. However, if the integrated circuit device can be removed from the circuit, then the logic comparator can be used to determine whether the device is good.

15–10 GLOSSARY

clock delay A feature of a logic analyzer which permits data storage after a given number of data words following the trigger signal.

event delay A circuit that delays the output of the logic analyzer's trigger circuit by a fixed number of trigger events.

glomper clip See **logic clip.**

logic analyzer An instrument for digital system analysis, troubleshooting, and maintenance. It works by sampling system voltages and displaying them on a CRT as a sequence of logic levels versus time.

logic clip Also called a *glomper clip,* a device that clips onto the top of DIP ICs and continuously shows the logic levels of each pin of the IC.

logic comparator An instrument that tests digital ICs by comparing them to a good IC of the same type.

logic probe A hand-held probe that monitors a single digital signal, indicating whether it is a logic 0 or 1 or a pulse, via one or more LEDs.

logic pulser A hand-held device used to inject logic-level pulse signals into a digital circuit, for troubleshooting.

qualifier lines Data lines combined with the trigger data word, to establish triggering requirements in a logic analyzer.

signature analyzer An instrument that detects problems in a faulty piece of equipment by comparing the data stream, or signature, at a node against the signature at the same node in a good piece of equipment.

trigger data word A user-selected word compared against incoming data, which will trigger the logic analyzer as the starting point for the storage of data.

trigger events The conditions or protocols that trigger the logic analyzer.

16

INSTRUMENTATION SYSTEMS

16–1 INSTRUCTIONAL OBJECTIVES

After completing this chapter, you will be able to

- Define the characteristics that must be considered in assembling any measurement system.
- Explain signal conditioning in terms of buffering, linearization, filtering, and offset/level conversion.
- Explain the advantage of using a voltage-to current converter.

- Describe the components of a typical analog-to-digital measurement system.
- Describe the operation of the IEEE-488 bus structure.
- Describe the RS-232C serial standard.
- Explain the operation of current loops.
- Describe how data can be transmitted over long distances using modems.

16–2 INTRODUCTION

When two or more instruments are used in a coordinated manner to perform an overall measuring function, the resulting arrangement is referred to as a **measurement system.** To completely characterize many complex processes and systems, it is often necessary to measure a number of parameters simultaneously and to store or output the results in a coherent format. This is the function of the instrumentation system.

The interconnection of instruments and devices is called **interfacing.** With the variety of instruments and data recorders now available, this subject is becoming very complex. Basically, however, there are three types of system interfaces:

1. analog-to-analog
2. analog-to-digital
3. digital-to-digital

These are discussed in the following sections.

16–3 ANALOG SYSTEMS

A purely **analog system** measures, transmits, displays, and stores data in analog form. Analog signals are characterized by *continuous* levels of voltage or current, rather than discrete or digital levels. Figure 16–1 shows the block diagram of an analog system used to measure and store pressure. First, the pressure is converted to a voltage signal by a pressure-to-voltage transducer. The signal is **conditioned** to remove any unwanted components, while presenting the proper level to the measuring instruments. The oscilloscope displays high-speed changes and noise, while both the recorder and analog meter respectively store and display the short-term average values. Although analog systems are characterized by low cost and wide bandwidth, their accuracy and resolution tend to be lower than that of other configurations.

To be compatible, the parameters of the individual system components must coordinate with the parameters of the other components. For example, a transducer with a slow response time cannot produce high-speed data, regard-

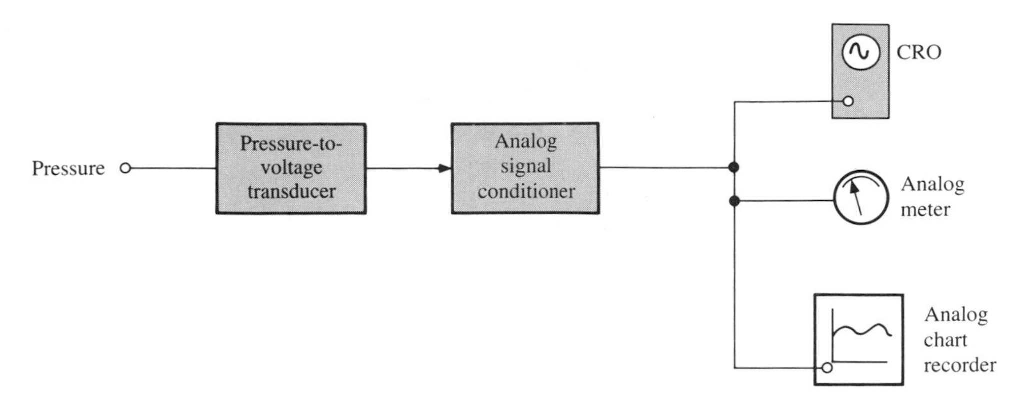

Figure 16–1 Block diagram of a typical analog measurement system.

less of the high-speed characteristics of the measuring and recording instruments that follow it. In this case, the transducer's speed is the limiting factor of the overall system. Similarly, a high-speed transducer cannot reach its full potential if the devices following it are not able to follow the signal faithfully. In this case, the recording and measuring instruments following the transducer are the limiting factors on the speed of the overall system.

Besides cost, the following characteristics must be considered when assembling any measurement system:

- signal level compatibility
- impedance matching
- susceptibility to noise
- proper grounding
- proper shielding
- power line conditioning
- signal conditioning
- ambient temperature

16–4 SIGNAL CONDITIONING

Signal conditioning usually involves one or more of the following four functions:

1. filtering
2. offset/level conversion
3. linearization
4. buffering

Filtering is used to reject unwanted components of a signal or the components of a spurious signal that would cause a measurement error. By using a filter to narrow the bandwidth of the signal that can be passed, broadband noise energy is reduced and signals outside of the passband are mostly rejected. By using a narrow band-reject, or *notch,* filter, individual frequencies such as 60-Hz power-line interference (hum) can be greatly suppressed or eliminated. Since the frequency response of many transducers is relatively low (less than 10 Hz), a simple low-pass filter is often all that is required to reject noise and power-line interference.

Offset/level conversion either changes the output signal of the transducer from one voltage or current level to another, or from one form to another. This is frequently done for compatibility with either the transmission medium in use or the levels required by the instruments being used.

In process-instrumentation systems, signals are frequently transmitted as a current level (i.e., 4 to 20 mA) rather than a voltage level, as shown in Figure 16–2. Using a current level technique eliminates the effects of voltage drop. Consequently, such a system requires that transducer output signals be converted to the proper current level at the sending end and back to a compatible form (e.g., a voltage) with the instruments at the receiving end.

Transmission of signals based on current offers the following three important features:

1. Impedance changes within the loop do not affect the signal level as long as the total impedance of the loop does not exceed the capabilities of the components to maintain the current level.
2. By using a standard current signal level, the instruments and controllers used in the loop are easily interchanged, added, or removed.
3. Generally, the current loop conductors are also used to deliver operating power to the transducers and signal conditioning equipment.

Figure 16–3 shows a circuit used to convert a voltage signal to a current signal.

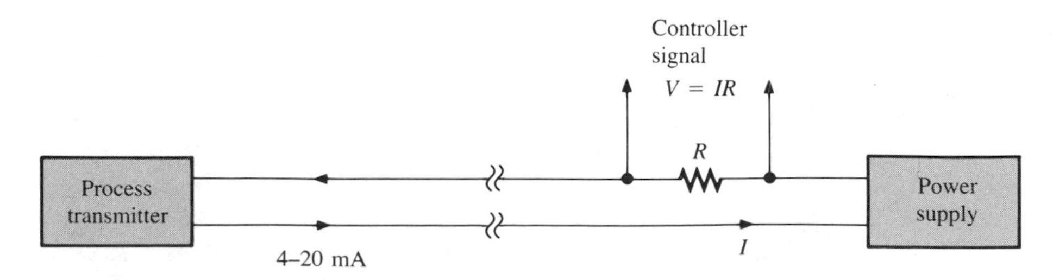

Figure 16–2 Signal transmission using a current loop.

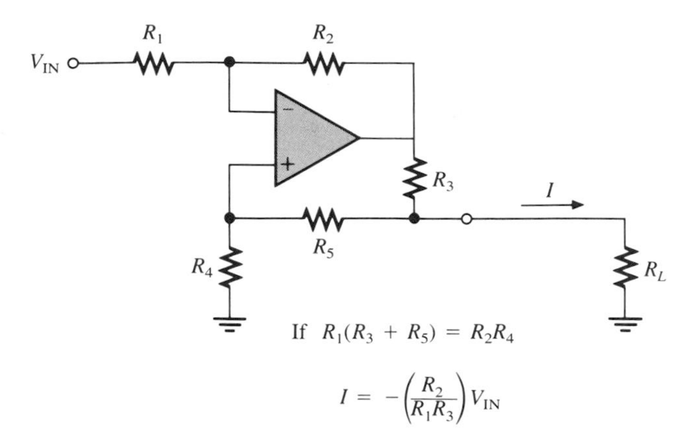

Figure 16–3 Voltage-to-current converter.

When the output signal from a transducer is not a linear function of the measured variable, it is often desirable to *linearize* this relationship before it is passed on to instruments and recorders. This can be done either with analog circuitry or by a computer. In *analog linearization,* the signal is passed through a circuit that has a response that is the inverse of the transducer. For example, if the transducer has an exponential response, its output signal might be passed through an amplifier circuit that has a logarithmic response, producing an output that is a linear function of the measured variable.

Alternatively, the signal could be digitized, after which a digital computer could be used to generate the logarithm of the input signal. Although the result is the same, there is a considerable difference in both cost and complexity. The system size and requirements dictate whether the computer is justified.

Finally, **buffering** isolates the signal source from its load. This serves both to make the transducer output independent of load changes and to provide a form of impedance matching.

16–5 ANALOG-TO-DIGITAL SYSTEMS

In an *analog-to-digital (A/D) system,* the original data is acquired in analog form and converted to digital before being passed on to other parts of the system. The A/D conversion technique is particularly good where the measured variable changes slowly with time and where high accuracy and resolution are required. The digital voltmeter discussed in Chapter 7 is an excellent example of a simple analog-to-digital system.

The addition of an analog **multiplexer** ahead of the A/D converter permits the sequential reading of a number of transducers. The analog multiplexer, such as that shown in Figure 16–4, is the electronic equivalent of a

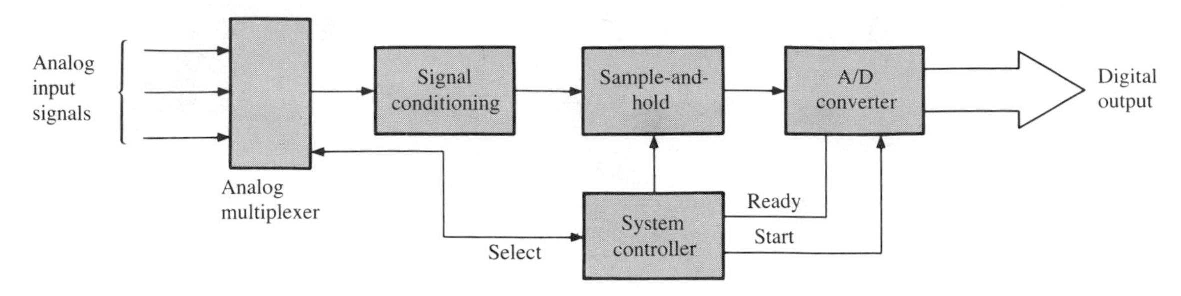

Figure 16–4 Block diagram of a multichannel A/D temperature measurement system.

selector switch that permits a number of inputs to be sequentially connected to a single output, one at a time, in a repeatable pattern.

Analog-to-digital systems are particularly useful when mathematical operations must be performed on the measured values before they are stored or displayed. Furthermore, digital signals offer better noise immunity than their analog counterparts. Analog-to-digital instrumentation systems usually contain some or all of the following components:

1. A transducer or other signal source that converts a measured physical parameter to a proportional voltage or current (Chapter 10).
2. A multiplexer used in multichannel systems to sequentially connect the input signals to the system. This is actually a form of an electronic multiposition selector switch operated by the system controller.
3. A sample-and-hold circuit, which samples the input signal and holds that value while the A/D converter makes its conversion. This is necessary when the conversion time is long compared to the time for the signal to go through a significant change in value. With a sample-and-hold circuit, it is possible to read the voltage at a particular point in a rapidly changing waveform, even if the conversion time is relatively long.
4. Signal conditioning, which includes the functions for filtering, buffering/amplification, linearization, and offset/level shifting.
5. An analog-to-digital converter, to convert the conditioned signal to a digital form (Chapter 7).
6. A system control section that can be a computer, a calculator, or a special logic section, which controls the functions of the other elements.

16–6 DIGITAL INTERFACING

Very simple systems can transmit data in straight serial or parallel binary form through simple data ports. However, as systems become more complex, there is a definite advantage to standardizing both the hardware and the format used to communicate between the various members of the system. *Standardization* means that off-the-shelf instruments and accessories from various manufacturers can be easily added to a system and can be expected to be compatible with the other system components. With standardization, upgrading and addition of system components becomes a relatively simple task, at the same time permitting instruments to be easily moved from system to system. Currently, two standards enjoy widespread use: the IEEE-488 standard and the RS-232C standard.

The IEEE-488 Standard and Bus

Many instruments are now manufactured that include an output socket, or **port,** to pass measurement data and various control signals in digital form. To facilitate interfacing, an industry standard establishing a truly general purpose interface bus, or **GPIB,** was first published in 1975, and was revised in 1978. Originally, this system was developed by Hewlett-Packard so that they could interconnect their own instruments, and was called the **HPIB.** The Institute for Electrical and Electronic Engineers (IEEE) subsequently adopted the HPIB system as the basis of its own standard. This standard is now known as the **IEEE-488 standard** and is used extensively by instrument manufacturers. The IEEE-488 *standard* is a set of rules, specifications, and characteristics of the interface. The IEEE-488 **bus** is the hardware that is used to implement the standard.

The development and worldwide acceptance of the IEEE-488 standard makes possible the automation of conventional manual measurements without extensive cost or engineering time. Besides having the ability to make a number of measurements virtually simultaneously, an automated measurement system offers the following advantages over a manual system:

- Repeatable results—operator skill and fatigue factors are eliminated.
- Automatic error correction—correction factors and system errors can be corrected immediately.
- Greater throughput—the automated system is usually faster than manual measurements.
- More complete testing—greater throughput allows more parameters to be tested in a shorter time.
- Better test records—the test records can be automatically stored and retrieved at a later time for analysis.

The IEEE-488 interface bus provides the capability to interconnect up to 15 devices, using a standard passive multiconductor cable. The cable's only purpose is to interconnect all devices in parallel. Once connected together, any one device can transfer data to one or more other devices on the bus. Every device connected to the bus must be able to perform at least one of the following roles: talker, listener, and controller.

A **talker** transmits data to, while the **listener** receives data from, other devices via the bus. Some devices perform both functions, i.e., "listen" to receive their control instructions and "talk" to send their data. A **controller** manages the operation of the system by designating which devices are to send and receive data, as well as commanding specific actions within other devices.

A minimum system consists of one talker and one listener without a controller. With this arrangement, data transfer is limited to direct transfer between one device manually set to "talk only" and one device to "listen only." An example is a measuring instrument talking to a printer. A controller increases the flexibility and power of the system. The role of the controller is usually filled by a computer or programmable calculator. A controller participates by being programmed to configure individual devices to perform certain measurements, schedule the measurements, monitor the progress of a measurement as it proceeds, and interpret the results of the measurement.

The IEEE-488 standard specifies logic levels that are similar to those exhibited by the TTL (transistor-transistor logic) family of integrated circuit devices. A logic 0 is any level 0.8 V or less. A logic 1 is any level greater than 2.0 V. If an instrument load is placed every 2 m, the overall length of the cable connecting the IEEE-488 devices cannot exceed 20 m. The maximum unrestricted data rate is 250 kilobytes/second.

The bus structure specified by the IEEE-488 standard is a form of party line in which all devices are connected in parallel. Sixteen signal lines are grouped into three clusters, according to their function:

1. Data bus (eight signal lines)
2. Data byte transfer control bus (three signal lines)
3. General interface management bus (five signal lines)

Table 16–1 summarizes the arrangement of these 16 lines, using a standard 24-pin connector.

The eight DATA IN signal lines (DI01–DI08) of the data bus carry data in *bit-parallel, byte-serial* format across the interface. These signal lines can carry addresses, program data, measurement data, universal commands, and status bytes to and from devices interconnected in a system.

The three signal lines control transfer of each byte on the databus:

- DAV (data valid)
- NDAC (not data accepted)
- NRFD (not ready for data)

These lines operate in an interlocked handshake mode. Two of these signal lines (NRFD and NDAC) are connected to each device via a "wired or" circuit. Active listeners not ready to accept data hold the NRFD line low. With this

Table 16–1 IEEE-488 (GPIB) pin connections

Pin number	Signal line	Pin number	Signal line
1	DIO1	13	DIO5
2	DIO2	14	DIO6
3	DIO3	15	DIO7
4	DIO4	16	DIO8
5	EOI	17	REN
6	DAV	18	ground
7	NRFD	19	ground
8	NDAC	20	ground
9	IFC	21	ground
10	SRQ	22	ground
11	ATN	23	ground
12	shield	24	logic ground

scheme, the active talker cannot send more data until the slowest listener is ready, releasing the NRFD line to go high.

The talker controls the DAV line to indicate when data is ready to transmit, and the listeners control the NRFD and the NDAC lines to indicate both their readiness to receive data and the receipt of data, respectively. As shown in the bus timing diagram of Figure 16–5, the active talker outputs eight bits of data to the data bus and, 2 μs later, pulls the DAV line low. This 2-μs delay provides settling time for the data to reach valid levels. After the DAV line goes low, the listeners respond by pulling the NRFD line low to prevent any

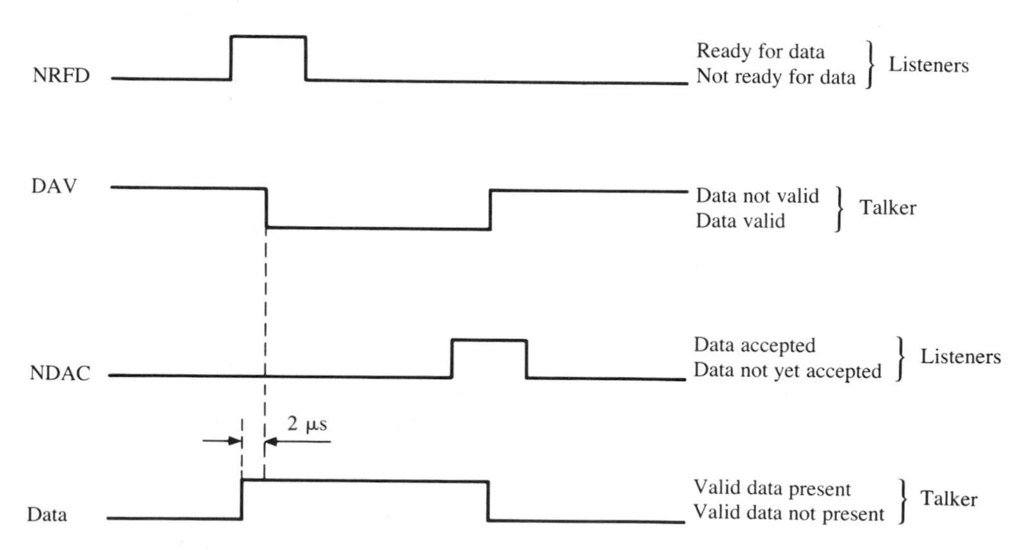

Figure 16–5 IEEE-488 bus timing diagram.

additional data transfer. The listeners accept the data byte at their individual rates. When each listener has accepted the data, it releases the NDAC line. When the last listener releases the NDAC line, it goes high. When the active talker sees the NDAC line go high, it stops driving the data bus and releases the DAV line. The listeners pull the NDAC line low. Once the controller has configured the talker and listeners, it takes no part in the data transfer.

The remaining five interface management bus lines are used to help control the data lines:

- ATN (attention)
- EOI (end or identify)
- IFC (interface clear)
- REN (remote enable)
- SRQ (service request)

The ATN line determines how many data lines are processed. When high, the data lines are interpreted as either containing addresses or commands by all the devices on the bus. When low, only those devices addressed are able to use the data lines. Consequently, transmitted data is *device-dependent.* The IFC line clears the bus to a quiescent state. In practice, the IFC line is frequently used to clear or initialize the bus lines either when an error has occurred or something has gone wrong.

The REN permits a device to be controlled by the bus. In effect, a given instrument can be *programmed* by the bus rather than being controlled by front-panel controls of the instrument. The SRQ line can be used by any of the devices on the bus to command the attention of the controller when it has data to send, where it functions as a talker. When a device has data to receive, it functions as a listener. Finally, the EOI line can be used either by the active talker device to designate the end of message, or by the controller as a polling line.

The RS-232C Standard

The **RS-232C standard** is issued by the Electronic Industries Association (EIA). It applies specifically to the interconnection of data terminal equipment (DTE) and data communication equipment (DCE) employing serial binary data interchange. Data terminal equipment includes such devices as digital instruments, computer terminals, computers, and teleprinters. Data communications equipment includes **modems** (acronym for *modulator-demodulator*), devices that convert between the digital signals and the corresponding tone signals that are carried by the telephone lines.

The RS-232C standard defines the following four characteristics:

1. electrical signal characteristics
2. mechanical interface characteristics

3. the description of interchange circuits
4. standard interfaces for selected communications system configurations

The RS-232C standard applies to both synchronous and nonsynchronous communications systems operating at speeds up to 19,200 bits per second, or 19.2 kBd (kilobaud), where a **baud** equals the number of bits per second. For example, 110 Bd equals 110 bits/second. The standard is general in its application, in that it assigns serial signals to specific pins on a standard "DB-25" series connector, but does not restrict the type of data that can be sent. Any character length, bit code, and bit sequence can be used. This interface standard is intended for fairly short cable runs, of about 50 feet or less between devices.

The RS-232C standard specifies the following four types of lines:

1. data signals
2. control signals
3. timing signals
4. signal grounds (returns)

Any voltage between −3 V and −25 V represents a logic 1, and any voltage between +3 V and +25 V represents a logic 0. Consequently, any voltage between −3 V and +3 V is *undefined,* and should be avoided. Control signals are "on" if the voltage is between +3 V and +25 V and "off" if the voltage level is between −3 V and −25 V.

Table 16–2 lists the RS-232C standard DB-25 pin configuration. Although there are 25 possible lines, the present standard specifies functions for only 20 lines, and the remaining five are either unassigned or reserved. The convention is such that the pins are named from the viewpoint of the DTE. Consequently, the DTE transmits data via pin 2 to a DCE, but a DCE transmits via pin 3 to the DTE.

Figure 16–6 shows a block diagram of a complete system using the RS–232C interface. The standard interface can also be used in systems that do not contain an actual DCE—systems in which no telephone line is involved. Here, confusion may arise as to which devices assume the role of DTE and DCE. That is, it must be carefully defined which connector pins serve to transmit data and which to receive. This is called the **null modem** connection.

16–7 THE CURRENT LOOP

The **current loop** used for digital communication is similar to the analog current loop used in process instrumentation systems, except that this current has only two values: on and off. Originally, the current loop was designed specifically for use with teletype machines, but it has since been adapted to a variety of other uses, including that of communications between digital instruments. A

Table 16–2 RS-232C pin connections

Pin number	Signal line	Pin number	Signal line
1	protective ground	14	secondary transmitted data (to DCE)
2	transmitted data (to DCE)	15	transmit clock (to DTE)
3	received data (to DTE)	16	secondary received data (to DTE)
4	request to send (to DCE)	17	receiver clock (to DTE)
5	clear to send (to DTE)	18	unassigned
6	data set ready (to DTE)	19	secondary request to send (to DCE)
7	logic ground	20	data terminal ready (to DCE)
8	carrier detect (to DTE)	21	signal quality detect (to DTE)
9	reserved	22	ring detect (to DCE)
10	reserved	23	data rate select (to DCE)
11	unassigned	24	transmit clock (to DCE)
12	secondary carrier detect (to DTE)	25	unassigned
13	secondary clear to send (to DTE)		

20-mA current flowing in the continuous loop is called a **mark,** and usually represents a binary 1. The absence of a current in the loop is called a **space,** and usually represents a binary 0.

Figure 16–7 shows a circuit used to generate a 20-mA current loop from a TTL logic system. As shown, the current loop is electrically isolated with an opto-isolator, the output of which controls the current flowing in the loop. A TTL high level turns on the LED. The light-sensitive transistor inside the opto-isolator, responding to the lit LED, saturates, or turns on, thus permitting

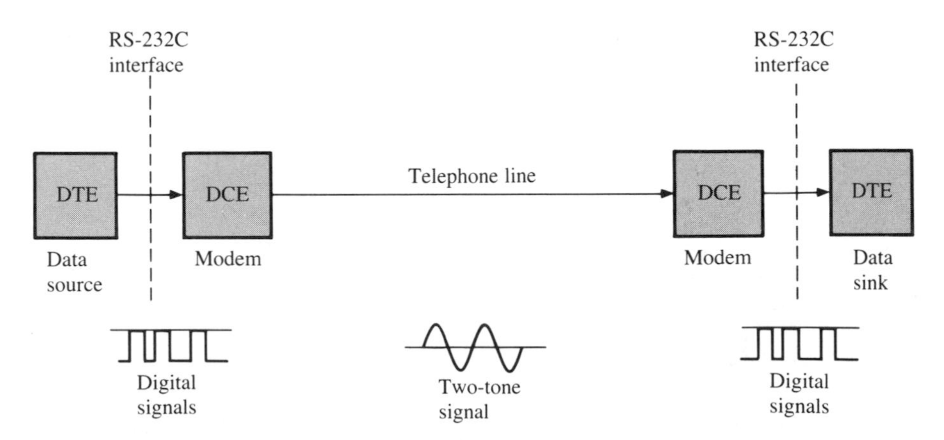

Figure 16–6 Block diagram of a complete data transmission system using the RS-232C interface.

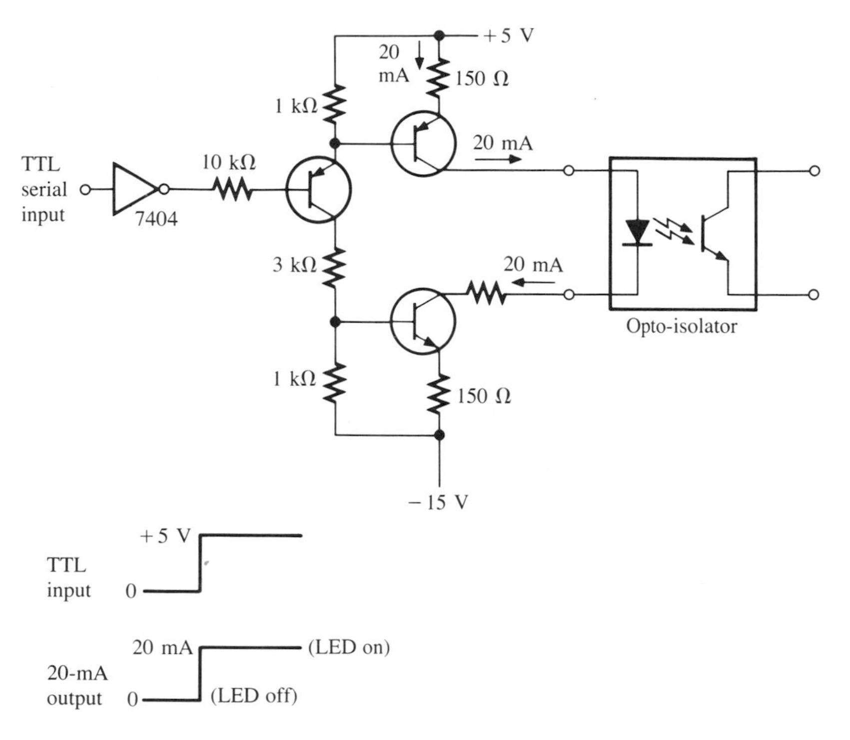

Figure 16–7 Circuit for generating a 20-mA current loop.

current to flow from collector to emitter, which completes the loop. Depending on the load of the receiver, the voltage source can be adjusted to give an "on" current of 20 mA. When the TTL level is low, no current flows through the LED, which cuts off the transistor and causes the loop current to be zero.

The current loop is a simple, low-impedance path that is resistant to electrical noise, yet requires only two wires to transmit serial data up to 5000 feet. However, the two wires are prone to crosstalk interference from nearby circuits. As shown by Figure 16–8, a typical low-speed, 20-mA current loop connected to a teletype set uses an 11-bit format to serially transmit a single character. In order, these bits are

- one start bit
- eight data bits
- two stop bits

The transmission rate is 110, 150, or 300 Bd. It is common to use the 7- or 8-bit **ASCII codes** for the transmitted character codes. Although 20-mA "current on" values are normally used, older systems using 60-mA levels persist.

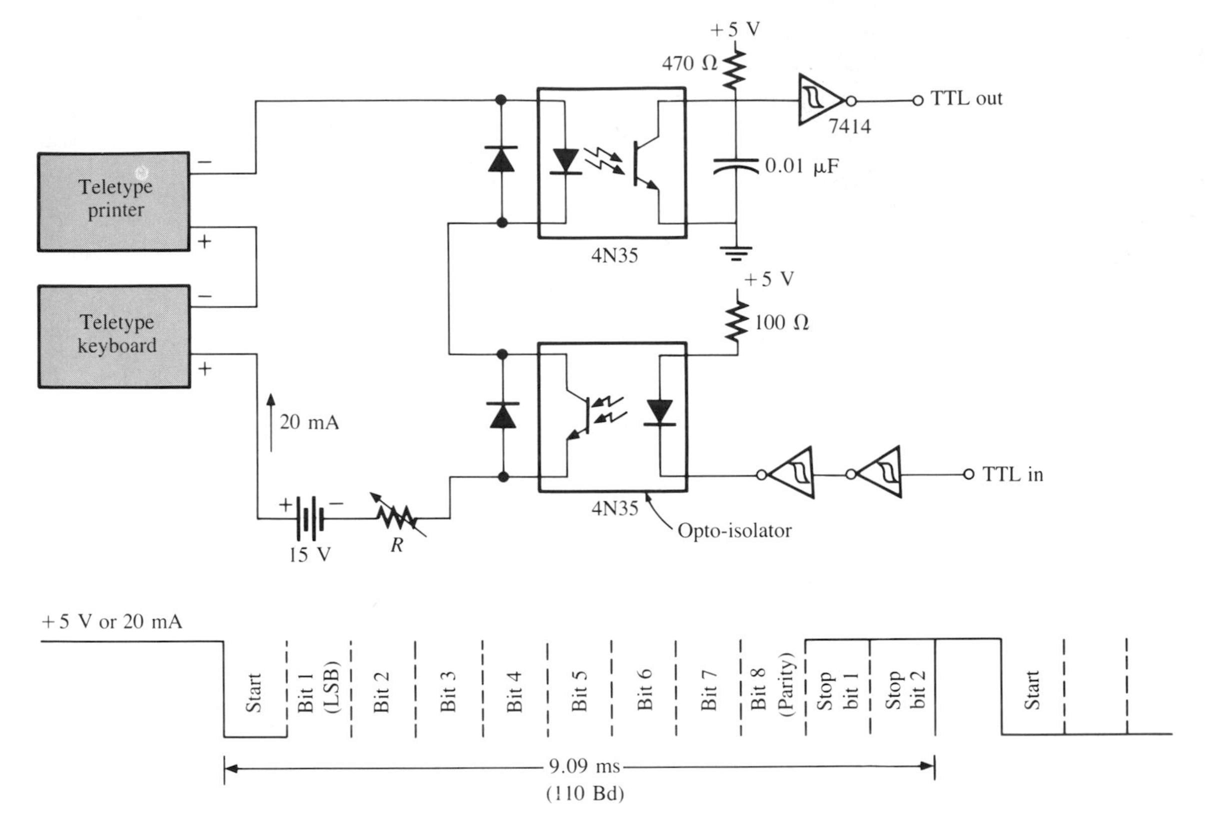

Figure 16–8 Typical low-speed 20-mA current loop connected to a teletype set.

16–8 LONG-DISTANCE DATA TRANSMISSION

When data must be transmitted over distances greater than those permitted by the RS-232C (50 feet) or the current loop (5000 feet), telephone lines are usually used. Since the telephone system cannot handle digital signals directly, the digital signals must be converted to audio tones at the sending end and back to digital signals at the receiving end. This conversion is the function of a device called a **modem.** Most modems used for medium-speed (1200 Bd) asynchronous data transmission use **frequency-shift keyed (FSK)** modulation to generate the tones. In FSK, mark and space signals are converted to one of two harmonically unrelated frequencies by the modem. Summarized in Table 16–3 are the popular Bell System FSK-type modem characteristics. For a Bell 202-type system, a mark (binary 1) uses a 1200-Hz tone, and a space (binary 0) uses a 2200-Hz tone (Figure 16–9). These frequencies are compatible with the fre-

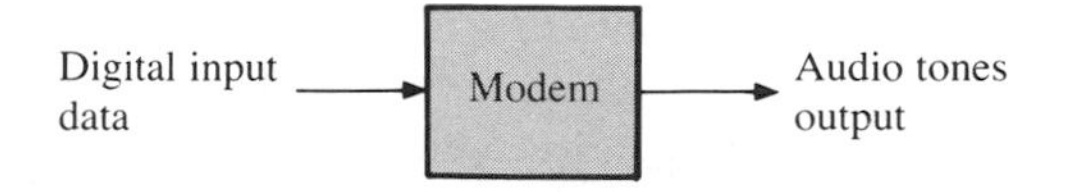

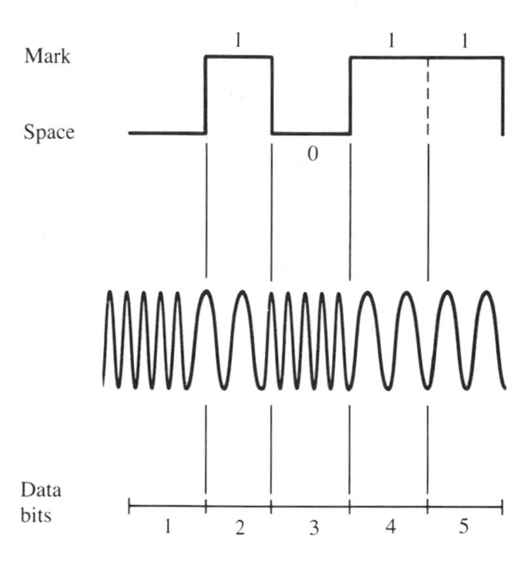

Figure 16–9 Asynchronous data transmission using frequency-shift keying (FSK).

Table 16–3 Operating characteristics of Bell System FSK MODEMs

	103	113D	202S	202T
Maximum data rate (baud)	300	1200	1200	1800
Mode	FDX	FDX	HDX	HDX–2W FDX–4W
Originate				
space frequency (Hz)	1070	1070	2200	2200
mark frequency (Hz)	1270	1270	1200	1200
Answer				
space frequency (Hz)	2025	2025	2200	2200
mark frequency (Hz)	2225	2225	1200	1200
Frequency shift (Hz)	200	200	1000	1000

Abbreviations:
FDX = full duplex
HDX = half duplex
2W = two wires
4W = four wires

quency response of a typical telephone line, which is about 300 Hz to 3 kHz. For all FSK modems, different mark and space frequencies can be used to identify the device which initiates the transmission of data, called an **originator,** and the device which initially receives data, or the **answerer.**

Either the familiar dial-up or dedicated leased phone lines can be used for the communications channel. Dial-up lines require that a person or a computer dial the phone number of the receiving system and verify that a link has been established. With dedicated leased lines, the link is permanently established.

The digital equipment interfaces with the modem, using one of the digital exchange protocols: EIA-RS-232C, IBM, MIL-STD-188C, ITT Consultative Committee (CCITT), and Bell Telephone. Currently, there are three classes of exchange interfaces: simplex, half-duplex, and full-duplex. A **simplex** interface is one that can transmit data in only one direction. A **half-duplex** interface is one where both ends can transmit, but not simultaneously. Simplex and half-duplex operation require only two wires. **Full-duplex (FDX)** interface, however, permits simultaneous two-way transmissions, but requires four wires between the transmitter and receiver.

Two very popular modem standards are the Bell 103/113 and the Bell 202 types. The Bell 103/113 modem can handle data up to 300 Bd full-duplex, and the 202-style modem to 1200 Bd full-duplex. Higher speeds, up to 9600 Bd, are possible, but require synchronous data transmission, more sophisticated modulation, and a special "data grade" telephone line. For Bell 103/113-type formats, the space frequency is *lower* than the mark frequency; while in the Bell 202 format, the space frequency is *higher* than the mark frequency.

The frequency separation between mark and space is called the **frequency shift.** A figure of merit (h), quantifying the degree of modulation achieved in an FSK system, is the h factor. It relates the mark (f_M) and space (f_S) frequencies for a given baud rate (f_B), and is given by

$$\boxed{h = \frac{|f_M - f_S|}{f_B}} \tag{16–1}$$

The h factor should be less than 1.0, and a value of 0.7 is considered optimum. As an example, a Bell 103 modem in the answer mode operating at 300 Bd has a value of

$$h = \frac{200 \text{ Hz}}{300 \text{ Bd}}$$

$$= 0.66$$

16–9 SUMMARY

This final chapter discussed various types of measurement systems, in which two or more instruments and other components are interconnected in a coordinated manner. These systems can be either analog or digital, or a hybrid system of both digital and analog devices. Among digital systems, two interface standards enjoy widespread use, the IEEE-488 and RS-232C standards and associated bus structures, which are fast replacing the older 20-mA and 60-mA current loop systems. Though the IEEE-488 and RS-232C systems have limitations on the maximum distance between receive and transmit devices, the distances can be extended worldwide by the use of modems and telephone lines.

16–10 GLOSSARY

analog system A system whose signals are represented by continuously varying voltages or currents.

answerer A communications device which initially receives data.

ASCII Acronym for American Standard Code for Information Interchange.

ASCII code A seven-bit digital communications code representing letters, numbers, symbols, and printer controls.

baud (Bd) A unit of signaling speed equal to 1 bit/second.

buffering The process providing electrical isolation between two other circuits.

bus One or more conductors used as a path for the transmission of signals.

conditioning A process whereby a signal can be amplified, buffered, filtered, level-converted, or linearized, to be compatible with other stages of a system.

controller A device, such as a computer, which manages the operation of an IEEE-488 bus system.

current loop A continuous loop whereby the absence of current flow represents a binary 0 and the presence of a current represents a binary 1.

frequency shift Represented by h; the difference between mark (f_M) and space (f_S) frequencies.

frequency shift keying (FSK) A form of frequency modulation in which a digital signal is converted into corresponding audio tones that are not harmonically related.

full-duplex (FDX) A four-wire interface involving simultaneous two-way transmission of data between two devices.

General Purpose Interface Bus (GPIB) See **IEEE-488 standard.**

half-duplex (HDX) A two-wire interface between two devices, whereby both devices can transmit data, but not simultaneously.

HPIB Acronym for Hewlett-Packard Interface Bus. Also referred to as **General Purpose Interface Bus (GPIB).**

IEEE-488 standard A standard and 24-pin bus structure designed to interface various instruments.

interfacing The interconnection of instruments and devices.

listener The receiving device on an IEEE-488 bus system.

mark Corresponding to a logic (or binary) 1.

measurement system An arrangement whereby two or more instruments are used in a coordinated manner to perform an overall function.

modem Short for modulator-demodulator, a device that translates between digital logic levels and corresponding specific audio frequencies.

multiplexer The equivalent of a single-pole, multiposition switch in which a number of inputs are sequentially connected to a single output.

null modem A method of interconnecting two computers or two computer-controlled devices so that both devices will not attempt to transmit or receive simultaneously on the same wire pair.

originator A communications device which initiates the transmission of data.

port Any input or output data or control signal connection.

RS-232C standard A standard 25-pin bus structure used for the interconnection of data communication (DCE) and data terminal (DTE) equipment.

simplex A two-wire interface between two devices, where data is transmitted in only one direction.

space Corresponding to a logic (or binary) 0.

talker The transmitting device on an IEEE-488 bus system.

APPENDIX A

PROPER SPELLING OF COMBINED FORMS OF MULTIPLIER PREFIXES AND UNITS

In general, the spelling of combined forms of multiplier prefixes and a unit of measure is straightforward. For example, the prefix *kilo-* and the unit *volt* are combined as *kilovolt*. No letters are dropped or added. However, like most rules of English spelling, there are exceptions. Multiplier prefixes end in the vowels *a*, *i*, or *o*. Examples of the combination of units that start with the vowels *a*, *e*, or *o* are:

1. *mega* + *ampere* = *megampere* (*MA*).
2. *mega* + *electronvolt* = *megaelectronvolt* (*MeV*).
3. *mega* + *ohm* = *megohm* ($M\Omega$).
4. *milli* + *ampere* = *milliampere* (*mA*).
5. *milli* + *electronvolt* = *millielectronvolt* (*meV*).
6. *milli* + *ohm* = *milliohm* ($m\Omega$).
7. *pico* + *ampere* = *picoampere* (*pA*).
8. *kilo* + *electronvolt* = *kiloelectronvolt* (*keV*).
9. *kilo* + *ohm* = *kilohm* ($k\Omega$).

APPENDIX B

STANDARD QUANTITY SYMBOLS

The following quantity symbols are those frequently used in the study of electricity and electronics. Where two symbols are separated by three dots, the second is to be used only where there is a specific need to avoid conflict.

Quantity	Symbol	Unit
active power	P	watt
admittance	Y	siemen
angle (phase angle)	θ, ϕ	radian, degree
angular frequency	ω	radian per second
apparent power	$S \ldots P_s$	voltampere
capacitance	C	farad
capacivity (permittivity)	ε	farad per meter
conductance	G	siemen
conductivity	Γ, σ	siemen per meter
coupling coefficient	κ	dimensionless
dissipation factor	D	dimensionless
efficiency	η	dimensionless
elastance	S	reciprocal farad, daraf
electric charge	Q	coulomb
electric field strength	E	volt per meter
electric flux	Ψ	coulomb
electrostatic potential	$V \ldots \phi$	volt
energy	E	joules, watt-hour
force	F	newton
frequency	$f \ldots \nu$	hertz
gain (ordinary)	A	dimensionless
gain (logarithmic)	A	bel, neper
illuminance (illumination)	E	lux

Quantity	Symbol	Unit
impedance	Z	ohm
inductance	L	henry
irradiance	E	watt per square meter
leakage coefficient	σ	dimensionless
luminance	L	candela per square meter, lambert, nit
luminous intensity	I	candela
luminous flux	Φ	lumen
magnetic field strength	H	ampere per meter
magnetic flux	Φ	weber
magnetic flux density	B	tesla
magnetic flux linkage	Ψ	weber
magnetization	H, M	ampere per meter
magnetomotive force	F, F_m	ampere
period	T	second
permeability	μ	henry per meter
permeance	P, P_m, γ	henry
power	P	watt
power factor	PF	dimensionless
quality factor	Q	dimensionless
reactance	X	ohm
reactive power	$Q \ldots P_q$	var
reluctance	R, R_m	reciprocal henry, ampere-turn per maxwell
reluctivity	ν	meter per henry
resistance	R	ohm
resistivity (specific resistance)	ρ	ohm-meter
susceptance	B	siemen
susceptibility	χ, κ	dimensionless
temperature (customary)	t	degree
temperature (thermodynamic)	$T \ldots \theta$	kelvin
time constant	$\tau \ldots T$	second
transmittance	τ	dimensionless
turns ratio	N	dimensionless
voltage (potential difference)	$V, E \ldots U$	volt
wavelength	Λ, λ	meter
work	W	joule

APPENDIX C

COMMON CONVERSION FACTORS

To convert from	to	multiply by
acre	sq. meter	4047
ampere/hour	coulomb	3600
BTU	joule	1055
BTU/second	watt	1055
centimeter Hg (0 °C)	pascal	1333
centimeter of water (4 °C)	pascal	98.06
day	second	86400
degree Celsius	kelvin	°C + 273.15
degree Fahrenheit	kelvin	1.8 × (°F + 459.67)
degree Fahrenheit	degree Celsius	(°F − 32)/1.8
degree Rankine	kelvin	°R/1.8
dyne	newton	1×10^{-5}
erg	joule	1×10^{-7}
foot	meter	0.3048
cu. foot	cu. meter	0.02831
sq. foot	sq. meter	0.09293
footcandle	lumen/sq. meter	10.76
footcandle	lux	10.76
gauss	tesla	1×10^{9}
gilbert	ampere-turn	0.7958
horsepower	watt	746
inch	meter	0.0254
cu. inch	cu. meter	1.639×10^{-5}
sq. inch	sq. meter	6.452×10^{-4}
maxwell	weber	1×10^{-8}
mile	meter	1.609×10^{3}
mile/hour	meter/second	0.4470

To convert from	to	multiply by
oersted	ampere/meter	79.58
ounce	cu. meter	2.957×10^{-5}
pound/sq. inch	pascal	6895
watt-second	joule	1.000
yard	meter	0.9144

APPENDIX D

SI UNIT SYMBOLS DERIVED FROM NAMES OF INDIVIDUALS

Unit	Symbol	Person
ampere	A	Andre Marie Ampere (1775–1836)
bell	B	Alexander Graham Bell (1847–1922)
coulomb	C	Charles Augustin de Coulomb (1736–1806)
curie	Ci	Marie Sklodowska Curie (1867–1934)
faraday	F	Michael Faraday (1791–1867)
gauss	G	Karl Friedrich Gauss (1777–1855)
gilbert	Gb	William Gilbert (c. 1540–1603)
henry	H	Joseph Henry (1798–1878)
hertz	Hz	Heinrich Rudolf Hertz (1857–1894)
joule	J	James Prescott Joule (1818–1889)
kelvin	K	William Thompson Kelvin (1824–1907)
lambert	L	Johann Heinrich Lambert (1728–1777)
maxwell	Mx	James Clerk Maxwell (1831–1879)
neper	Np	John Napier (1550–1617)
oersted	Oe	Hans Christian Oersted (1777–1851)
ohm	Ω	Georg Simon Ohm (1787–1854)
roentgen	R	Wilhelm Conrad Roentgen (1843–1923)
siemen	S	Ernst Werner Siemens (1816–1892)
tesla	T	Nikola Tesla (1856–1943)
volt	V	Alessandro Volta (1745–1827)
watt	W	James Watt (1736–1819)
weber	Wb	William Eduard Weber (1804–1891)

APPENDIX E

GREEK LETTER SYMBOLS

The following Greek letter symbols are those frequently used in the study of electricity and electronics. Unless otherwise indicated, the small letter is used.

Letter			
Capital	Small	Name	Designates
Α	α	alpha	transistor forward current gain
Β	β	beta	transistor short circuit current gain
Γ	γ	gamma	permeance, conductivity
Ε	ε	epsilon	capacivity, permittivity
Η	η	eta	efficiency, intrinsic standoff ratio
Θ	θ	theta	temperature, angle
Κ	κ	kappa	coupling coefficient, susceptibility
Λ	λ	lambda	wavelength
Μ	μ	mu	permeability, micro (prefix, 10^{-6})
Ν	ν	nu	frequency, reluctivity
Π	π	pi	constant equal to 3.14159 . . . .
Ρ	ρ	rho	resistivity, volume density
Σ	σ	sigma	conductivity, Stefan-Boltzmann constant, leakage coefficient, summation (Σ)
Τ	τ	tau	time constant, transmittance
Φ	φ	phi	angle, potential, luminous flux (Φ), magnetic flux (Φ)
Χ	χ	chi	susceptibility
Ψ	ψ	psi	electric flux (Ψ), magnetic flux linkage (Ψ)
Ω	ω	omega	angular frequency, ohm (Ω)

ANSWERS TO ODD-NUMBERED PROBLEMS

Chapter 1

1. 0.2 mA, 1.4%
3. (a) 28.8 (b) 28.9 (c) 28.8 (d) 28.6 (e) 28.5 (f) 28.6
5. (a) 3200 ±320 Ω (b) 678.5 ±206.3 Ω
7. 60.5 Ω, percent voltmeter error = 2.63%, percent ammeter error = 3.18%, total limit error = 5.81%.
9. (a) 35.6 ppm (b) ±0.00355%
11. Measurement 2 (650 Hz), precision = 0.9991
13. (a) −1.043 mA/V, 10.41 mA (b) 5.195 mA (c) −0.989, indirect correlation (d) 97.8%

Chapter 2

1. 0.502 V
3. 5.74 VU
5. 3.5526×10^{12} ft/hr
7. 5.597 Mx/m^2
9. 1394 pF

Chapter 4

1. 2.85 μV
3. 2.41 μV
5. 13.5 nA
7. 12.3 dB
9. 316 pW
11. 15 dB
13. 1.426 V RMS
15. 52.5 dB
17. 101.98 dB

Chapter 5

1. (a) 10 mA (b) 5 mA (c) 1 mA
3. (a) 5.56 Ω (b) 1.02 Ω (c) 0.336 Ω
5. 0.2%
7. Range A = 10 mA, range B = 100 mA, range C = 1 A
9. (a) 49,950 Ω (b) 1 V = 20 μA, 2 V = 40 μA, 3 V = 60 μA, 4 V = 80 μA, 5 V = 100 μA
11. Meter B (5 kΩ/V vs. 2 kΩ/V)
13. (a) $Y_{\text{expected}} = -0.0613 + 1.01\ X_{\text{observed}}$ (b) 5.393 V

Chapter 6

1. 24.8 μA
3. (a) 2 kΩ/V (b) 900 Ω/V (c) 13.35 kΩ
5. (a) 5.29 V RMS (b) 11.8%
7. 3.33 V RMS
9. (a) 5.53 V RMS (b) 9.584 V

11. 2.5 A

13. (a) 13.7 W (b) 0.86

Chapter 7

1. 14 kΩ

3. (a) 5 kΩ (b) 6750 Ω

5. (a) 0.391% (b) 0.0244% (c) 0.0122%

7. 3 μs

9. 9 bits

Chapter 8

1. 246 mV

3. $V_{true} = 0.133 + 0.962\ V_{measured}$ ($r = 0.997$)

5. 66.57 Ω

Chapter 9

1. 127.9 Ω

3. 0.0098 Ω

5. 51.45 m

7. 0.186 μF

9. (a) 0.183 μF (b) 63.3 Ω (c) 0.073

11. (a) 0.058 μF (b) 798 Ω (c) 3.44

13. (a) 157 mH (b) 591 Ω (c) 1.67

15. (a) 0.104 μH (b) 1.04 Ω (c) 0.0006

Chapter 10

1. 168°

3. 2.035 V

5. 77.4 °C

7. 1.023 mV

9. 2.86

11. 24.6 mV

13. 78.4 °C

Chapter 11

1. 12.9 MHz

3. 8 pF

5. (a) 263 mV peak-to-peak (b) 0.18 V

7. 2 μs/division

9. 53.1°

11. 19.4°

13. 0.29 μs

Chapter 12

1. (a) ±10 Hz, 99.99999 MHz (b) ±1 Hz, 9,999,999 Hz (c) ±0.1 Hz, 999,999.9 Hz

3. (a) ±0.104% (b) ±0.00051%

5. ±0.044%

7. 1.4652071 GHz

9. 7.025 MHz

11. 27.7°

13. ±0.016 Hz

Chapter 13

1. 3339 Hz

3. upper sideband frequencies: 902, 904, 906, 908 kHz, etc.
lower sideband frequencies: 898, 896, 894, 892 kHz, etc.

5. (a) 3.5 kHz (b) 0.093 (c) 3.89

7. 87%

Chapter 14

1. 2.08:1

3. 0.16

5. (a) 38.7% (b) 61.3%

7. 4.21 dB

9. (a) 50 mV (b) 0.63 V

INDEX